STUDENT
SOLUTIONS MANUAL

VOLUME 1: CHAPTERS 1–14

Essential COLLEGE PHYSICS

Andrew F. Rex
University of Puget Sound

Richard Wolfson
Middlebury College

Addison-Wesley

Boston Columbus Indianapolis New York San Francisco Upper Saddle River
Amsterdam Cape Town Dubai London Madrid Milan Munich Paris Montréal Toronto
Delhi Mexico City São Paulo Sydney Hong Kong Seoul Singapore Taipei Tokyo

We would like to thank the following people for their contributions to the
Student Solutions Manual for Essential College Physics:

Freddy Hansen

Brett Kraabel

Sen-Ben Liao

Michael Schirber

Larry Stookey

Dirk Stueber

Robert White

Publisher: Jim Smith

Executive Editor: Nancy Whilton

Director of Development: Michael Gillespie

Senior Project Editor: Martha Steele

Editorial Assistant: Dyan Menezes

Managing Editor: Corinne Benson

Senior Production Supervisor: Shannon Tozier

Production Service: Pre-Press PMG

Illustrations: Rolin Graphics

Text Design: Pre Press-PMG

Cover Design: Derek Bacchus

Cover Production: Seventeenth Street Studios

Manufacturing Manager: Jeff Sargent

Text and Cover Printer: Edwards Brothers Inc.

Cover Image: Mark Madeo Photography. "Formation", four men jumping over a wall at the beach in San Francisco

ISBN-13: 978-0-321-61120-8

ISBN-10: 0-321-61120-9

Addison-Wesley
is an imprint of

www.pearsonhighered.com

1 2 3 4 5 6 7 8 9 10—EB—13 12 11 10 09

Manufactured in the United States of America.

Contents

MEASUREMENTS IN PHYSICS

CONCEPTUAL QUESTIONS

1. **SOLVE** An advantage of using AU is that astronomical distances can be expressed as simple multiples of AUs.
   REFLECT An astronomical unit (abbreviated as AU) is a unit of length equal to the mean distance between the Earth and the Sun. One AU is 149.6 million kilometers (approximately 93 million miles). Some astronomical distances in AU are given below:
   The Earth is 1.00 ± 0.02 AU from the Sun.
   The Moon is 0.0026 ± 0.0001 AU from the Earth.
   Mars is 1.52 ± 0.14 AU from the Sun.
   Jupiter is 5.20 ± 0.05 AU from the Sun.
   Pluto is 39.5 ± 9.8 AU from the Sun.
   108 AU: As of November 16, 2008, Voyager 1 is the farthest of any human-made objects from the Sun.

3. **SOLVE** The disadvantage of using a prototype, a standard piece of metal for the standard kilogram is the fact that the mass of the piece could change its mass over time.
   REFLECT Alternative options for a standard of mass are under consideration.

5. **SOLVE** Different units can describe a quantity with the same dimension. For example speed has the dimensions of distance per time, and can be expressed as m/s, miles/hour, or any other units of distance per time. Units reveal the dimensions of physical quantities.
   REFLECT A dimension analysis of a calculated physical property is very helpful for identifying errors in a calculation. For example, if an area was supposed to be calculated, and the units of the answer are m or m^3, it's clear that a mistake was made.

7. **SOLVE** Physical quantities with different units have to be converted to the same units in order to properly add and subtract them.
   REFLECT In 1999, the Mars climate orbiter spacecraft was lost due a navigation error, which was tracked to different scientific teams using different unit systems.

MULTIPLE-CHOICE PROBLEMS

9. **SOLVE** The answer is (b). Ten million kg are $10 \times 10^6 \ \text{kg} \times \left(\dfrac{1000 \ \text{g}}{1 \ \text{kg}} \right) = 10^{10}$ g.

   REFLECT The aircraft carrier USS Ronald Reagan has an overall length of 1,092 ft or 333 m and weighs around 20628 tons (2.0628×10^{10} g!). It can carry more than 80 aircraft and 5500 sailors at one time.

11. **SOLVE** The answer is (c).

$$13.7 \ \text{Gy} = 13.7 \times 10^9 \ \text{y} = 13.7 \times 10^9 \ \text{y} \times \frac{3600 \ \text{s}}{1 \ \text{h}} \times \frac{24 \ \text{h}}{1 \ \text{day}} \times \frac{365 \ \text{d}}{1 \ \text{y}} = 4.32 \times 10^{17} \ \text{s}$$

   REFLECT Expressing the time in years gives us a better feel for the magnitude of the immense time period, whereas it is difficult to comprehend that large number of seconds, the SI units of time. For everyday events, it is sometimes more convenient to use non-SI units.

13.   SOLVE   The answer is (a).

$$85\frac{\text{mi}}{\text{h}} = 85\frac{\text{mi}}{\text{h}} \times \frac{1602 \text{ m}}{1 \text{ mi}} \times \frac{1 \text{ h}}{3600 \text{ s}} = 38\frac{\text{m}}{\text{s}}$$

REFLECT   A very important point for any driver to remember is that if you double your speed–say from 30 mph to 60 mph–your braking distance does not become twice as long, it becomes *four times* as far.

15.   ORGANIZE AND PLAN   The two surface areas are given by: $A_1 = 4\pi r_1^2$ and $A_2 = 4\pi r_2^2$

Setting $A_1 = 4 \times A_2$ $A_1 = 4 \times A_2$ we obtain for A_1:

$$4A_2 = 4\pi r_1^2$$
$$A_2 = \pi r_1^2$$

Therefore, we obtain for the relationship between the two radii:

$$\pi r_1^2 = 4\pi r_2^2$$
$$r_1 = 2r_2$$

SOLVE   And then for the volume ratio:

$$V_1 = \frac{4}{3}\pi r_1^3 \text{ and } V_2 = \frac{4}{3}\pi r_2^3$$

$$\frac{V_1}{V_2} = \frac{(2r_2)^3}{r_2^3} = 2^3 = 8$$

The answer is (c).

REFLECT   The exponent of 3 that the doubling of the radius has to be raised with to calculate the ratio of the volumes reflects the dimension of volume, i.e., distance cubed.

17.   SOLVE   The answer is (a).

REFLECT   Zeroes before a decimal point are not significant.

PROBLEMS

19.   SOLVE   Expressing in scientific notation: (a) 1.3950×10^4 m (b) 2.46×10^{-5} kg (c) 3.49×10^{-8} s (d) 1.28×10^9 s

REFLECT   Scientific notation helps you to manage very large and small numbers.

21.   SOLVE   We convert from megatons to kg:

$$1 \text{ megaton} = 10^6 \text{ tons} = \frac{1000 \text{ kg}}{1 \text{ ton}} = 10^9 \text{ kg}$$

23.   ORGANIZE AND PLAN   In part (a), we use the volume of a sphere, which is given by $V(\text{sphere}) = \frac{4}{3}\pi r^3$, where r is the radius of the sphere. In part (b), we consider that density is defined as mass per volume.

SOLVE   (a) For Earth we obtain:

$$V(\text{Earth}) = \frac{4}{3}\pi \left(6.371 \times 10^6 \text{ m}\right)^3 = 1.083 \times 10^{21} \text{ m}^3$$

(b) We use the volume from part (a) and the mass given to calculate the average density of Earth, $\overline{\rho}$ (Earth) :

$$\overline{\rho}\,(\text{Earth}) = \frac{m(\text{Earth})}{V(\text{Earth})} = \frac{\left(5.97 \times 10^{24} \text{ kg}\right)}{\left(1.083 \times 10^{21} \text{ m}^3\right)} = 5511.44 \frac{\text{kg}}{\text{m}^3}$$

The average density of Earth is approximately 5.5 times the density of water.

REFLECT   The interior of the Earth, like that of the other terrestrial planets, is divided into layers by their chemical or physical (rheological) properties. The outer layer of the Earth is a chemically-distinct silicate solid crust, which is underlain by a highly viscous solid mantle. The crust is separated from the mantle by the

Mohorovičić discontinuity, and the thickness of the crust varies: averaging 6 km under the oceans and 30–50 km on the continents. The crust and the cold, rigid, top of the upper mantle are collectively known as the lithosphere, and it is of the lithosphere that the tectonic plates are comprised. Beneath the lithosphere is the asthenosphere, a relatively low-viscosity layer on which the lithosphere rides.

25.   **SOLVE**   The ratio of Earth's radius and the height of Mount Everest's summit is:

$$\frac{h(\text{Mount Everest})}{r(\text{Earth})} = \frac{(8847 \text{ m})}{(6.371 \times 10^6 \text{ m})} = 1.39 \times 10^{-3}$$

REFLECT   The height of Mount Everest makes up only about 0.1% of Earth's radius.

SOLVE   (a) The number of Earth's diameters to make up the distance from Earth to the Moon is:

$$\frac{d(\text{Earth to Moon})}{d(\text{Earth diameter})} = \frac{d(\text{Earth to Moon})}{2 \times r(\text{Earth radius})} = \frac{(3.85 \times 10^8 \text{ m})}{2 \times (6.371 \times 10^6 \text{ m})} = 30.2$$

(b) The number of Earth's diameters to make up the distance from Earth to the Sun is:

$$\frac{d(\text{Earth to Sun})}{d(\text{Earth diameter})} = \frac{d(\text{Earth to Sun})}{2 \times r(\text{Earth radius})} = \frac{(1.496 \times 10^{11} \text{ m})}{2 \times (6.371 \times 10^6 \text{ m})} = 11740$$

REFLECT   Earth is the third planet from the Sun, and the largest of the terrestrial planets in the Solar System in terms of diameter, mass, and density.

27.   **SOLVE**   We need to convert mi to m and h to s:

$$v(\text{cheetah}) = \left(70 \frac{\text{mi}}{\text{h}}\right) \times \left(\frac{\text{h}}{3600 \text{ s}}\right) \times \left(\frac{1609 \text{ m}}{\text{mi}}\right) = 31.3 \frac{\text{m}}{\text{s}}$$

REFLECT   Below is a list with the next 4 fastest animals:
Pronghorn Antelope 61 mph (98 km per hour)
Wildebeest 50 mph (80 km per hour)
Lion 50 mph (80 km per hour)
Thomson's Gazelle 50 mph (80 km per hour)

29.   **SOLVE**   We need to convert feet and inches to m and add the lengths:

$$\text{height (Yao)} = 7 \text{ feet} + 6 \text{ inches} = \left(7 \text{ feet} \times \frac{0.3048 \text{ m}}{\text{feet}}\right) + \left(6 \text{ inches} \times \frac{0.0254 \text{ m}}{\text{inches}}\right) = 2.286 \text{ m}$$

REFLECT   Robert Wadlow is confirmed as the tallest male person by the *Guinness Book of World Records* at 8'11.1" (2.72 m).

31.   **SOLVE**   Converting inches to meters gives for the average annual rainfall:

$$\text{average rainfall} = \left(200 \text{ inches} \times \frac{0.0254 \text{ m}}{\text{inch}}\right) = 5.08 \text{ m}$$

REFLECT   The most annual rainfall on Earth is in the Amazon at about 20 m per year!

33.   **SOLVE**   (a) The number of Earth's diameters to make up the distance from Earth to the Moon is:

$$\frac{d(\text{Earth to Moon})}{d(\text{Earth diameter})} = \frac{d(\text{Earth to Moon})}{2 \times r(\text{Earth radius})} = \frac{(3.85 \times 10^8 \text{ m})}{2 \times (6.371 \times 10^6 \text{ m})} = 30.2$$

(b) The number of Earth's diameters to make up the distance from Earth to the Sun is:

$$\frac{d(\text{Earth to Sun})}{d(\text{Earth diameter})} = \frac{d(\text{Earth to Sun})}{2 \times r(\text{Earth radius})} = \frac{(1.496 \times 10^{11} \text{ m})}{2 \times (6.371 \times 10^6 \text{ m})} = 11740$$

REFLECT   Earth is the third planet from the Sun, and the largest of the terrestrial planets in the Solar System in terms of diameter, mass, and density.

35.   **SOLVE**   The mass for a ^{12}C atom is given by:

$$m\left(^{12}\text{C atom}\right) = 12\frac{\text{g}}{\text{mol}} \times \frac{\text{mol}}{6.02 \times 10^{23}} = 1.99 \times 10^{-23}\text{g}$$

REFLECT   The mass of one ^{238}U atom, the heaviest naturally occurring atom, is 3.95×10^{-22} g.

37.   **SOLVE**   (a) To convert from mi to km we use the fact that 1609 m are 1 mile and 1000 m are in 1 km:

$$\left(\text{factor mi} \rightarrow \text{km}\right) = \frac{\left(1602 \text{ m mi}^{-1}\right)}{\left(1000 \text{ m km}^{-1}\right)} = 1.602 \text{ km mi}^{-1}$$

(b) To convert from kg to μg we use the fact that 1000 g are 1 kg and 10^{-6} g are in 1 μg :

$$\left(\text{factor kg} \rightarrow \mu\text{g}\right) = \left(1000\frac{\text{g}}{\text{kg}}\right) \times \left(\frac{1 \, \mu\text{g}}{10^{-1} \text{ g}}\right) = 10^{-1} \, \mu\text{g kg}^{-1}$$

(c) To convert from km/h to m/s we have to convert km to m and h to s:

$$\left(\text{factor } \frac{\text{km}}{\text{h}} \rightarrow \frac{\text{m}}{\text{s}}\right) = \left(1000\frac{\text{m}}{\text{km}}\right) \times \left(\frac{1 \text{ h}}{3600 \text{ s}}\right) = 0.2778 \text{ m h}^{-1} \text{ km}^{-1}$$

(d) To convert from ft^3 to m^3 we use that 0.3048 m is 1 foot:

$$\left(\text{factor ft}^3 \rightarrow \text{m}^3\right) = \left(\frac{0.3048 \text{ m}}{1 \text{ ft}}\right)^3 = 0.0283 \text{ m}^3 \text{ ft}^{-3}$$

REFLECT   Always convert all physical properties to the same unit system before you use the corresponding numbers in your calculations.

39.   **ORGANIZE AND PLAN**   To solve the problem we use the fact the speed has the dimensions of distance per time. Furthermore, the circumference, Cf, of a circle with radius, r, is given by Cf = 2 π r.

SOLVE   Therefore, we obtain for the speed of the spacecraft with converting the units appropriately:

$$v = \frac{d}{t} = \frac{2 \, \pi \, \left(6378 \text{ km} + 100 \text{ km}\right) \times 10^3 \left(\dfrac{\text{m}}{\text{km}}\right)}{86.5 \text{ min} \times \left(\dfrac{60 \text{ s}}{\text{min}}\right)} = 7842.5\frac{\text{m}}{\text{s}}$$

REFLECT   The speed of the spacecraft corresponds to an incredible 17542 miles/hour!

41.   **SOLVE**   To convert the astronomic distances, d, to AU we use the definition of 1 AU = 1.496×10^{11} m:
(a) Mercury

$$d = 5.76 \times 10^{10} \times \left(\frac{1 \text{ AU}}{1.496 \times 10^{11} \text{ m}}\right) = 0.385 \text{ AU}$$

(b) Mars

$$d = 2.28 \times 10^{10} \times \left(\frac{1 \text{ AU}}{1.496 \times 10^{11} \text{ m}}\right) = 1.524 \text{ AU}$$

(c) Jupiter

$$d = 7.78 \times 10^{11} \times \left(\frac{1 \text{ AU}}{1.496 \times 10^{11} \text{ m}}\right) = 5.200 \text{ AU}$$

(d) Neptune

$$d = 4.50 \times 10^{12} \times \left(\frac{1 \text{ AU}}{1.496 \times 10^{11} \text{ m}}\right) = 30.080 \text{ AU}$$

REFLECT   Using AUs, astronomical distances can easily be expressed as multiples and fractions of the average distance between the Earth and Sun.

43. **SOLVE** (a) We set up two equations for the surface areas of planets A and B, and replace the radius of planet A, r_A with 2 $r_B r_B$:

$$A_A = 4\pi r_A{}^2 = 4\pi \left(2r_B\right)^2 = 16\pi r_B{}^2$$

$$A_B = 4\pi r_B{}^2$$

We now Solve both equations for $r_B{}^2$ and set them equal to obtain the ratio of the surface areas:

$$r_B{}^2 = \frac{A_A}{16\pi} \quad \text{and} \quad r_B{}^2 = \frac{A_B}{4\pi}$$

Then:

$$\frac{A_A}{16\pi} = \frac{A_B}{4\pi}$$
$$A_A = 4A_B$$

(b) We use an equivalent scheme as in a) to obtain the ratio of the volumes, using the equation for the volume of a sphere:

$$V_A = \frac{4}{3}\pi r_A{}^3 = \frac{4}{3}\pi \left(2\, r_B\right)^3 = \frac{32}{3}\pi r_B{}^3$$

$$V_B = \frac{4}{3}\pi r_B{}^3$$

Then:

$$r_B{}^3 = \frac{3V_A}{32\pi} \quad \text{and} \quad r_B{}^3 = \frac{3V_B}{4\pi}$$

$$\frac{3V_A}{32\pi} = \frac{3V_B}{4\pi}$$
$$V_A = 8V_B$$

REFLECT The ratio of the planet's volumes is twice the ratio of their surfaces.

45. **ORGANIZE AND PLAN** To obtain the travel times we use the equation for speed, v, defined as distance per time, and solve for time:

$$v = \frac{d}{t}$$

$$t = \frac{d}{v}$$

We use the speed of light given in the text.

SOLVE (a) The distance between the Moon and Earth is given in Problem 24. With converting the units properly we obtain:

$$t = \frac{d}{v} = \frac{385{,}000 \text{ km} \times \left(\dfrac{1000 \text{ m}}{\text{km}}\right)}{299{,}792{,}458 \text{ m s}^{-1}} = 1.28 \text{ s}$$

(b) The distance from the Sun to Earth is given in Table 1.1. With converting the units properly we obtain:

$$t = \frac{d}{v} = \frac{1.5\times10^{11} \text{ m}}{299{,}792{,}458 \text{ m s}^{-1}} = 500.3 \text{ s}$$

(c) The distance from the Sun to Neptune is given in Problem 41. With converting the units properly we obtain:

$$t = \frac{d}{v} = \frac{4.50\times10^{12} \text{ m}}{299{,}792{,}458 \text{ m s}^{-1}} = 15{,}010.4 \text{ s} = 15{,}010.4 \text{ s}\times\left(\frac{\text{h}}{3600 \text{ s}}\right) = 4.2 \text{ h}$$

REFLECT Considering the incredible speed of light, the 4.2 h of travel time of light from Earth to Neptune indicates the immense size of our solar system.

47. **SOLVE** To answer the question we analyze the dimensions of the constant and solve for time to obtain:

$$k = \frac{m}{t^2}$$

$$t = \sqrt{\frac{m}{k}}$$

This means that the period depends on the square root of the ratio of mass over the constant.

REFLECT In mechanics, and physics, Hooke's law of elasticity is an approximation that states that the extension of a spring is in direct proportion with the load added to it as long as this load does not exceed the elastic limit. Materials for which Hooke's law is a useful approximation are known as linear-elastic or "Hookean" materials.

49. **ORGANIZE AND PLAN** We know that velocity has dimensions of length per time:

$$v = \left[\frac{L}{T} \right]$$

Therefore, we must combine the height and gravitational constant to produce the dimensions of length per time.

SOLVE

$$v = \left[\frac{L}{T} \right] = \left[\sqrt{gh} \right] = \left[\sqrt{\frac{L}{T^2} L} \right] = \left[\frac{L}{T} \right]$$

REFLECT Gravitation is a natural phenomenon by which objects with mass attract one another. In everyday life, gravitation is most commonly thought of as the agency which lends weight to objects with mass. Gravitation compels dispersed matter to coalesce, thus it accounts for the very existence of the Earth, the Sun, and most of the macroscopic objects in the universe. It is responsible for keeping the Earth and the other planets in their orbits around the Sun; for keeping the Moon in its orbit around the Earth, for the formation of tides; for convection (by which fluid flow occurs under the influence of a temperature gradient and gravity); for heating the interiors of forming stars and planets to very high temperatures; and for various other phenomena that we observe. Modern physics describes gravitation using the general theory of relativity, in which gravitation is a consequence of the curvature of spacetime which governs the motion of inertial objects. The simpler Newton's law of universal gravitation provides an excellent approximation for most calculations.

51. **SOLVE** (a) one, (b) three, (c) three, and (d) five.

REFLECT The final answer to a given problem should be rounded to the appropriate number of figures.

53. **ORGANIZE AND PLAN** Labeling the longest side of the right triangle with 25.0 cm of length, c, and the other two shorter sides with lengths 15.0 cm and 20.0 cm, a and b, we obtain for the area of the right triangle:

$$A_{\text{right triangle}} = \frac{1}{2}(a \times b)$$

SOLVE $A_{\text{room}} = \frac{1}{2}(15.0 \text{ cm} \times 20.0 \text{ cm}) = 150 \text{ m}$

REFLECT We apply the same rules for rounding to significant figures when multiplying numbers as described in Problem 52.

55. **ORGANIZE AND PLAN** The volume of a cylinder is given by: $V = \pi r^2 h$.

SOLVE We expand the equation for the density by introducing the equation for the volume of a cylinder:

$$\rho = \frac{m}{V} = \frac{m}{(\pi r^2 h)} = \frac{m}{\pi \left(\frac{d}{2} \right)^2 h} = \frac{27.13 \text{ g}}{8.625 \text{ cm } \pi \left(\frac{1.218 \text{ cm}}{2} \right)^2} = 2.6996 \text{ g cm}^{-3}$$

REFLECT A dimension analysis of our answer indicates that the equation to solve the problem was set up correctly.

57.   **ORGANIZE AND PLAN**   We use the average number of heart beats per minute of 70 from Problem 68, and then use a series of conversion factors to estimate the number of heart beats per lifetime. We further assume an average lifetime of 80 years.

SOLVE $70\dfrac{\text{beats}}{\text{min}} \times \dfrac{60\ \text{min}}{\text{h}} \times \dfrac{24\ \text{h}}{\text{day}} \times \dfrac{365\ \text{days}}{\text{yr}} \times \dfrac{80\ \text{yr}}{\text{lifetime}} = 2.94 \times 10^9\ \dfrac{\text{yr}}{\text{lifetime}}$

The number of heart beats is in the order of 10^9.

REFLECT   Humans live on average 31.99 years in Swaziland and on average 81 years in Japan (2008 est.). The oldest confirmed recorded age for any human is 122 years (see Jeanne Calment), though some people are reported to have lived longer. This is referred to as the life span, which is the upper boundary of life, the maximum number of years an individual can live. The following information is derived from the *Encyclopaedia Britannica*, 1961, as well as other sources and represents estimates of the life expectancies of the population as a whole. In many instances life expectancy varied considerably according to class and gender. It is important to note that life expectancy rises sharply in all cases for those who reach puberty. A pre-20th Century individual who lived past the teenage years could expect to live to an age close to the life expectancy of today. The ages listed below are an average that includes infant mortalities, but not miscarriage or abortion. This table also rejects certain beliefs that the ancient humans had life expectancy of hundreds of years.

59.   **ORGANIZE AND PLAN**   We use the thickness of the entire book and divide by the number of pages to estimate the thickness of one page.

SOLVE   thickness of page $= \dfrac{\text{thickness of book}}{\text{number of pages}} = \dfrac{2.5\ \text{cm}}{400} = 6.25 \times 10^{-3}\ \text{cm}$

The thickness of a page in a 400-page book with a total thickness of 2.5 cm is in the order of 10^{-3} cm.

REFLECT   The term *e-book* is a contraction of "electronic book"; it refers to a digital version of a conventional print book. An e-book is usually made available through the internet, but also on CD-ROM and other forms. E-books are read by means of a physical book display device known as an e-book reader, such as the Sony Reader or the Amazon Kindle. These devices attempt to mimic the experience of reading a print book.

61.   **ORGANIZE AND PLAN**   We use the equation for density and substitute the ratios given for plant's masses and radii.
SOLVE

$$\rho = \frac{m}{V} \text{ with } m_E = 1.23 \times m_E \text{ and } r_E = 1.05 \times r_V$$

$$\frac{\rho_E}{\rho_V} = \frac{\left(\dfrac{m_E}{V_E}\right)}{\left(\dfrac{m_V}{V_V}\right)} = \frac{\left(\dfrac{m_E}{\frac{4}{3}\pi r_E^3}\right)}{\left(\dfrac{m_V}{\frac{4}{3}\pi r_V^3}\right)} = \frac{\left(\dfrac{1.23 \times m_V}{\frac{4}{3}\pi (1.05 \times r_V)^3}\right)}{\left(\dfrac{m_V}{\frac{4}{3}\pi r_V^3}\right)} = \frac{1.23}{1.05} = 1.06$$

REFLECT   The Solar System consists of the Sun and those celestial objects bound to it by gravity, all of which formed from the collapse of a giant molecular cloud approximately 4.5 billion years ago. The Sun's retinue of objects circle it in a nearly flat disc called the ecliptic plane, in which most of the mass is contained within eight relatively solitary planets whose orbits are nearly circular. The four smaller inner planets, Mercury, Venus, Earth and Mars, also called the terrestrial planets, are primarily composed of rock and metal. The four outer planets, Jupiter, Saturn, Uranus and Neptune, also called the gas giants, are composed largely of hydrogen and helium and are far more massive than the terrestrials.

63.   **SOLVE**   (a) To calculate the average density of Saturn we use the equation for density:

$$\rho_S = \frac{m_S}{\frac{4}{3}\pi r_S^3} = \frac{\left(5.69 \times 10^{26}\ \text{kg}\right)}{\frac{4}{3}\pi \left(6.03 \times 10^7\ \text{m}\right)^3} = 619.543\ \text{kg m}^{-3}$$

(b) The density of Saturn is only about 60% of the density of water, reflecting the predominantly gaseous nature of the planet.

REFLECT   The planet Saturn is composed of hydrogen, with small proportions of helium and trace elements. The interior consists of a small core of rock and ice, surrounded by a thick layer of metallic hydrogen and a gaseous outer layer. The outer atmosphere is generally bland in appearance, although long-lived features can appear. Wind speeds on Saturn can reach 1,800 km/h, significantly faster than those on Jupiter. Saturn has a planetary magnetic field intermediate in strength between that of Earth and the more powerful field around Jupiter.

65.   **ORGANIZE AND PLAN**   The average speed is distance per time.

SOLVE   (a) We convert the units given to the SI units of distance and time, m and s, and obtain:

$$v = \frac{s}{t} = \frac{1.5 \text{ mil} \times \left(\dfrac{1\,609.344 \text{ m}}{1 \text{ mil}}\right)}{\left(2 \text{ min} \times \left(\dfrac{60 \text{ s}}{1 \text{ min}}\right) + 24 \text{ s}\right)} = 16.76 \text{ m/s}^{-1}$$

Secretariat ran the race at an amazingly average speed of 16.76 m/s^{-1}.

(b) Repeating the average speed calculation for a human sprinter yields:

$$v = \frac{s}{t} = \frac{100 \text{ m}}{9.8 \text{ s}} = 10.20 \text{ m/s}^{-1}$$

Therefore, the ratio of Secretariat's speed and the speed of a human sprinter is:

$$\frac{v_{\text{Secretariat}}}{v_{\text{humnan sprinter}}} = \frac{16.76 \text{ m/s}^{-1}}{10.20 \text{ m/s}^{-1}} = 1.64$$

REFLECT   Secretariat (March 30, 1970 – October 4, 1989) was an American thoroughbred racehorse. When Secretariat won the 1973 Triple Crown, he became the first Triple Crown winner in 25 years, and set still-standing track records in two of the three races in the Series, the Kentucky Derby (1:59 2/5), and the Belmont Stakes (2:24). Like the famous Man o' War, Secretariat was a large chestnut colt and was given the same nickname, "Big Red."

67.   **SOLVE**   To get the number of water molecules in a water bottle containing 0.500 L we combine the equations for density and molecular mass:

$$\rho = \frac{m}{V} \quad \Rightarrow \quad m = \rho V$$

$$M = \frac{m}{n} \quad \Rightarrow \quad m = nM$$

Therefore:

$$V\rho = nM$$
$$n = \frac{V\rho}{M}$$

The units for density and volume have to be made equivalent, and then finally, to convert from moles to actual water molecules we need to multiply with Avogadro's number:

$$\text{number of molecules} = \frac{3V\rho N_\Lambda}{M}$$

$$\text{number of molecules} = \frac{3(0.500 \text{ L}) \times \left(\dfrac{0.001 \text{ m}^3}{1 \text{ L}}\right) \times (1000 \text{ kg m}^{-3}) \times (6.022 \times 10^{23} \text{ mol}^{-1})}{0.01802 \text{ kg mol}^{-1}} = 5.0127 \times 10^{25}$$

REFLECT   More than 83 moles of water molecules are in the 0.5 L bottle.

69.   **ORGANIZE AND PLAN**   (a) We use the definition of density and solve for mass:

$$\rho = \frac{m}{V} \implies m = \rho V$$

Then we convert units appropriately, include the rate of 15 breaths per minute, and consider that air contains 23% O_2. (b) To get the number of O_2 molecules we use:

$$M = \frac{m}{n} \implies n = \frac{m}{M}$$

$$\text{number of molecules} = nN_A = \frac{m}{M}N_A$$

SOLVE   (a)

$$\text{mass } O_2 \text{ per day} = 1.29 \text{ kg m}^{-3} \times 400 \frac{\text{mL}}{\text{breath}} \times \left(\frac{1 \text{ m}^3}{10^6 \text{ mL}}\right) \times 15 \frac{\text{breaths}}{\text{min}} \times \left(\frac{60 \text{ min}}{1 \text{ h}}\right) \times \left(\frac{24 \text{ h}}{1 \text{ d}}\right) \times 0.23$$

$$\text{mass } O_2 \text{ per day} = 2.5635 \text{ kg d}^{-1}$$

(b) $\text{number of molecules} = \dfrac{\left(2.5635 \text{ kg d}^{-1}\right)}{\left(0.032 \text{ kg mol}^{-1}\right)} \times \left(6.022 \times 10^{23} \text{ mol}^{-1}\right) = 4.824 \times 10^{25} \text{ d}^{-1}$

REFLECT   In nature, free oxygen is produced by the light-driven splitting of water during oxygenic photosynthesis. Green algae and cyanobacteria in marine environments provide about 70% of the free oxygen produced on earth and the rest is produced by terrestrial plants.

2

MOTION IN ONE DIMENSION

CONCEPTUAL QUESTIONS

1. **SOLVE** In one-dimensional motion, displacement is the same as distance traveled when the entire distance traveled is in the same direction, which will also be the direction of the displacement.

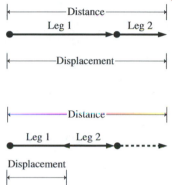

   REFLECT If an object reverses its direction of motion, the distance traveled continues to add up. Since the object is returning to its starting point, the displacement decreases.

3. **SOLVE** Yes. We define a coordinate system along a driveway so that the positive direction is toward the garage and the negative direction is toward the street. Suppose you back your car out of the garage in a straight line at a constant velocity. The velocity is negative. However, velocity is not changing with time, so acceleration is zero.
   REFLECT Since acceleration is the change in velocity with time, $a_x = \dfrac{v_x - v_{x0}}{\Delta t} = 0$. Velocity can be positive, negative or zero. Acceleration can be zero in any of these cases. An object with zero velocity and zero acceleration is at rest.

5. **SOLVE** Yes. If you throw the apple of Question 4 straight upward, it slows down to a velocity of zero at its highest point. The acceleration due to gravity is still negative at this highest point.
   REFLECT Think about a drag race with the positive *x*-direction toward the finish line. At the instant the race starts, the vehicles have zero velocity. At the same instant, the acceleration has a large positive value. As a result, the velocities of the vehicles increase rapidly as they move down the track.

7. **SOLVE**

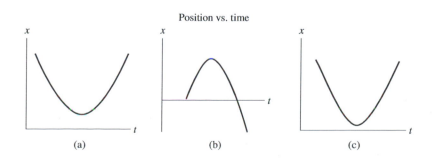

Position vs. time

(a)   (b)   (c)

Acceleration vs. time

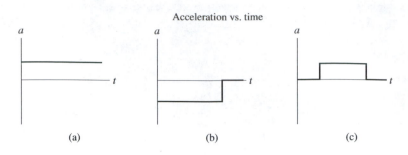

(a)   (b)   (c)

REFLECT   Look at a point on a velocity-versus-time graph. This point gives the slope of the line on the corresponding position-versus-time graph at that same time. The slope of the velocity-versus-time graph gives the location of a point on the acceleration-versus-time graph at that same time.

9.   **SOLVE**   We cannot make any determination about the accelerations of the two cars.

REFLECT   The two cars might both have zero acceleration and the faster car has just caught up to the slower car. Another alternative is that the faster car has zero acceleration while the slower car has a positive acceleration. The faster car passes the slower car, but in a short time, the car that was slower will accelerate to a speed faster than 25 m/s and will pass the first car. This is what happens when you pass a stationary police car while you are driving at constant speed over the speed limit! In fact, there are $3^2 = 9$ combinations of positive, negative and zero acceleration with the two cars.

11.   **SOLVE**   If an object's average velocity is zero, then its displacement must be zero.

$$\Delta x = \overline{v}_x \Delta t$$

For $\overline{v}_x = 0$, $\Delta x = 0 \times \Delta t = 0$ m

If the object's acceleration is zero, then the displacement might be zero, positive, or negative, depending on the value and sign of the velocity.

REFLECT   Even if the object's instantaneous velocity has not been zero throughout the time interval, a velocity of zero implies that it has returned to its starting point, so $\Delta x = 0$. If acceleration is zero and velocity is zero, then the object is at rest and has not moved. If velocity is positive or negative, then displacement constantly increases in the same direction as velocity.

MULTIPLE-CHOICE PROBLEMS

13.   **ORGANIZE AND PLAN**   We are to find average speed. We use the definition of average speed, $\overline{v} = \Delta x / \Delta t$. The responses are all in m/s, so we must convert distance to meters and time to seconds.

Known: $\Delta x = 385{,}000$ km; $\Delta t = 2.5$ day.

SOLVE   Using the definition of speed,

$$\overline{v}_x = \frac{\Delta x}{\Delta t}$$

$$\overline{v}_x = \frac{385{,}000 \text{ km} \times 1000 \text{ m/km}}{2.5 \text{ day} \times 24 \text{ h/day} \times 3600 \text{ s/h}} = 1800 \text{ m/s}$$

The answer is response (C).

REFLECT   The escape velocity from Earth is about 11,000 m/s. A spacecraft traveling toward the moon initially achieves escape velocity, and then slows down due to Earth's gravity. If the intention is to make a "soft" landing on the moon with $v_x = 0$ m/s, it makes sense that the average speed must be less than the escape velocity.

15.   **ORGANIZE AND PLAN**   In this problem, the runner has already run part of the race with known quantities. We have to figure out what's going to happen during the rest of the race. We'll use subscript "1" for the first part of the race, subscript "2" for the remainder of the race, and no subscript if the variable applies to the entire race. We need to find the speed v_2 the runner must maintain until the finish line. We know that it's a 1500 m race,

with 1200 m already run. Since the runner must finish the race in under 4 minutes, our strategy is to find the time already elapsed and then the allowed time remaining. From that, we can calculate the necessary speed.

Known: $\Delta x = 1500$ m; $\Delta x_1 = 1200$ m; $v_1 = 6.14$ m/s; $\Delta t = 4$ min.

SOLVE   For the part of the race already completed, we use

$$\overline{v}_1 = \frac{\Delta x_1}{\Delta t_1}$$

Solving for Δt_1, we get

$$\Delta t_1 = \frac{\Delta x_1}{\overline{v}_1} = \frac{1200 \text{ m}}{6.14 \text{ m/s}} = 195.44 \text{ s}$$

$$\Delta t_2 = 4 \text{ min} \times 60 \text{ s/min} - 195.44 \text{ s} = 44.56 \text{ s}$$

$$\Delta x_2 = 1500 \text{ m} - 1200 \text{ m} = 300 \text{ m}$$

$$\overline{v}_2 = \frac{\Delta x_2}{\Delta t_2} = \frac{300 \text{ m}}{44.56 \text{ s}} = 6.73 \text{ m/s}$$

The answer is response (a).

REFLECT   This is nearly a one-mile race. It is normal for runners to sprint, or run faster, at the end of a long race. The required average speed to finish under 4 minutes is only a little faster than the average speed in the first part of the race, which is reasonable.

17. **ORGANIZE AND PLAN**   This problem is about the properties of displacement in one-dimensional motion and comparison of displacement with distance.

   SOLVE   The definition of displacement in one-dimensional motion is the difference between final and initial positions in a coordinate system:

$$\Delta x = x - x_0$$

   A one-dimensional coordinate system extends without bound in both the positive and negative directions from the origin. Therefore both x and x_0 can have values that are positive, negative, or zero, independently of each other. Since this is the case, the difference between the two values can also be positive, negative or zero.
   The answer is response (b).

   REFLECT   Distance is the sum of the lengths of all the straight-line segments of a trip, without regard to sign. Distance is always positive. The displacement value may be equal to the distance value, but does not have to be. Displacement is never greater than distance.

19. **ORGANIZE AND PLAN**   We must find average acceleration from initial velocity, final velocity and elapsed time. This means we use the definition of average acceleration, $\overline{a}_x = \Delta v_x / \Delta t$.

   Known: $v_{x0} = 1250$ m/s; $v_x = 1870$ m/s; $\Delta t = 35$ s.

   SOLVE   We use the definition of average acceleration

$$\overline{a}_x = \frac{\Delta v_x}{\Delta t} = \frac{v_x - v_{x0}}{\Delta t} = \frac{1870 \text{ m/s} - 1250 \text{ m/s}}{35 \text{ s}} = 17.7 \text{ m/s}^2$$

   The answer is response (c).

   REFLECT   The acceleration of the spacecraft may seem modest, about a 50% increase in velocity. But the velocity values themselves are not important. Only their difference matters, Δv_x. In this case the acceleration is about 180% of that due to gravity. If this spacecraft has human occupants, acceleration of this magnitude would certainly be noticeable. By comparison, a sports car accelerating from zero to 60 miles per hour in four seconds only accelerates at about 68% of that due to gravity.

21. **ORGANIZE AND PLAN**   This problem requires us to find acceleration from initial velocity, final velocity and displacement.

   Known: $v_{x0} = 21.4$ m/s; $v_x = 0$ m/s $\Delta x = 3.75$ cm.

   SOLVE   Since we don't know time, we will make use of the kinematic equation

$$v_x^2 = v_{x0}^2 + 2a_x \Delta x$$

Since Δx is given in centimeters, we need to convert it to meters.

Rearranging to find a_x,

$$\frac{v_x^2 - v_{x0}^2}{2\Delta x} = a_x = \frac{(0 \text{ m/s})^2 - (21.4 \text{ m/s})^2}{2 \times 3.75 \text{ cm} \times (1 \text{ m}/100 \text{ cm})} = -6100 \text{ m/s}^2$$

The answer is response (c).

REFLECT   The arrow has a relatively high speed and stops in a very short distance. This is like an automobile traveling 50 mi/h stopping in about 1.5 inches. This requires a very large negative acceleration. The sign on the acceleration is negative because the initial velocity of the arrow is positive, and it slows down as it travels into the target.

23.   **ORGANIZE AND PLAN**   This is a free-fall problem. Notice that only an algebraic solution is needed. We need to find the effect on final velocity if we double the distance an object falls. We won't use a subscript for the original height, but we'll use the subscript "2" for the doubled height. Since time is not included in the problem, we'll use the kinematic equation $v_y^2 = v_{y0}^2 + 2a_y\Delta y$.

Known: $\Delta y = h$; $v_y = 0$ m/s; $v_y = v$; $a_y = -g$.

SOLVE   Substituting the variables given in the problem, we get

$$v^2 = v_0^2 + 2a_y h$$

Since $v_{y0} = 0$ for both the original height and the doubled height,

$$v^2 = 2ah$$

$$v = \sqrt{2ah}$$

If we now double Δy to the value $2h$, then the new velocity is

$$v_2 = \sqrt{2a_y(2h)} = \sqrt{2}\sqrt{2ah} = \sqrt{2}v$$

The answer is response (c).

REFLECT   Displacement in the y-direction is a function of the square of velocity. It makes sense that velocity is a function of the square root of height during free-fall.

25.   **ORGANIZE AND PLAN**   We are asked to find the shape of the graph of velocity versus time for a moving object under constant acceleration. Since the object is moving, we know that velocity can be positive or negative, but it cannot be zero. No calculation is needed.

Known: $v \neq 0$ m/s; $a_x \neq 0$ m/s^2.

SOLVE   We know that velocity is a function of time and can be expressed as

$$v_x = v_{x0} + a_x\Delta t$$

We rearrange this formula to put it in the slope-intercept form of the equation of a straight line.

$$v_x = a_x\Delta t + v_{x0}$$

We see that a_x is the slope and v_{x0} is the vertical intercept (the intercept on the velocity axis). We are told that $a_x \neq 0$. The slope of the line must be either positive, sloping upward, or negative, sloping downward. Either of these possibilities gives a diagonal line on the graph.

The answer is response (b).

REFLECT   An acceleration of zero would give us a horizontal line (slope = zero). In order for the graph to have the shape of a parabola, velocity would have to be a function of t^2. Rather, it is a function of the first power of t, a linear function.

PROBLEMS

27.   **ORGANIZE AND PLAN**   In this problem we must show the difference between the distance and the displacement for a round trip between two points. We'll use "1" as the subscript for the first part of the trip and "2" as the subscript for the second part of the trip.

Known: $d_1 = 200$ m.

SOLVE   The distance between two points is always positive regardless of the direction traveled. For a round trip to the video store, the distance from your friend's house is $d_1 = 200$ m. The distance from the video store back to your friend's house is also $d_2 = 200$ m . So the total distance for the round trip is

$$d = d_1 + d_2 = 200 \text{ m} + 200 \text{ m} = 400 \text{ m}$$

But for displacement, we must take into account the sign of the direction of travel for each part of the trip. Traveling in the positive direction, from your friend's house to the video store,

$$\Delta x_1 = 200 \text{ m}$$

Returning from the video store in the negative direction,

$$\Delta x_2 = -200 \text{ m}$$

The total displacement for the round trip is

$$\Delta x = \Delta x_1 + \Delta x_2 = 200 \text{ m} + (-200 \text{ m}) = 0 \text{ m}$$

REFLECT   A round trip, with the ending position the same as the starting position, always has a positive distance and a zero displacement.

29.   **ORGANIZE AND PLAN**   In this problem, we have to consider what effect a fractional round trip has on distance and displacement. We set up a coordinate system with the positive direction to the east. We start at Grand Island, so we can declare this position to be x_0. The position of Lincoln will be x.

SOLVE   (a) Here we have 3 round trips from Grand Island to Lincoln. The displacement Δx_1 traveling from Grand Island to Lincoln is

$$\Delta x_1 = x - x_0 = 160 \text{ km} - 0 \text{ km} = 160 \text{ km}$$

Traveling from Lincoln back to Grand Island, the displacement is

$$\Delta x_2 = -160 \text{ km}$$

The total displacement for one round trip is

$$\Delta x = 160 \text{ km} + (-160 \text{ km}) = 0 \text{ km}$$

Therefore, the displacement for three round trips is

$$3\Delta x = 3 \times 0 \text{ km} = 0 \text{ km}$$

The distance traveled from Grand Island to Lincoln is 160km. Since distance is always positive, the distance traveled from Lincoln back to Grand Island is also 160km. The distance traveled in three round trips is

$$d = 3 \times (160 \text{ km} + 160 \text{ km}) = 960 \text{ km}$$

(b) Here we have $3\frac{1}{2}$ round trips. This means three round trips plus one last trip from Grand Island to Lincoln. Since the displacement from (a) for exactly 3 round trips was 0km , the displacement here is the same as one-half of a round trip, or the displacement from Grand Island to Lincoln, so

$$\Delta x = x - x_0 = 160 \text{ km} - 0 \text{ km} = 160 \text{ km}$$

However, the distance for $3\frac{1}{2}$ round trips from Grand Island to Lincoln is

$$d = 3 \times (160 \text{ km} + 160 \text{ km}) + 160 \text{ km} = 7 \times 160 \text{ km} = 1120 \text{ km}$$

(c) This part asks about displacement and distance for $3\frac{3}{4}$ round trips. From (b), the displacement for 3 round trips is zero. During the remaining $\frac{3}{4}$ round trip, we travel from Grand Island to Lincoln in the positive direction, and then halfway back to Grand Island in the negative direction. The displacement for this trip is

$$\Delta x = 160 \text{ km} + \tfrac{1}{2} \times (-160 \text{ km}) = 80 \text{ km}$$

The distance for each leg of the trip is always positive, so

$$d = 7 \times 160 \text{ km} + \tfrac{1}{2} \times 160 \text{ km} = 1200 \text{ km}$$

REFLECT   The displacement for a round trip is always zero. The distance for each leg of a trip is positive, regardless of direction. Think about how often a car's owner must fill the fuel tank just from driving round trips to school!

31. **ORGANIZE AND PLAN**   To calculate elapsed time, we need to know average distance between the Earth and the sun, and the speed of light. We use 3.00×10^8 m/s for the speed of light and 1.50×10^{11} m for the orbital radius of Earth.

Known: distance traveled $=1.50\times10^{11}$ m; $\bar{v}=3.00\times10^8$ m/s.

SOLVE   Using the definition of average speed,

$$\bar{v}=\frac{\text{distance traveled}}{\Delta t}$$

and rearranging for Δt,

$$\Delta t=\frac{\text{distance traveled}}{\bar{v}}=\frac{1.50\times10^{11}\text{ m}}{3.00\times10^8\text{ m/s}}=500\text{ s}$$

REFLECT

One occasionally hears that if the sun suddenly stopped shining, it would take the inhabitants of Earth about eight minutes to realize this. Checking this value,

$$(500\text{ s})\left(\frac{1\text{ min}}{60\text{ s}}\right)=8.33\text{ min or about eight minutes and 20 seconds.}$$

33. **ORGANIZE AND PLAN**   In this problem there are two parts to the motion, at different speeds. We do not simply average the speeds, even though he distances traveled are both 100.m. We must go back to the definition of

average speed, $\bar{v}=\dfrac{\text{distance traveled}}{\Delta t}$. We'll use subscript (1) for the 4.0 m/s run and subscript (2) for the

5.0 m/s run.

Known: $\bar{v}_1=4.0$ m/s; distance traveled$_1=100$.m; $\bar{v}_2=5.0$ m/s; distance traveled$_2=100$.m.

SOLVE   First we find the distance traveled

$$\text{distance traveled}=\text{distance traveled}_1+\text{distance traveled}_2$$

Then we find total time

$$\Delta t=\Delta t_1+\Delta t_2=\frac{\text{distance traveled}_1}{v_1}+\frac{\text{distance traveled}_2}{v_2}$$

Finally we substitute known values into the definition of average speed

$$\bar{v}=\frac{\text{distance traveled}}{\Delta t}=\frac{100.\text{m}+100.\text{m}}{25\text{ s}+20\text{ s}}=4.4\text{ m/s}$$

REFLECT   To understand why we can't just average the speeds, we see that this is a rate problem. We have to add the reciprocals of the individual speeds to get the reciprocal of the average speed:

$$\frac{\Delta t}{200.\text{m}}=\frac{20.\text{s}}{200\text{ m}}+\frac{25\text{ s}}{200\text{ m}}=\frac{45\text{ s}}{200\text{ m}}=\frac{1}{\bar{v}}$$

$$\bar{v}=\frac{200\text{ m}}{45\text{ s}}=4.4\text{ m/s}$$

35. **ORGANIZE AND PLAN**   We are given the distances traveled and the speeds for two legs of a flight, and the layover time between legs. We calculate the times for each leg using the definition of velocity. Then we add the layover time to find Δt. From the total time and total distance, we find average speed. We'll use subscript (1) for the first leg of the flight and subscript (2) for the second leg.

Known: distance traveled$_1=1100$ km; distance traveled$_2=550$ km; $v=800$.km/h; $\Delta t_{\text{layover}}=80$.min.

SOLVE

(a) First we find the time for the first leg of the flight. Notice that distance is in kilometers and time is in hours.

$$\Delta t_1 = \frac{\text{distance traveled}_1}{v_1} = \frac{1100 \text{ km}}{800. \text{km/h}} = 1.375 \text{ h}$$

$$\Delta t_2 = \frac{\text{distance traveled}_2}{v_2} = \frac{550 \text{ km}}{800. \text{km/h}} = 0.688 \text{ h}$$

Then we convert the layover time to hours:

$$\Delta t_{\text{layover}} = (80. \text{min})(1 \text{ h}/60 \text{ min}) = 1.333 \text{ h}$$

$$\Delta t = \Delta t_1 + \Delta t_2 + \Delta t_{\text{layover}} = 1.375 \text{ h} + 0.688 \text{ h} + 1.333 \text{ h} = 3.4 \text{ h}$$

(b) Average speed is total distance divided by total time, so

$$\overline{v} = \frac{\text{distance traveled}_1 + \text{distance traveled}_2}{\Delta t_1 + \Delta t_2 + \Delta t_{layover}} = \frac{1100 \text{km} + 550 \text{km}}{1.375 \text{h} + 0.688 \text{h} + 1.333 \text{h}} = 490 \text{ km/h}$$

REFLECT   This is one kind of problem in physics where it is not necessary to convert all units to SI base or derived units. Since the problem does not ask for specific units, we are free to express speed in km/h and time in hours. These units allow us to use values of reasonable magnitude for travel by air.

37.   **ORGANIZE AND PLAN**   We are to find both speed and velocity. First we must find both distance and displacement. To find total distance, we use the definition of speed to find the distance traveled on each leg of the trip. To find displacement, we take into account the signs of the distances. For this purpose, our coordinate system will establish east as positive and west as negative. We'll use subscript (1) for the first leg and subscript (2) for the second leg.
Known: $v_1 = 210$ km/h east; $\Delta t_1 = 3.0$ h; $v_2 = 170$ km/h west, or -170 km/h; $\Delta t_2 = 2.0$ h.
SOLVE   First we find the displacement for each leg

$$\Delta x_1 = 630 \text{ km}$$

The second leg of the trip is in the negative direction:

$$\Delta x_2 = (-170 \text{ km/h})(2.0 \text{ h}) = -340 \text{ km}$$

$$\Delta x = \Delta x_1 + \Delta x_2 = 630 \text{ km} + (-340 \text{ km}) = 290 \text{ km}$$

Dividing displacement by time to obtain velocity,

$$\overline{v}_x = \frac{\Delta x}{\Delta t} = \frac{290 \text{ km}}{5.0 \text{ h}} = 58 \text{ km/h}$$

Then we find the distance for each leg:

$$\text{distance traveled}_1 = v_1 \Delta t_1 = (210 \text{ km/h}) \times (3.0 \text{ h}) = 630 \text{ km}$$

Remember that distance is always positive, so we have to use the absolute value of velocity

$$\text{distance traveled}_2 = |v_2| \Delta t_2 = (170 \text{ km/h}) \times (2.0 \text{ h}) = 340 \text{ km}$$

The total distance is

$$\text{distance traveled} = 630 \text{ km} + 340 \text{ km} = 970 \text{ km}$$

So the average speed is

$$\overline{v} = \frac{\text{distance traveled}}{\Delta t} = \frac{970 \text{ km}}{5.0 \text{ h}} = 194 \text{ km/h}$$

REFLECT   Since the plane reverses its path and flies back toward its starting point, the distance is great than the displacement and the speed is greater than the velocity.

39.   **ORGANIZE AND PLAN**   This problem emphasizes that velocity takes into account all the elapsed time, not just the time an object is in motion. Here we will use the definition of velocity. The dogsled goes "straight" so we are free to establish our own coordinate system, with the "straight" direction of travel being in the positive *x*-direction. We'll use the subscript (1) for the time the dogsled is in motion, and no subscript for the variables pertaining to the entire 24-hour period.

Known: $\Delta t = 24$ h; $\Delta t_1 = 10$ h; $v_{x1} = 9.5$ m/s.

SOLVE   First, we convert the two known times to seconds.

$$\Delta t = 24 \text{ h} \times 3600 \text{ s/h} = 86,400 \text{ s}$$

$$\Delta t_1 = 10 \text{ h} \times 3600 \text{ s/h} = 36,000 \text{ s}$$

Then we find the displacement Δx_1 during the 10-hour period when the dogsled is moving.

$$\Delta x_1 = v_1 \Delta t_1 = (9.5 \text{ m/s})(36,000 \text{ s}) = 342,000 \text{ m}$$

This gives us velocity.

$$\overline{v}_x = \frac{\Delta x_1}{\Delta t} = \frac{342,000 \text{ m}}{86,400 \text{ s}} = 4.0 \text{ m/s}$$

REFLECT   A velocity of 9.5 m/s is about the highest velocity a human can achieve for short periods of time. No wonder that humans in snowy regions of the Earth use dogsleds for transportation.

41.   **ORGANIZE AND PLAN**   In this problem, we have to find the error in an observation. Error is observed value − true value . The observed value is the speedometer reading, 60.0 mi/h . We must calculate the true speed from the true distance between highway mileposts and the true elapsed time (measured by your clock).

Known: $\overline{v}_{observed} = 60.0$ mi/h; distance traveled$_{true} = 5.00$ mi; (a): $\Delta t_{true} = 4$ min 45 s; (b): $\overline{v}_{true} = 65.0$ mi/h.

SOLVE   (a) First we convert time to hours:

$$\Delta t_{true} = 4.75 \text{ min} \times (1 \text{ h}/60 \text{ min}) = 0.792 \text{ h}$$

Then we calculate our true speed

$$\overline{v} = \frac{\text{distance traveled}_{true}}{\Delta t_{true}} = \frac{5.00 \text{ mi}}{0.0792 \text{ h}} = 63.2 \text{ mi/h}$$

$$\text{error} = \overline{v}_{observed} - \overline{v}_{actual} = 60.0 \text{ mi/h} - 63.2 \text{ mi/h} = -3.2 \text{ mi/h}$$

(b) If our true speed is 65.0 mi/h, then

$$\Delta t = \frac{\text{distance traveled}_{true}}{\overline{v}_{true}} = \frac{5.00 \text{ mi}}{65.0 \text{ mi/h}} = 0.769 \text{ h} = 4 \text{ min } 37 \text{ s}$$

REFLECT   Error is one of many statistical functions that we use as tools to compare data from a limited number of observations to data from a large population.

43.   **ORGANIZE AND PLAN**   In this problem, two moving objects start at different positions. Here we have to use the full form of a kinematic equation for position, not just displacement. We'll establish a coordinate system with both animals traveling in the positive x-direction. The cheetah starts at the origin and the zebra starts at a position of 35 m. Since both animals end up at the same spot and at the same time, we can set the equations for each animal's position equal to each other and solve for Δt. We'll use subscript (1) for the cheetah and subscript (2) for the zebra.

Known: $x_{10} = 0$ m; $x_{20} = 35$ m; $v_{x1} = 30.$m/s; $v_{x2} = 14$ m/s.

SOLVE   This is a constant velocity problem. For the cheetah,

$$x = x_{10} + v_{x1}\Delta t$$

For the zebra,

$$x = x_{20} + v_{x2}\Delta t$$

Setting these equal,

$$x_{10} + v_{x1}\Delta t = x_{20} + v_{x2}\Delta t$$

Solving for Δt ,

$$x_{10} - x_{20} = (v_{x2} - v_{x1})\Delta t$$

$$\Delta t = \frac{(x_{10} - x_{20})}{(v_{x2} - v_{x1})} = \frac{0 \text{ m} - 35 \text{ m}}{14 \text{ m/s} - 30 \text{ m/s}} = 2.2 \text{ s}$$

REFLECT   The answer is just the zebra's initial lead divided by the difference in velocities. Think of the animals on a treadmill that is moving at the zebra's velocity of 14 m/s. To a stationary observer standing on the ground next to the treadmill, the zebra appears to be standing still while the cheetah approaches at 30. m/s − 14 m/s = 16 m/s. The time it takes the cheetah to reach the zebra under these conditions is given by solving the definition of velocity $\Delta x = v\Delta t$. This gives the exact answer we obtained above. We'll discuss this notion of relative motion in a later chapter.

45.   **ORGANIZE AND PLAN**   We are to find an experimental speed of light and compare it with today's accepted value of 3.00×10^8 m/s. We must convert the experimental distance from kilometers to meters and the experimental time from minutes to seconds.

Known: distance traveled=299,000,000 km; $\Delta t = 22$ min.

SOLVE

$$\bar{v} = \frac{\text{distance traveled}}{\Delta t} = \frac{(2.99 \times 10^8 \text{ km})(10^3 \text{ m/km})}{(22 \text{ min})(60 \text{ s/min})} = 2.3 \times 10^8 \text{ m/s}$$

REFLECT   Today we find Römer's value to be significantly in error, low by 70,000,000 m/s, or −23%. We must remember that only shortly before his work, scientists were still considering the speed of light to be infinite. Only about 65 years earlier, Galileo Galilei had developed the modern telescope. Before this, Jupiter's moons had not been seen at all. In 1675, timekeeping devices were still not accurate. Sixty years later, in 1735, John Harrison started his work on the first accurate marine chronometer, which was not completed until a voyage to Jamaica in 1761. Römer's value was quite a scientific feat for his time.

47.   **ORGANIZE AND PLAN**   Here we are to construct a graph of velocity versus time for the 8-second time interval. From Problem 46, we already have the average velocity for each time interval. Velocity is the slope if a line tangent to curve of the position-versus-time graph when Δt is small.

SOLVE   The table below summarizes the data from Problem 46 that we need to construct our graph.

Time interval, s	Midpoint values, s	$\bar{v}_x$
0.0-2.0	1.0	-2.8
2.0-4.0	3.0	0.2
4.0-6.0	5.0	3.8
6.0-8.0	7.0	6.2

Using the values from the table below, we plot the following:

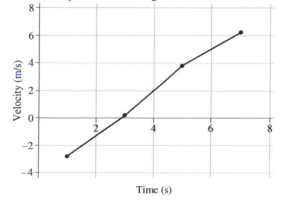

REFLECT   Velocity increases steadily with time, from an initial negative (downward) value to positive values. The original path of the object (from Problem 46) is a parabola, concave upward. It shows an initial downward velocity, decreasing to zero, and then increasing as Δt increases.

49.   **ORGANIZE AND PLAN**   Here we have to draw a graph of instantaneous acceleration. We'll need to look at the graph in Problem 48. This graph has three regions of constant slope. We'll find the slope for each region using $a_x = \dfrac{\Delta v_x}{\Delta t}$.

Known: From $t = 0$ s to $t = 6$ s velocity changes from 0 m/s to 15 m/s. From $t = 8$ s to $t = 12$ s velocity is constant at 18 m/s. From $t = 16$ s to $t = 20$ s velocity decreases from 12 m/s to –3 m/s.

SOLVE   We'll indicate the three line segments using the subscripts 1, 2 and 3. The acceleration values for the three segments are

$$a_{x1} = \frac{\Delta v_x}{\Delta t} = \frac{15 \text{ m/s} - 0 \text{ m/s}}{6 \text{ s}} = 2.5 \text{ m/s}^2$$

$$a_{x2} = \frac{\Delta v_x}{\Delta t} = \frac{18 \text{ m/s} - 18 \text{ m/s}}{12\text{s} - 8 \text{ s}} = 0 \text{ m/s}^2$$

$$a_{x3} = \frac{\Delta v_x}{\Delta t} = \frac{-3 \text{ m/s} - 12 \text{ m/s}}{20 \text{ s} - 16 \text{ s}} = -3.8 \text{ m/s}^2$$

REFLECT   Acceleration is the slope of the graph of velocity versus time. The graph of instantaneous acceleration consists of three horizontal line segments, each indicating a constant acceleration during that time interval.

51.   **ORGANIZE AND PLAN**   Acceleration is the value of the slope of a graph of velocity versus time. We are to consider the graph in problem 48. The greatest acceleration occurs when the slope of the graph is most positive. The least acceleration occurs when the slope is most negative. Acceleration is zero where the graph is a horizontal line. We use the equation $\bar{a}_x = \dfrac{v_x - v_{x0}}{\Delta t}$ to calculate acceleration.

Known: From $t = 0$ s to $t = 6$ s velocity changes from 0 m/s to 15 m/s. From $t = 8$ s to $t = 12$ s velocity is constant at 18 m/s. From $t = 16$ s to $t = 20$ s velocity decreases from 12 m/s to –3 m/s.

SOLVE   (a) Acceleration is greatest between 0s and about 6 s.

(b) Acceleration is least between about 16 s and 20 s.

(c) Acceleration is zero between about 8 s and 12 s.

(d) For the greatest acceleration,

$$a_x = \frac{v_x - v_{x0}}{\Delta t} = \frac{12.5 \text{ m/s} - 0 \text{ m/s}}{5.0 \text{ s}} = 2.5 \text{ m/s}^2.$$

For the least acceleration,

$$a_x = \frac{v_x - v_{x0}}{\Delta t} = \frac{-2.5 \text{ m/s} - 10.0 \text{ m/s}}{20.0\text{s} - 16.5 \text{ s}} = -3.6 \text{ m/s}^2.$$

REFLECT   This motion is like starting from rest in your car, accelerating to some constant speed, realizing that you should have turned at the corner, then slowing down and backing up toward the corner. Actually turning the corner is not included in this problem.

53.   **ORGANIZE AND PLAN**   The stock car's velocity is related to time by the equation $v_x = 1.4t^2 + 1.1t$. We are asked to find v_x after 4.0 s. This means we have to evaluate the given equation at 4.0 s. For the units to cancel properly the value 1.4 must have units of m/s^3 and the value 1.1 must have units of m/s^2.

Known: $t_0 = 0$ s; $t = 4.0$ s.

SOLVE   Substituting 4.0s for t, we get

$$v_x = 1.4t^2 + 1.1t = \left(1.4 \text{ m/s}^3\right)\left(4.0 \text{ s}\right)^2 + \left(1.1 \text{ m/s}^2\right)\left(4.0 \text{ s}\right) = 26.8 \text{ m/s}$$

REFLECT   We know the car starts from rest at time $t = 0$ s since substituting this value into the given equation gives

$$v_x = 1.4t^2 + 1.1t = \left(1.4 \ \text{m/s}^3\right)\left(0 \ \text{s}\right)^2 + \left(1.1 \ \text{m/s}^2\right)\left(0 \ \text{s}\right) = 0 \ \text{m/s}$$

During the time interval of this problem, velocity increases as a quadratic function of time. This situation can't last. The engine's ability to accelerate the car will decrease as engine speed increases past a certain point. Air resistance also reduces acceleration as the car moves faster.

55.   **ORGANIZE AND PLAN**   The instantaneous acceleration is the slope of a line tangent to the graph of velocity versus time. This is the graph we constructed in Problem 54. Taking another look at this graph, we choose the point $t = 2.0$ s to draw the tangent. Then we can find the slope of the tangent using $\dfrac{\Delta v_x}{\Delta t} = \dfrac{v_x - v_{x0}}{t - t_0}$.
Known: $t = 2.0$ s.

SOLVE   Carefully drawing a tangent to the curve at $t = 2.0$ s , we get

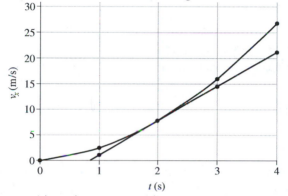

From the ordered pairs $\left(1.0 \ \text{s}, 1.1 \ \text{m/s}\right)$ and $\left(4.0 \ \text{s}, 21.2 \ \text{m/s}\right)$ we calculate the tangent line to have the slope

$$\frac{\Delta v_x}{\Delta t} = \frac{v_x - v_{x0}}{t - t_0} = \frac{21.2 \ \text{m/s} - 1.1 \ \text{m/s}}{4.0 \ \text{s} - 1.0 \ \text{s}} = 6.7 \ \text{m/s}^2$$

REFLECT   This value is the acceleration at the point in time $t = 2.0$ s. We can see from the graph that for any later time up to $t = 4.0$ s the slope, and therefore the acceleration, will be greater.

57.   **ORGANIZE AND PLAN**   We are to draw a motion diagram based on Figure 2.15(b) in the text, which is also shown below. The car speeds up in the negative direction during the first equal time interval. Then it moves with constant velocity during the second equal time interval. Finally the car slows down and stops during the final equal time interval.
Known:

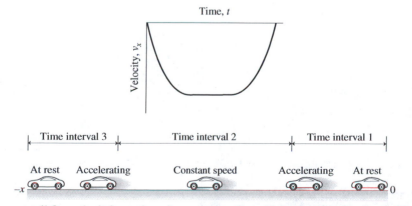

SOLVE

REFLECT   We cannot tell from the information given whether the vehicle is pointed in the positive or negative direction. It doesn't matter, because we model the car as a point, as shown in Section 2.1. We only know that the car is moving in the negative direction, regardless of whether it is in a forward gear or reverse gear.

59.   **ORGANIZE AND PLAN**   We know our initial velocity $v_{x0} = 50.\,\text{km/h}$ and the distance to the stoplight, $\Delta x = 40.\,\text{m}$. We see that we need to convert the initial velocity to m/s to agree with the units of Δx. We need to find both acceleration and how long it takes to stop. We can use $v_x^2 - v_{x0}^2 = 2\bar{a}_x \Delta x$ to find the acceleration. Then, knowing $\bar{a}_x$, we use $\bar{a}_x = \dfrac{v_x - v_{x0}}{\Delta t}$ to find the stopping time, Δt.

Known: $\Delta x = 40.\,\text{m};\ v_{x0} = 50\,\text{km/h}$.

SOLVE   First, convert the initial velocity to m/s

$$50.\,\text{km/h}\left(\frac{1000\text{m}}{1\text{km}}\right)\left(\frac{1\text{h}}{3600\text{s}}\right) = 13.9\,\text{m/s}$$

(a) Then rearranging

$$v_x^2 - v_{x0}^2 = 2\bar{a}_x \Delta x$$

We get

$$\bar{a}_x = \frac{v_x^2 - v_{x0}^2}{2\Delta x} = \frac{0\ \text{m/s} - 13.9\ \text{m/s}}{2(40.\text{m})} = -2.4\ \text{m/s}^2$$

(b) To find the time it takes to stop, we rearrange the definition of acceleration

$$\bar{a}_x = \frac{v_x - v_{x0}}{\Delta t}$$

$$\Delta t = \frac{v_x - v_{x0}}{\bar{a}_x} = \frac{0 - 13.9\ \text{m/s}}{-2.4\ \text{m/s}^2} = 5.8\ \text{s}$$

The acceleration is about the same as that of a car accelerating from 0 to 60 mi/h in 11 seconds. This seems to be a reasonable acceleration.

REFLECT   According to the sign of our initially velocity, we are traveling in the positive x-direction. We are slowing down, so the sign of acceleration must be negative, as we calculated.

61.   **ORGANIZE AND PLAN**   This is a comparison problem. We are given the distance to the hole. Now we have to calculate how far the ball goes, and compare that value to the distance to the hole. We assume for the moment that the ball comes to rest and use $v_x^2 - v_{x0}^2 = 2\bar{a}_x \Delta x$ with a final velocity of zero. If we find that the ball makes it to the hole, then we can recalculate using the same formula, but instead using $\Delta x = 4.0$ m and solving for final velocity v_x at the position of the hole.

Known: $v_{x0} = 2.52\ \text{m/s};\ a_x = -0.65\ \text{m/s}^2;$ distance to the hole $= 4.0$ m.

SOLVE   (a) To find whether the ball makes it to the hole,

$$v_x^2 - v_{x0}^2 = 2\bar{a}_x \Delta x$$

Rearranging,

$$\Delta x = \frac{v_x^2 - v_{x0}^2}{2\bar{a}_x} = \frac{(0\ \text{m/s})^2 - (2.52\ \text{m/s})^2}{2(-0.65\ \text{m/s}^2)} = 4.9\ \text{m}$$

So the ball does reach the hole!

(b) Now, at the hole,

$$v_x^2 - v_{x0}^2 = 2\bar{a}_x \Delta x$$

$$v_x = \sqrt{2\bar{a}_x \Delta x + v_{x0}^2} = \sqrt{2(-0.65\ \text{m/s}^2)(4.0\ \text{m}) + (2.52\ \text{m/s})^2}$$

$v_x = 1.1\,\text{m/s}$ as the ball passes the hole.

REFLECT   Since the problem tells us that the golfer putts straight toward the hole, we have confidence that the ball goes into the hole at this modest speed. If the ball's speed were too great, it could strike the far edge of the cup and bounce out. When we study Chapters 3 and 6, we'll learn about projectile motion and collisions, which will help us understand why the ball might not end up in the cup!

63.   **ORGANIZE AND PLAN**   We have a car that speeds up from some initial speed, then slows down and stops in two distinct time intervals. We need to find the car's maximum speed, and the total time and distance it travels. We'll use the subscripts 1 and 2 to represent what happens during the first and second time intervals. Between the time intervals, acceleration changes from a_{x1} to a_{x2}. We can use the formulas $v = v_0 + a_x \Delta t$ and distance traveled $= v_0 \Delta t + \frac{1}{2} a_x (\Delta t)^2$. We have to be careful because the car slows down to less than its original speed.

Known: $\Delta t_1 = 6.2\text{s}$; $a_{x1} = 1.9 \text{ m/s}^2$; $a_{x2} = -1.2 \text{ m/s}^2$; $v_0 = 13.5 \text{ m/s}$; $v_2 = 0 \text{ m/s}$.

SOLVE   (a) The car's maximum speed occurs at the end of the first time interval, that is, after the maximum time with positive acceleration:

$$v_1 = v_0 + a_x \Delta t = 13.5 \text{ m/s} + \left(1.9 \text{ m/s}^2\right)(6.2 \text{ s}) = 25.3 \text{ m/s}$$

(b) We use the answer to (a) to find the length of the second time interval:

$$\Delta t = \frac{v_2 - v_1}{a_{x2}} = \frac{0 \text{ m/s} - 25.3 \text{ m/s}}{-1.2 \text{ m/s}^2} = 21.1 \text{ s}$$

$$\Delta t_{\text{total}} = \Delta t_1 + \Delta t_2 = 6.2 \text{ s} + 21.1 \text{ s} = 27.3 \text{ s}$$

(c) Since we now know both time intervals, we can calculate the distance traveled for both intervals:

$$\text{distance traveled}_1 = v_0 \Delta t_1 \frac{1}{2} a_{x1} (\Delta t_1)^2 = (13.5 \text{ m/s})(6.2 \text{ s}) + \frac{1}{2}\left(1.9 \text{ m/s}^2\right)(6.2 \text{ s})^2 = 120.1 \text{ m}$$

$$\text{distance traveled}_2 = v_1 \Delta t_2 \frac{1}{2} a_{x2} (\Delta t_2)^2 = (25.3 \text{ m/s})(21.1 \text{ s}) + \frac{1}{2}\left(-1.2 \text{ m/s}^2\right)(21.1 \text{ s})^2 = 266.7 \text{ m}$$

$$\text{distance traveled} = \text{distance traveled}_1 + \text{distance traveled}_2 = 120.1 \text{ m} + 266.7 \text{ m} = 387 \text{ m}$$

REFLECT   To solve graphically, we could plot distance traveled versus time, and find distance as the area under the line. We'd divide the graph into a trapezoid and a triangle and add the areas of each. The numeric equivalent of this is to multiply the average speed for each time interval by the length of each interval, and add the two values.

65.   **ORGANIZE AND PLAN**   We'll model the bullet as a point at the front of the bullet. Since we know displacement but not time, we can use $v^2 - v_0^2 = 2a_x \Delta x$ for both parts of this problem.

Known: $v_{x0} = 310 \text{ m/s}$; $\Delta x = 5.0 \text{ cm}$; (a) $v_x = 0 \text{ m/s}$; (b) $v_x = 50.\text{m/s}$.

SOLVE   First, we convert the displacement to meters.

$$\Delta x = 5.0 \text{ cm}\left(\frac{1 \text{ m}}{100 \text{ cm}}\right) = 0.050 \text{ m}$$

(a) For the bullet stopping in the target,

$$a_x = \frac{v_x^2 - v_{x0}^2}{2\Delta x} = \frac{(0 \text{ m/s})^2 - (310 \text{ m/s})^2}{2(0.050 \text{ m})} = -9.6 \times 10^5 \text{ m/s}^2$$

(b) For the bullet leaving the target at $50.\text{m/s}$,

$$a_x = \frac{v_x^2 - v_{x0}^2}{2\Delta x} = \frac{(50 \text{ m/s})^2 - (310 \text{ m/s})^2}{2(0.050 \text{ m})} = -9.4 \times 10^5 \text{ m/s}^2$$

REFLECT   These are very large values! Since the acceleration is proportional to the difference of the squares of the velocities, the velocity of the bullet leaving the target would have to be a lot greater to make a significant difference in the acceleration. We have simplified this problem by specifying constant acceleration and modeling the bullet as a point. In fact, the bullet begins its negative acceleration at the moment the front of the bullet enters the target. Its acceleration does not stop until either the bullet stops as in (a) or until the bullet exits the target completely, as in (b). [We can leave out these last three sentences if they are too far afield. The bullet is subsonic, and I'm also leaving out any mention of compressing the bow wave as the bullet approaches the target! -LLS]

67.   **ORGANIZE AND PLAN**   The police cruiser and the speeder must end up at the same position at the same time. The speeder will travel 1.2km. The cruiser will have to travel the 1.2km plus 100m, or 1.3km. We'll use subscript 1 for the speeder and subscript 2 for the cruiser. Since the speeder travels at constant speed, we can directly calculate his time to the state line, Δt , which will be the same for the cruiser. Once we know the time, we can use $\Delta x = v_{x0}\Delta t + \frac{1}{2}a_x(\Delta t)^2$ to find the cruiser's acceleration.

Known: $v_{x1} = 75\,\text{km/h}$; $\Delta x_1 = 1.2\,\text{km}$; $v_{x02} = 0$ m/s; $\Delta x_2 = 1.3$ km.

SOLVE   First, convert 1.2 km and 1.3 km to meters, and 75 km/h to m/s.

$$\Delta x_1 = 1.2\,\text{km}\left(\frac{1000\,\text{m}}{1\,\text{km}}\right) = 1200\,\text{m}$$

Likewise, $\Delta x_2 = 1.3\,\text{km} = 1300\,\text{m}$

$$v_{x1} = 75\ \text{km/h}\left(\frac{1\ \text{m/s}}{3.6\ \text{km/h}}\right) = 20.8\ \text{m/s}$$

Then find the time it takes the speeder to get to the state line:

$$\Delta x_1 = v_{x1}\Delta t$$

$$\Delta t = \frac{\Delta x_1}{v_{x1}} = \frac{1200\ \text{m}}{20.8\ \text{m/s}} = 57.6\ \text{s}$$

Finally, find the acceleration of the cruiser:

$$\Delta x_2 = v_{x0}\Delta t + \frac{1}{2}a_x(\Delta t)^2$$

Since $v_{x2} = 0$ m/s,

$$a_x = \frac{2\Delta x_2}{(\Delta t)^2} = \frac{2(1300\ \text{m})}{(57.6\text{s})^2} = 0.78\ \text{m/s}^2$$

REFLECT   This is a relatively modest acceleration, and the police cruiser should reasonably be able to overtake the speeder at the state line.

69.   **ORGANIZE AND PLAN**   This is a constant acceleration problem. We know the values of three variables, so we can find a solution: Initial velocity, final velocity and displacement. We have the equation $v_x^2 - v_{x0}^2 = 2a_x\Delta x$ to directly find acceleration.

Known: $v_x = 0$ m/s; $v_{x0} = 86.1$ m/s; $\Delta x = 1.00 \times 10^3$ m.

SOLVE

$$v_x^2 - v_{x0}^2 = 2a_x\Delta x$$

Since $v_x = 0$,

$$a_x = \frac{-v_{x0}^2}{2\Delta x} = \frac{-(86.1\ \text{m/s})^2}{2(1.00 \times 10^3\ \text{m})} = -3.71\ \text{m/s}^2$$

REFLECT   Commercial aircraft approach and land at a comparatively high rate of speed due to the size and shape the their wings and also so they can gain altitude and "go around" quickly if something unexpected happens on the runway ahead of them. The wheel brakes would wear out very quickly stopping a plane at this speed. Pilots use a combination of resistance on parts of the wing and reverse thrust by the engines to slow the aircraft down on the runway before applying the brakes.

71.   **ORGANIZE AND PLAN**   This problem requires conversion from English units to SI units. It is a constant acceleration problem. We know initial and final velocity, and elapsed time, so we can use the definition of acceleration to find acceleration. We must also find its displacement. We can do that independently using the average velocity formula $\Delta x = \frac{1}{2}(v_x + v_{x0})\Delta t$.

Known: $v_{x0} = 0$ m/s; $v_x = 60.\text{mi/h}$; $\Delta t = 2.8$ s.

SOLVE   (a) First, convert the final velocity to SI units:

$$v_x = 60.\text{mi/h}\left(\frac{5280\ \text{ft}}{1\ \text{mi}}\right)\left(\frac{0.3048\ \text{m}}{1\ \text{ft}}\right)\left(\frac{1\ \text{h}}{3600\ \text{s}}\right) = 26.8\ \text{m/s}$$

Then,

$$a_x = \frac{v_x - v_{x0}}{\Delta t} = \frac{26.8 \text{ m/s} - 0 \text{ m/s}}{2.8 \text{ s}} = 9.6 \text{ m/s}^2$$

(b) Finding the displacement,

$$\Delta x = \tfrac{1}{2}(v_x + v_{x0})\Delta t = \tfrac{1}{2}(0 \text{ m/s} + 26.8 \text{ m/s})(2.8 \text{ s}) = 38 \text{ m}$$

REFLECT   The acceleration is very nearly that due to gravity, so the effect is somewhat like falling sideways. There are custom-built "performance" automobiles that can accelerate from rest to 60 mi/h in 2.8 s or less!

73.   **ORGANIZE AND PLAN**   We know three of the five possible variables in this constant acceleration problem, allowing us to solve it. We solve for final velocity using $v_y^2 - v_{y0}^2 = 2a_y\Delta y$. We establish a coordinate system in which "down" is negative.
Known: $v_{y0} = 0 \text{ m/s};\ a_y = -g = -9.80 \text{ m/s}^2;\ \Delta y = -9.5 \text{ m}$.
SOLVE   Using the chosen equation,

$$v_y^2 - v_{y0}^2 = 2a_y\Delta y$$

Since $v_{y0} = 0 \text{ m/s}$,

$$v_y = \pm\sqrt{2a_y\Delta y} = \pm\sqrt{2(-9.80 \text{ m/s}^2)(-9.5 \text{ m})}$$

We know that the flowerpot is traveling down, so we choose the negative root:

$$v_y = -14 \text{ m/s}$$

REFLECT   Notice that the displacement and the acceleration are both negative. This gives us a positive number under the radical. In free-fall problems, we must remember that the sign of the displacement is negative if the object falls from a higher to a lower position.

75.   **ORGANIZE AND PLAN**   We know the initial velocity, acceleration and displacement. We'll use the quadratic formula to find time, Δt. We start with $\Delta y = v_{y0}\Delta t + \tfrac{1}{2}a_y(\Delta t)^2$.
Known: $v_{y0} = 0 \text{ m/s};\ a_y = -g = -9.80 \text{ m/s}^2;\ \Delta y = -25.6 \text{ m}$.
SOLVE   Beginning with the kinematic equation

$$\Delta y = v_{y0}\Delta t + \tfrac{1}{2}a_y(\Delta t)^2$$

We rearrange it to the general form of a quadratic equation,

$$\left(\tfrac{1}{2}a_y\right)(\Delta t)^2 + \left(v_{y0}\right)(\Delta t) - \Delta y = 0$$

This is analogous to the commonly taught form of the quadratic formula,

$$ax^2 + bx + c = 0$$

Where $a = \tfrac{1}{2}a_y$, $b = v_{y0}$, and $c = -y$.
The quadratic formula solved for Δt is then

$$\Delta t = \frac{-v_{y0} \pm \sqrt{\left(v_{y0}\right)^2 - 4\left(\tfrac{1}{2}a_y\right)(-y)}}{2\left(\tfrac{1}{2}a_y\right)}$$

Which simplifies to

$$\Delta t = \frac{-v_{y0} \pm \sqrt{v_{y0}^2 + 2a_y y}}{a_y}$$

Choosing the negative root so that time is positive,

$$\Delta t = \frac{0 - \sqrt{(0 \text{ m/s})^2 + 2(-9.80 \text{ m/s}^2)(-25.6 \text{ m})}}{-9.80 \text{ m/s}^2} = 2.29 \text{ s}$$

REFLECT   A time of 2.29s to drop a distance of 84 feet seems reasonable. The answer is easily shown to be correct using $\Delta y = \frac{1}{2} a_y (\Delta t)^2$ knowing that $v_{y0} = 0$ m/s.

77.   **ORGANIZE AND PLAN**   We are to graph the position of the rock and its velocity versus time. We need to know two pieces of information. First, in order to proper scale the horizontal (time) axis, how long will it take the rock to hit the ground? Second, to perhaps simplify our task, what are the shapes of the two graphs? We know initial velocity and acceleration. For any given value of Δt, we can find vertical displacement using $\Delta y = v_{y0} \Delta t + \frac{1}{2} a_y (\Delta t)^2$. We see that this is a quadratic equation that will have the shape of a parabola concave downward. We'll construct a table of values to plot y-position versus time. For velocity, the equation we'll use is $v_y = v_{y0} + a_y \Delta t$. We see that velocity is a linear function of time, and we only need two points to graph this line. *Known:* $v_{y0} = 16.5$ m/s; $a_y = -g = -9.80$ m/s^2.

SOLVE   First, let's find out how long it takes for the rock to return to the ground. Ignoring air resistance, we see that final velocity is just the opposite of initial velocity, and

$$\Delta t = \frac{v_y - v_{y0}}{a_y} = \frac{-16.5 \text{ m/s} - 16.5 \text{ m/s}}{-9.80 \text{ m/s}^2} = 3.37 \text{ s}$$

So a range of 0-4 seconds on the horizontal will produce an acceptable graph for either position or velocity. First, we complete a table of values for Δy and Δt using

$$\Delta y = v_{y0} \Delta t + \frac{1}{2} a_y (\Delta t)^2$$

Δt , s	Δy , m
0	0
0.5	7.03
1.0	11.60
1.5	13.73
2.0	13.40
2.5	10.63
3.0	5.40
3.37	0

Plotting these points and drawing a smooth curve through them, we obtain the graph shown in the figure below.

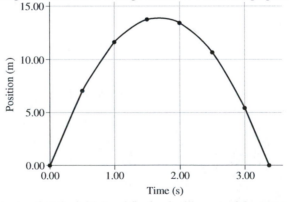

From our calculation of Δt, we see that the initial and final velocities are of the same order of magnitude as the positions. It would be possible to plot both position and velocity on the same graph, but for clarity we will make two separate graphs this time. From that same calculation, we have two points on this linear graph, the ordered pairs $(0s, 16.5 \text{ m/s})$ and $(3.37s, -16.5 \text{ m/s})$. Plotting these points and connecting them with a straight line, we obtain the graph shown in the figure below. We check the linearity with a third point, $(1.50 \text{ s}, 1.80 \text{ m/s})$.

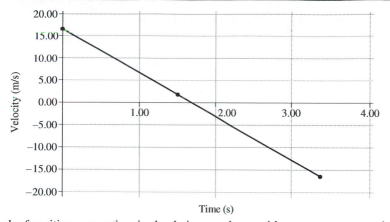

Time (s)

REFLECT The graph of position versus time is clearly in accordance with our common experience. The object goes up, stops momentarily, and descends. The graph of velocity versus time is also reasonable. The rock starts out going up quite fast, slows to a stop at the highest point in its path, and then speeds up as it comes back down.

79. **ORGANIZE AND PLAN** Here we have to find the value of g on a different planet by experimentation! Since we know time and displacement, we can use $\Delta y = v_{y0}\Delta t + \frac{1}{2}(\Delta t)^2$. We'll substitute $-g_{Mars}$ for a_y and solve for g_{Mars}.
Known: $\Delta t = 5.01$ s ; $\Delta y = -45.2$ m.
SOLVE

$$\Delta y = v_{y0}\Delta t + \frac{1}{2}a_y(\Delta t)^2$$

Since $a_y = g_{Mars}$,

$$\Delta y = v_{y0}\Delta t - \frac{1}{2}g_{Mars}(\Delta t)^2$$

The value of $v_{y0} = 0$ because the rock starts from rest:

$$2\Delta y = -g_{Mars}(\Delta t)^2$$

$$g_{Mars} = \frac{2\Delta y}{(\Delta t)^2} = \frac{2(-45.2 \text{ m})}{(5.0 \text{ s})^2} = -3.60 \text{ m/s}^2$$

REFLECT Since g_{Mars} is smaller than g_{Earth}, we expect the rock to take longer to fall than it would on Earth.

The units come out to be m/s^2 so we are confident of our answer.

81. **ORGANIZE AND PLAN** In this problem we have to use the player's maximum height above the floor to calculate time in the air to reach that point. Then we can show that the time interval to fall back to the floor is the same. Adding these gives us total time in the air. Knowing Δt, we graph Δy versus time. We calculate 10 points and complete a table of ordered points $(\Delta t, \Delta y)$.
Known: $\Delta y = -1.1$ m; $v_y = 0$ m/s; $a_y = -g = -9.80$ m/s^2.
SOLVE (a) For the athlete traveling from the playing surface to the highest point in the path,

$$v_y^2 = v_{y0}^2 + 2a_y\Delta y$$

$$v_y^2 - 2a_y\Delta y = v_{y0}^2$$

$$v_{y0} = \sqrt{-2a_y\Delta y} = \sqrt{-2(-9.80 \text{ m/s}^2)(1.1 \text{ m})} = 4.64 \text{ m/s}$$

On the way back down, just before the player touches the floor

$$v_{y0} = -\sqrt{2(-9.80 \text{ m/s}^2)(-1.1 \text{ m})} = -4.64 \text{ m/s}$$

Considering the entire path of the player and using the definition of acceleration,

$$\Delta t = \frac{v_y - v_{y0}}{a_y} = \frac{(-4.64 \text{ m/s}) - (4.64 \text{ m/s})}{-9.80 \text{ m/s}^2} = 0.95 \text{ s}$$

(b) Now that we know Δt for the entire path of the player, we construct a table of values:

$\Delta t, s$	$\Delta y, m$
0	0
0.1	0.42
0.2	0.73
0.3	0.95
0.4	1.02
0.5	1.07
0.6	1.10
0.7	0.85
0.8	0.58
0.9	0.21
0.95	0

Graphing these points, we get the graph we need.

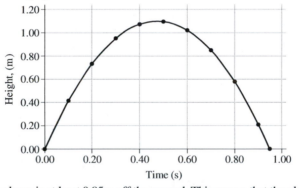

(c) Between 0.3s and 0.7s, the player is at least 0.85m off the ground. This means that the player is at least 77% of maximum height for 43% of the time. This is the reason the player appears to "hang" in the air near the top of the jump.

REFLECT If an object starts and ends at the same y-position, the graph of position versus time is symmetrical about the vertical line $\Delta t = \frac{1}{2}\Delta t_{total}$. Mathematically, we can use $v_y^2 = v_{y0}^2 + 2a_y\Delta y$ to show that $v_y = -v_{y0}$ if $\Delta y = 0$.

83. **ORGANIZE AND PLAN** Each of the two parts here is a constant acceleration problem. The first is a powered ascent for which we'll use the subscript 1. The second is a period of free fall. We'll use the subscript 2 for the period of free fall until maximum height, and the subscript 3 for the descent to the ground. In the first part we'll use $v_y = v_{y0} + 2a_y\Delta t$ to find final velocity when the engine cuts off. We'll also find the height from $v_y^2 = v_{y0}^2 + 2a_y\Delta y$. During free fall, we'll reset $v_y = 0$ m/s and use the same formula to find the change in height between the time the engine cuts off and the rocket reaches maximum height. We'll find Δt for both the powered ascent and free-fall descent. Adding these gives the total time to reach the ground.

Known: For powered ascent: $y_0 = 0$ m; $a_{y1} = 12.6 \text{ m/s}^2$; $\Delta t_1 = 11.0$ s; $v_{y0} = 0$ m/s. For free fall: $a_{y2} = -g = -9.80 \text{ m/s}^2$; $v_{y2} = 0$ m/s.

SOLVE (a) For the powered ascent,

$$v_{y1} = v_{y0} + a_{y1}\Delta t_1 = 0 \text{ m/s} + (12.6 \text{ m/s})(11.0 \text{ s}) = 139 \text{ m/s}$$

(b) For the free fall to maximum height, first we find the change in height after the engine cuts off,

$$v_{y2}^2 = v_{y1}^2 + 2a_{y2}\Delta y$$

$$\Delta y_2 = \frac{-v_{y1}^2}{2a_{y2}} = \frac{-(139 \text{ m/s})^2}{2(-9.80 \text{ m/s}^2)} = 985.8 \text{ m}$$

Now we need the height the rocket reached during powered ascent, Δy_1.

$$\Delta y_1 = v_{y0}\Delta t + \tfrac{1}{2}a_{y1}(\Delta t)^2 = (0 \ \text{m/s})(11.0 \ \text{s}) + \tfrac{1}{2}(12.6 \ \text{m/s}^2)(11.0 \ \text{s})^2 = 762.3 \ \text{m}$$

$$y = y_0 + \Delta y_1 + \Delta y_2 = 0 \ \text{m} + 762.3 \ \text{m} + 985.8 \ \text{m} = 1750 \ \text{m}$$

(c) Now considering the entire free fall, we want to find the velocity when the rocket strikes the Earth.

$$v_{y3}^2 = v_{y1}^2 + 2a_{y2}\Delta y$$

Choosing the negative root,

$$v_{y3} = -\sqrt{(v_{y1})^2 + 2a_{y2}\Delta y_1} = -\sqrt{(139 \ \text{m/s}) + 2(-9.80 \ \text{m/s}^2)(762.3 \ \text{m})} = -185 \ \text{m/s}$$

(d) In order to calculate total time of flight, we need the time the rocket was in free fall.

$$\Delta t_{\text{free-fall}} = \frac{v_{y3} - v_{y1}}{a_{y2}} = \frac{-185 \ \text{m/s} - 139 \ \text{m/s}}{-9.80 \ \text{m/s}^2} = 33.1 \ \text{s}$$

$$\Delta t_{\text{total}} = \Delta t_1 + \Delta t_{\text{free-fall}} = 11.0 \ \text{s} + 33.1 \ \text{s} = 44.1 \ \text{s}$$

REFLECT   In (c) we could have just as easily considered only the descent from maximum height. The velocity as the rocket strikes the Earth is negative since it is traveling in the negative y-direction.

85.   **ORGANIZE AND PLAN**   We aren't told what the height is, so we'll call it y, taking the horizontal surface to be zero on the coordinate system. The vertical distance the ball travels is then Δy. We'll use trigonometry to find the distances the ball rolls down the 30°incline and up the 45°incline, calling them d_1 and d_2. Since the ball rolling uphill is analogous to it rolling downhill, the acceleration a_2 is related to the same quantities, angle θ and Δy, that the initial acceleration is. We'll find an expression for the velocity of the ball at the bottom of the 30° incline and find an expression for the acceleration required to reduce that velocity to zero at the top of the 45° incline. To do this, we'll use $v^2 = v_0^2 + 2ad$ where d is the distance the ball rolls.
Known: $\theta_1 = 30°$; $\theta_2 = 45°$; $a_1 = 3.50 \ \text{m/s}^2$; $v_2 = v_0 = 0 \ \text{m/s}$.
SOLVE   Rolling downhill,

$$v_1^2 = v_0^2 + 2a_1d$$

$$v_1^2 = v_0^2 + 2a_1\frac{\Delta y}{\sin\theta_1}$$

Rolling uphill,

$$v_2^2 = v_1^2 + 2a_2\frac{\Delta y}{\sin\theta_2} \quad \text{and}$$

$$v_1^2 = v_2^2 - 2a_2\left(\frac{\Delta y}{\sin\theta_2}\right)$$

Setting the two expressions for v_1^2 equal,

$$v_0^2 + 2a_1\left(\frac{\Delta x}{\sin\theta_1}\right) = v_2^2 - 2a_2\left(\frac{\Delta x}{\sin\theta_2}\right)$$

Since $v_2 = v_0 = 0 \ \text{m/s}$,

$$\frac{a_1\Delta y}{\sin\theta_1} = \frac{-a_2\Delta y}{\sin\theta_2}$$

Canceling and rearranging for a_2,

$$a_2 = -a_1\left(\frac{\sin\theta_2}{\sin\theta_1}\right) = -3.50 \ \text{m/s}^2\left(\frac{\sin 45°}{\sin 30°}\right) = -4.95 \ \text{m/s}^2$$

REFLECT   The magnitude of the acceleration is greater rolling up the steeper slope. It is also negative, because the ball is slowing down. In this problem, neither the initial nor final motion is parallel to either the *x*- or *y*-axis. Since the angles aren't complementary, there's no advantage to tilting the axes as in Figure 2.2. If it appears that there is not enough information given to find the answer to a problem, the best strategy is to set the problem up and solve it algebraically before trying to substitute known values. The unknown variables will cancel, such as v_0, v_2, and Δy did in this problem. You will be left with a simplified and more easily soluble problem.

87.   ORGANIZE AND PLAN   Now that we know the initial velocity and the time to impact of the wrench, we can graph position versus time and velocity versus time.

Known: $y_{w0} = 20.\text{m}$; $v_{yw0} = 4.0\,\text{m/s}$; $a_{yw} = -g = -9.80 \ \text{m/s}^2$.

SOLVE   To find ordered pairs of position and time, we use $y_w = y_{w0} + v_{yw0}\Delta t + \frac{1}{2}a_{yw}(\Delta t)^2$ between $t = 0$ and $t = 2.5$ s. . To find ordered pairs of time and velocity, we use $v_{yw} = v_{yw0} + a_{yw}\Delta t$.

We construct a table with (arbitrarily) 12 ordered pairs:

$\Delta t, s$	y_w, m	$v_{yw}, \text{m/s}$
0	20.0	4.0
0.2	20.6	2.0
0.4	20.8	0.1
0.6	20.6	-1.9
0.8	20.1	-3.9
1.0	19.1	-5.8
1.2	17.7	-7.8
1.4	16.0	-9.7
1.6	13.9	-11.7
1.8	11.3	-13.6
2.0	8.4	-15.6
2.2	5.1	-17.6
2.5	0	-20.2

Now we can plot position versus time, shown in the first figure.

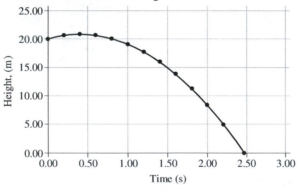

The graph of velocity versus time is shown in the second figure.

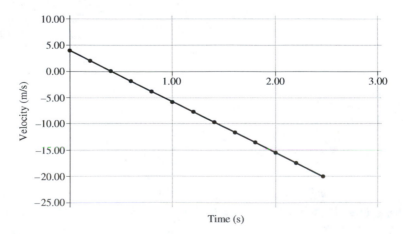

REFLECT   It makes sense that the wrench continues upward for a short time and then its velocity becomes negative, responding to the acceleration of gravity. We notice from the table and graphs that the wrench seems to "hang" in the air near its highest point, like the volleyball player in Problem 81.

89.   **ORGANIZE AND PLAN**   We'll find the final velocity of a cat falling through a distance of 6.4m . We know displacement, initial velocity and acceleration, so we'll use $v_y^2 = v_{y0}^2 + 2a_y\Delta y$. Then we'll use the same formula with the final velocity we just calculated to find the acceleration of the cat as it crouches to a stop.
Known: When the cat is in free fall: $v_{y0} = 0$ m/s; $\Delta y = -6.4$ m; $a_y = -g = -9.80$ m/s^2. As the cat crouches:
$$\Delta y = -14 \text{ cm.}$$

SOLVE   (a) To find final velocity in free fall,
$$v_y^2 = v_{y0}^2 + 2a_y\Delta y$$

Since $v_{y0} = 0$ m/s,
$$v_y = -\sqrt{2a_y\Delta y} = -\sqrt{2\left(-9.80 \text{ m/s}^2\right)\left(-6.4 \text{ m}\right)} = -11 \text{ m/s}$$

(b) First convert Δy to meters:
$$\Delta y = -14 \text{ cm}\left(\frac{1 \text{ m}}{100 \text{ cm}}\right) = -0.14 \text{ m}$$

Then as the cat crouches,
$$v_y^2 = v_{y0}^2 + 2a_y\Delta y$$

Using unrounded values,
$$a_y = \frac{v_y^2 - v_{y0}^2}{2\Delta y} = \frac{\left(0 \text{ m/s}\right)^2 - \left(11.2 \text{ m/s}\right)^2}{2\left(-0.14 \text{ m}\right)} = 448 \text{ m/s}$$

REFLECT   The velocity as the cat lands is negative, as we expect. The cat's acceleration as it crouches is positive, slowing the negative velocity. Cats can endure greater acceleration than humans (see Problem 72).

91.   **ORGANIZE AND PLAN**   Here we know initial and final velocity and displacement. We need to find acceleration. The problem is independent of time, so we use $v_x^2 = v_{x0}^2 + 2a_x\Delta x$. Once we have found a_x we find time using the definition of acceleration, $a_x = \frac{v_x - v_{x0}}{\Delta t}$. In the second part of the problem, we have a constant-velocity period before we accelerate. For this period we use $v_x = v_{x0} + a_x\Delta t$ to find the displacement, Δx_1. Finally we find the acceleration needed to stop in the remaining distance, Δx_2.
Known: (a) and (b) $v_{x0} = 13.4$ m/s; $v_x = 0$ m/s; $\Delta x = 15.0$ m. (c) Constant velocity: $\Delta t = 0.60$ s; $v_x = 0$ m/s.
SOLVE   (a) Rearranging for a_x,
$$v_x^2 = v_{x0}^2 + 2a_x\Delta x$$

$$a_x = \frac{v_x^2 - v_{x0}^2}{2\Delta x} = \frac{\left(0 \text{ m/s}\right)^2 - \left(13.4 \text{ m/s}\right)^2}{2\left(15.0 \text{ m}\right)} = -5.99 \text{ m/s}^2$$

(b) Rearranging for Δt,
$$a_x = \frac{v_x - v_{x0}}{\Delta t}$$

Using the unrounded value for a_x
$$\Delta t = \frac{v_x - v_{x0}}{a_x} = \frac{0 \text{ m/s} - 13.4 \text{ m/s}}{-5.98 \text{ m/s}^2} = 2.24 \text{ s}$$

(c) First we find how far we went in the 0.6s reaction time:
$$\Delta x_1 = v_x\Delta t = \left(13.4 \text{ m/s}\right)\left(0.6 \text{ s}\right) = 8.04 \text{ m}$$

This leaves the distance in which we have to stop as

$$\Delta x_2 = 15.0 \text{ m} - \Delta x_1 = 15.0 \text{ m} - 8.04 \text{ m} = 6.96 \text{ m}$$

Using the same formula as in (a),

$$a_x = \frac{v_x^2 - v_{x0}^2}{2\Delta x} = \frac{(0 \text{ m/s})^2 - (13.4 \text{ m/s})^2}{2(6.96 \text{ m})} = -13 \text{ m/s}^2$$

REFLECT With a positive velocity, we expect negative acceleration in order to slow down. The acceleration in (c) is a rather rapid acceleration, about one-third greater than that of gravity.

93.   **ORGANIZE AND PLAN** We know the values of three variables in this constant acceleration problem. In order to graph velocity and position versus time, we need to find acceleration. We'll use the definition of acceleration,

$a_x = \dfrac{v_x - v_{x0}}{\Delta t}$. Once we know a_x we can construct a table of values to graph, using $v_x^2 = v_{x0}^2 + 2a_x\Delta x$ to find position.

Known: $v_{x0} = 25 \text{ m/s}; \ v_x = 0 \text{ m/s}; \ \Delta t = 10.0 \text{ s}.$

SOLVE (a) Using the definition of acceleration,

$$a_x = \frac{v_x - v_{x0}}{\Delta t} = \frac{0 \text{ m/s} - 25 \text{ m/s}}{10.0 \text{ s}} = -2.5 \text{ m/s}^2$$

Rearranging the definition of acceleration to find velocity,

$$v_x = v_{x0} + a_x\Delta t$$

Calculating v_x at 2.0-s intervals,

Δt, s	$v_{x,\text{m/s}}$
0	25
2.0	20
4.0	15
6.0	10
8.0	5.0
10.0	0

Plotting ordered pairs of $(\Delta t, v_x)$ we get the graph shown in the figure below.

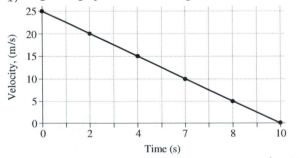

(b) Finding values of Δx at the end of each 2.0-s time interval using $\Delta x = v_{x0}\Delta t + \frac{1}{2}a_x(\Delta t)^2$,

Δt, s	x
0	0
2.0	45
4.0	80
6.0	105
8.0	120
10.	125

Plotting ordered pairs of $(\Delta t, x)$ we get the graph shown in the figure below.

(c) A motion diagram shows position at equal time intervals. We can use the data from the table in (a) to do this:

Displacement, (m)

REFLECT   The slope of the graph in (b) represents the acceleration. The slope is downward and to the right, corresponding to a negative acceleration. Looking at the graph in (c), we see that forward motion continues ever slower, until the vehicle stops.

95.   **ORGANIZE AND PLAN**   In this problem we're asked to find the time(s) when the ball is 5.20 m above its launch point. Those parentheses are important! The only case in which there would be only one time is if the ball is at its maximum height at that time. We can test this by using the quadratic formula to see if we get two positive roots, which are solutions for time. After that, we can use $v_x^2 = v_{x0}^2 + 2a_x\Delta x$ to find the velocities at each of these points in time.

Known: $v_{y0} = 12.1\,\text{m/s}$; $a_y = -g = -9.80\ \text{m/s}^2$; $\Delta y = 5.20$ m.

SOLVE   (a) We start with

$$\Delta x = v_{x0}\Delta t + \tfrac{1}{2}a_x\left(\Delta t\right)^2$$

Rearranging to the general form of a quadratic,

$$\tfrac{1}{2}a_y\left(\Delta t\right)^2 + v_{y0}\Delta t - \Delta y = 0$$

Substituting the coefficients into the quadratic formula,

$$\Delta t = \frac{-v_{y0} \pm \sqrt{\left(v_{y0}\right)^2 - 4\left(\tfrac{1}{2}a_y\right)\left(-\Delta y\right)}}{2\left(\tfrac{1}{2}a_y\right)}$$

Simplifying,

$$\Delta t = \frac{-v_{y0} \pm \sqrt{\left(v_{y0}\right)^2 - 2a_y\left(-\Delta y\right)}}{a_y}$$

Substituting known values,

$$\Delta t = \frac{-12.1\ \text{m/s} \pm \sqrt{\left(12.1\ \text{m/s}\right)^2 - \left(-9.80\ \text{m/s}^2\right)\left(-5.20\ \text{m}\right)}}{-9.80\ \text{m/s}^2}$$

The roots of this equation are

$$\Delta t = \left(\frac{-12.1 + 6.67}{-9.80}\right)\ \text{m/s} = 0.554\ \text{s and}$$

$$\Delta t = \left(\frac{-12.1 - 6.67}{-9.80}\right)\ \text{m/s} = 1.92\ \text{s}$$

(b) To find the final velocity, we use

$$v_y^2 = v_{y0}^2 + 2a_y\Delta y$$

$$v_y = \pm\sqrt{v_{y0}^2 + 2a_y\Delta y} = \pm\sqrt{(12.1 \text{ m/s})^2 + 2(-9.80 \text{ m/s})(5.20 \text{ m})} = \pm 6.67 \text{ m/s}$$

REFLECT   We notice that the expression for final velocity also appears in the numerator of the quadratic formula. We can deduce that on the way back down, the ball will strike the ground at $1.915 \text{ s} + 0.554 \text{ s} = 2.47 \text{ s}$. The 0.554s is the same time interval it took for the ball to reach a height of 5.20 m on the way up.

97. **ORGANIZE AND PLAN**   Here we have to describe the motions of the friends in a frame of reference that is moving with respect to the "stationary" fixed ends of the "moving sidewalk." We'll use subscripts 1 and 2 for the friends, and 3 for the sidewalk.

Known: $v_{x3} = 1.0 \text{ m/s}$ with respect to the fixed ends; $v_{x1} = 4.00 \text{ m/s}$ with respect to the sidewalk; $v_{x2} = -4.0 \text{ m/s}$ with respect to the sidewalk.

SOLVE   From the frame of reference of the sidewalk, each of the two friends, traveling at the same speed, will travel 25m, meeting in the middle of the sidewalk. For each person,

$$\Delta x = v_x \Delta t$$

$$\Delta t = \frac{\Delta x}{\Delta t} = \frac{25 \text{ m}}{4.0 \text{ m/s}} = 6.25 \text{ s}$$

In the same time period, the entire frame of reference of the sidewalk has moved

$$\Delta x_3 = v_{x3}\Delta t = (1.0 \text{ m/s})(6.25 \text{ s}) = 6.25 \text{ m}$$

So in the frame of reference of the fixed ends of the sidewalk, the friends meet at $25 \text{ m} + 6.25 \text{ m} = 31 \text{ m}$ from one of the fixed ends.

An alternative solution is to consider the relative velocities of the two friends with respect to the fixed ends. The first person will be traveling 5.0m/s while the second person will be traveling -3.0 m/s.

$$\Delta x_1 = (5.0 \text{ m/s})(\Delta t)$$

$$\Delta x_2 = (-3.0 \text{ m/s})(\Delta t)$$

$$\frac{\Delta x_1}{5.0 \text{ m/s}} = \Delta t = \frac{\Delta x_1 - 50.\text{m}}{-3.0 \text{ m/s}}$$

$$\Delta x_1 = \frac{(-5.0 \text{ m/s})(50.\text{m})}{-8.0 \text{ m/s}} = 31.25 \text{ m} \cong 31 \text{ m}$$

So the answer is 31m from one end.

REFLECT   This is an example of relationships between the motions of objects compared in two frames of references that are moving at constant velocity with respect to each other.

99. **ORGANIZE AND PLAN**   The runners meet at the same time, Δt. We can consider this a frame-of-reference problem. Suppose that Runner A is standing still and Runner B's initial velocity is 0 m/s. They are still separated by 85 m, but it has become easier to calculate the time interval in which Runner B has traveled the 85m.

Known: $v_{x0} = 0 \text{ m/s}$; $a_x = 0.10 \text{ m/s}^2$; $\Delta x = 85 \text{ m}$; for constant acceleration, $\Delta t = 10.\text{s}$.

SOLVE   First, we calculate how far Runner B travels during constant acceleration.

$$\Delta x = v_{x0}\Delta t + \frac{1}{2}a_x(\Delta t)^2 = (0 \text{ m/s})(10.\text{s}) + \frac{1}{2}(0.10 \text{ m/s}^2)(10.\text{s})^2 = 5.0 \text{ m}$$

After the time interval of $10.\text{s}$, Runner B is traveling

$$v_x = v_{x0} + a_x\Delta t = 0 \text{ m/s} + (0.10 \text{ m/s})(10.\text{s}) = 1.0 \text{ s}$$

Now Runner B has

$$\Delta x = 85 \text{ m} - 5.0 \text{ m} = 80.\text{m}$$

left to run. At $v_x = 1.0 \, \text{m/s}$,

$$\Delta t = \frac{\Delta x}{v_x} = \frac{80. \, \text{m}}{1.0 \, \text{m/s}} = 80. \, \text{s}$$

Runner B's total time to overtake Runner A is then

$$\Delta t_{\text{total}} = \Delta t_a + \Delta t_v = 10. \, \text{s} + 80. \, \text{s} = 90. \, \text{s}$$

REFLECT   We would get exactly the same result if we considered the velocities of the two runners with respect to the track. Notice that we have not taken into account the actual distance traveled with respect to the track. This would become important if the question asked for the winner of the race!

101.   **ORGANIZE AND PLAN**   Here the two trains must end up in the same place at the same time. We will use the kinematic equations $\Delta x = v_x \Delta t$ to find displacement at constant velocity and $\Delta x = v_{x0}\Delta t + \frac{1}{2}a_x\left(\Delta t\right)^2$ to find displacement at constant acceleration. We'll set these two equations equal to each other and solve for Δt. Then we can find where the two trains meet again by substituting Δt into one of the equations for position. We'll indicate variables for the two trains by using subscripts 1 and 2.
Known: $v_{x1} = 11 \, \text{m/s}$; $v_{x02} = 0 \, \text{m/s}$; $a_{x2} = 1.5 \, \text{m/s}^2$.
SOLVE   For the first train,

$$\Delta x_1 = v_{x1}\Delta t$$

$$\Delta t = \frac{\Delta x_1}{v_{x1}}$$

For the second train,

$$\Delta x = v_{x01}\Delta t + \frac{1}{2}a_x\left(\Delta t\right)^2$$

Since $v_{x01} = 0 \, \text{m/s}$,

$$\Delta x = \frac{1}{2}a_x\left(\Delta t\right)^2$$

Setting the two equations equal,

$$v_{x1}\Delta t = \frac{1}{2}a_y\left(\Delta t\right)^2$$

$$0 = \Delta t\left(\frac{1}{2}a_x\Delta t - v_{x1}\right)$$

$\Delta t = 0 \, \text{s}$ when the second train initially passes the first. Therefore, when the two trains meet again,

$$\frac{1}{2}a_x\Delta t = v_{x1}$$

$$\Delta t = \frac{2v_{x1}}{a_x} = \frac{2\left(11 \, \text{m/s}\right)}{1.5 \, \text{m/s}^2} = 14.7 \, \text{s} \cong 15 \, \text{s}$$

Now we substitute Δt into one of the original equations:

$$\Delta x = v_{x1}\Delta t = \left(11 \, \text{m/s}\right)\left(14.7 \, \text{s}\right) = 160 \, \text{m}$$

REFLECT   At an acceleration of $1.5 \, \text{m/s}^2$, the second train would have a velocity equal to that of the first train in a little over $7 \, \text{s}$, even though the second train would still be behind the first. Since the second train is still accelerating, it is reasonable to think that it will overtake the first train in about $15 \, \text{s}$.

103.   **ORGANIZE AND PLAN**   Since we are only given two known variables in this free-fall problem, we must consider the first half of the motion of the ball, where it reaches maximum height and a final velocity of $v_y = 0 \, \text{m/s}$. We'll use $v_y^2 = v_{y0}^2 + 2a_y\Delta y$ to find initial velocity. Then we can use the definition of acceleration to find the time to reach maximum height. From symmetry considerations, we can double the time to find when the ball is caught.
Known: $\Delta y = 17.2 \, \text{m} - 1.6 \, \text{m} = 15.6 \, \text{m}$; $v_y = 0 \, \text{m/s}$; $a_y = -g = -9.80 \, \text{m/s}^2$.

SOLVE   First we find initial velocity:

$$v_y^2 = v_{y0}^2 + 2a_y \Delta y$$

$$v_{y0} = \sqrt{-2a_y \Delta y} = \sqrt{-2\left(9.80 \text{ m/s}^2\right)\left(15.6 \text{ m}\right)} = 17.49 \text{ m/s}$$

Now, to find time to maximum height,

$$\Delta t = \frac{v_y - v_{y0}}{a_y} = \frac{0 \text{m/s} - 17.49 \text{ m/s}}{-9.80 \text{ m/s}^2} = 1.784 \text{ s}$$

Since the ball was caught at the same height at which it was thrown upward, we know from symmetry that the ball takes the same time interval to come back down, and

$$\Delta t_{\text{total}} = 2\left(1.784 \text{ s}\right) = 3.57 \text{ s}$$

With 4.8s on the clock, this leaves 1.2s for the opponent to take a shot.

REFLECT   Realistically, if the opposing player were facing the basket and could make the catch and release a shot in one smooth motion, there might be a chance of a basket. Otherwise the game is over.

3 MOTION IN TWO DIMENSION

CONCEPTUAL QUESTIONS

1. **SOLVE** (a) Surface area is a scalar; (b) position of point on earths surface is a vector; (c) centripetal acceleration is a vector; (d) the number of pages in this book is a scalar.

   REFLECT Vectors require at least two numbers to specify what they represent. Position in space requires the same number of coordinates as dimensions of the space (in the case of a position on the surface of the earth, 3). Centripetal acceleration requires a *magnitude* of the acceleration and a *direction* that "seeks" the center. Scalars require one, and only one number. Area and number of pages are perfect examples.

3. **SOLVE** The magnitude of the vector is a scalar. The direction of a vector in two dimensions is a scalar quantity.

   REFLECT If only one number is required to specify the quantity then the quantity is a scalar. The length of a vector is a scalar. In two dimensions the direction can be given with one angle (e.g., angle from the positive x axis). In other words, it is a scalar!

5. **SOLVE** Yes. Consider the vector $\vec{v} = v_x \hat{i} + v_y \hat{j}$. The magnitude of this vector is $v = \sqrt{v_x^2 + v_y^2}$. If each component is doubled $v_{new} = \sqrt{(2v_x)^2 + (2v_y)^2} = 2\sqrt{v_x^2 + v_y^2} = 2v$.

   REFLECT You see this all the time, for example, zooming into an image on a computer. If the ratio of the height to the width of the image remains fixed, scaling the height of the image by any factor results in every part of the image being scaled by that same factor. If that were not the case, pictures of faces would look very strange indeed when blown up or shrunk down from the original size.

7. **SOLVE** You cannot add a position vector to a velocity vector because they have different units.

   REFLECT Vectors follow the same rules as scalars regarding units. It makes no more sense to ask the question, "What is $4m + 2\frac{m}{s}$?" than that it does to ask "what is 4 gallons + 2 feet?".

9. **SOLVE** An object's average velocity over some time interval does not have to equal it's instantaneous velocity at some instant over the interval. We demonstrate by counter-example: Consider an object in uniform circular motion with period T. The average velocity over one period is $\langle \vec{v} \rangle = \frac{\Delta \vec{r}}{T} = 0$ since the object returned to where it began. Over this interval of time the instantaneous velocity has a constant magnitude greater than zero.

   REFLECT While it is generally false that an object's average velocity over some time interval has to equal it's instantaneous velocity at some instant over the interval, the statement is true for velocity in one dimension. The counter-example above does not work in one dimension because periodic motion in 1-d requires reversing direction which is impossible to do without reducing speed to zero.

11. **SOLVE** The horizontal distance that the projectile travels until it reaches the launch distance is called the range. We can calculate the range by first finding the time it takes to get to the range point given initial velocity in the y direction v_{0y} and noting that the velocity at the range point is $v_{Ty} = v_{0y}$. Using equation 3.19b $v_y = v_{0y} - gt$. we find $-2v_{0y} = -gT$. Solving for T yields $T = \frac{2v_{0y}}{g}$. (the time it takes to reach the range point).

    Given the time in transit to the range point we can calculate the range using eq. 3.18a

    $$x_R = v_{0x}T = \frac{2v_{0y}v_{0x}}{g}.$$

We assume that for any launch angle, the launch speed is the same. Encoding that sentence into an equation gives speed $S^2 = v_{0x}^2 + v_{0y}^2$, or equivalently: $v_{0y} = \sqrt{S^2 - v_{0x}^2}$. Finally we have an expression for the range in terms of one variable only:

$$x_R = \frac{2v_{0x}\sqrt{S^2 - v_{0x}^2}}{g}.$$

Now we can answer the question. What is the launch angle that produces the largest range? Plotting the expression for $x_r(v_{0x})$ produces the figure below which informs us that there is one value of v_{0x} that produces a extreme value of the range. That value corresponds to the value $v_{0x} = \frac{S}{\sqrt{2}}$, which fixes the value of the y-component of the launch velocity $v_{0y} = \frac{S}{\sqrt{2}}$. When $v_{0x} = v_{0y}$, the launch angle is $45°$.

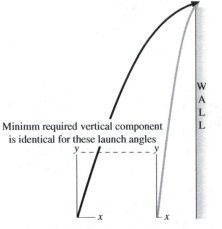

Minimm required vertical component is identical for these launch angles

Larger launch speed   Smaller launch speed

Therefore, the launch angle that maximizes the range is $45°$.

REFLECT   That was some work! A more qualitative explanation is that a large y-component of velocity results in longer flight time, but smaller velocity in the x direction. Small y-component results in larger velocity in the x-direction but less time of travel. The balance between these is when the x and y components of velocity are equal. Equal components result in a launch angle of $45°$.

13.   **SOLVE**   All points on the blade have the same period T, so the velocity on any point of the blade is proportional to the radius (Sec. 3.5). The centripetal acceleration is proportional to the square of the velocity and inversely proportional to the radius, which, combined leaves the centripetal acceleration proportional to the radius. Therefore $a_r(R) = 2a_r(\frac{R}{2})$, i.e., a_r is twice as large on the tip than it is half way out.

REFLECT   This qualitative relationship above is a sense memory for anyone who has played on a merry-go-round. It is harder to keep yourself from flying off if you are on the edge than if you move in toward the center.

MULTIPLE-CHOICE PROBLEMS

15.   **ORGANIZE AND PLAN**   Traveling east a distance x then turning left and heading north a distance y produces a net displacement from the origin. Since you made a $90°$ turn you can use the Pythagorean theorem to find the length of this displacement, i.e., the hypotenuse spanning the right angle formed by the trip yielding $x^2 + y^2 = h^2$ where h is the net displacement from the origin.

SOLVE   Replacing symbols with numbers given in the problem we have

$$h = \sqrt{(30 \text{ km})^2 + (25 \text{ km})^2} = 39 \text{ km}$$

Choice (b) 39 km

REFLECT   The total distance traveled was 55 km. Taking the most direct route possible (*as the crow flies* the saying goes) you would shave 16 km off the trip.

17.   **ORGANIZE AND PLAN**   When given a vector in component form $\vec{v} = v_x\hat{i} + v_y\hat{j}$, determining the angle the vector makes with the positive x-axis requires one step if the vector lies in the first or fourth quadrant, namely employing

the relationship $\theta = \tan^{-1}(\frac{v_y}{v_x})$. If the vector lies in the second or third quadrant then $180°$ must be added to the angle determined in the same manner as the QI and QIV vectors to obtain the correct angle.

The vector in this problem lies in the third quadrant since both components are negative, hence we will have to determine the QI,QIV angle then add $180°$.

SOLVE   The QI angle is $\theta = \tan^{-1}(\frac{4.0 \text{ m}}{2.3 \text{ m}}) = 60°$. Since the vector in the problem lies in the third quadrant we have the final angle as $60° + 180° = 240°$.

Choice (c) $240°$

REFLECT   This problem highlights dangers in applying equations without knowing where they come from. The formula for the angle a vector makes appears straight forward, but one must always remember the $180°$ addition that must be made in QII and QIII.

19.   **ORGANIZE AND PLAN**   The relationship between speed and velocity is straight forward. Velocity is a vector and speed is the length of the vector. The length of a vector (of any units) is given by the Pythagorean theorem:

$$v = |\vec{v}| = |v_x \hat{i} + v_y \hat{j}| = \sqrt{v_x^2 + v_y^2}$$

SOLVE   The instantaneous velocity of the car is given as $11.9 \text{ m/s}\hat{i} + 19.5 \text{ m/s}\hat{j}$. The speed S (magnitude of the velocity vector) is:

$$S = \sqrt{11.9 \text{ m/s}^2 + 19.5 \text{ m/s}^2} = 22.8 \text{ m/s}$$

Choice (a) 22.8 m/s

REFLECT   Selection c is the sum of the magnitudes of the components. This is as large a value as you can get by adding vectors together *and that can only occur when the vectors are in the same direction.* You can always check your vector magnitude solution for craziness by making sure it is larger than each of the components that compose it and smaller than the sum of the magnitudes of the components.

21.   **ORGANIZE AND PLAN**   To calculate the maximum range we shall require both the speed and the launch angle. The determination of the launch angle that produces the maximum range is shown in the Conceptual Questions section Question 10. We found that the range is given for all launch angles by

$$x_R = \frac{2v_{0x}\sqrt{S^2 - v_{0x}^2}}{g}$$

and that the maximum range is obtained for a launch angle of $45°$. For a launch angle of $45°$, $v_{0x} = \cos 45° S = \frac{S}{\sqrt{2}}$ where S is the launch speed. Replacing v_{0x} with $\frac{S}{\sqrt{2}}$ in the expression for the range yields:

$$x_R = \frac{\sqrt{2}S^2\sqrt{1 - 1/2}}{g} = \frac{S^2}{g}$$

SOLVE   Given the speed of 25.5 m/s the range is

$$x_R = \frac{(25.5 \text{ m/s})^2}{9.8 \text{ m/s}^2} = 66.4 \text{ m}$$

Choice (c) $x_R = 66.4$ m

REFLECT   Considering the effort that went into calculating the range, the final expression for the maximum range could be easily memorized. It is helpful to realize that the maximum range calculation in this text over-estimates the range that could be achieved by a ball due to air resistance, and under-estimates the range that could be achieved by an airfoil, for example a Frisbee or glider.

23.   **ORGANIZE AND PLAN**   Horizontally launched means that there is no initial component of the velocity in the vertical direction. We will use the vertical component of the motion to determine the time of flight and then use the time of flight to determine the total horizontal displacement of the motion.

Since the components of the velocity are independent from each other, we decouple the component in the direction of the acceleration from the horizontal velocity. The vertical component of the motion is free-fall from the building. We can determine the time it takes to hit the ground by using equation 3.19a noting that the initial velocity in the $\hat{j}$ direction is zero yielding:

$$y(t_b) = -h = -\frac{1}{2}gt_b^2$$

where h is the height of the building. Solving for $t_b = \sqrt{\dfrac{2h}{g}}$.

In the horizontal direction there is no acceleration so the kinematic equation of motion is $x = v_{0x}t$. Since the projectile was launched horizontally, $v_{0x} = S$ where S is the launch speed (given in the problem $S = 14.0$ m/s). It travels horizontally at this rate until it hits the ground, i.e. until $t = t_b$. Therefore the horizontal displacement will be:

$$x_{hit} = St_b$$

SOLVE   Solving for yields:

$$t_b = \sqrt{\frac{2 \times 12.4 \text{ m}}{9.80 \text{ m/s}^2}} = 1.59 \text{ s}$$

Feeding t_b into our expression for the total horizontal displacement:

$$x_{hit} = 14.0\frac{\text{m}}{\text{s}} \times 1.59 \text{ s} = 22.3 \text{ m}$$

THERE IS NO CORRECT SOLUTION GIVEN AS A CHOICE... CHANGE CHOICE B TO 22.3

REFLECT   A 12.4 m tall building is about a 40 ft tall building (the roof of a 3 story building). The drop time of 1.59 s is consistent with experience which means the horizontal displacement of 22.3 m makes sense. Choice a would correspond to a drop time of 2.5 seconds which seems a little too long. Choices c and d would correspond to drop times less than a second which seem a little too short.

25.   **ORGANIZE AND PLAN**   Centripetal acceleration as a function of radius of motion and the speed is given by equation 3.20: $a_r = \dfrac{v^2}{R}$. If we are given any two of the variables a_r, v or R we can find the remaining variable. In this case, we are given the radius of the track and the maximum centripetal acceleration and we are asked to find the velocity corresponding to the maximum acceleration. Faster than this speed and the tires lose their grip on the road and the car slides.

SOLVE   Solving the for yields:

$$v = \sqrt{Ra_r}$$

with given values:

$$v = \sqrt{875 \text{ m } 3.50 \text{ m/s}^2} = 55.3 \text{ m/s}$$

Choice (b) 55.3 m/s

REFLECT   The speed 55.3 m/s is approximately 200 km/hr which is a decent race car speed. What force would provide this acceleration? A combination of the frictional forces between tire and road as well as the bank (or tilt) of the racetrack.

PROBLEMS

27.   **ORGANIZE AND PLAN**   The best plan to solve this triangle is to work out the labels of the angles. Since the triangle is a right triangle, one angle is easy: 90°. The angle formed by the 4 m side and the 5 m we will call θ_1. The angle formed by the 3 m side and the 5 m will be θ_2.

There are a number of ways this problem can be solved. We shall use the tangent function solely. One could equally use the other trigonometric functions to solve the problem.

SOLVE   Recall that $\tan\theta = \dfrac{opposite}{adjacent}$. Angle θ_1 is adjacent to the 4 m side and opposite to the 3 m side.

Consequently,

$$\tan\theta_1 = \frac{3\text{ m}}{4\text{ m}}$$

In order to isolate the angle we need to apply the inverse tangent function to both sides which yields:

$$\theta_1 = \arctan\frac{3\text{ m}}{4\text{ m}} = 36.9°$$

Angle θ_2 is adjacent to the 3 m side and opposite to the 4 m side. Consequently,

$$\tan\theta_2 = \frac{4\text{ m}}{3\text{ m}}$$

Applying the inverse tangent function to both sides yields:

$$\theta_2 = \arctan\frac{4\text{ m}}{3\text{ m}} = 53.1°$$

REFLECT   As a method of checking our answer we note that adding all interior angles of a triangle must yield 180°. As we would expect 90° + 36.9° + 53.1° = 180°!

29.   **ORGANIZE AND PLAN**   We are given one angle of a right triangle ($\theta = 55°$), the angle adjacent to the shadow which has a length of 1.12 m. We need a relationship between between the angle, the given length and the length of the side opposite the given angle. We recall that $\tan\theta = \dfrac{opp}{adj}$. Solving this relationship for the unknown length we obtain:

$$opp = \tan\theta \times adj$$

SOLVE   Substituting values given in the problem into the relationship above yields:

$$opp = \tan\theta \times adj = 1.12\text{ m} \times \tan 55° = 1.60\text{ m}$$

Height of the person is 1.60 m.

REFLECT   This height corresponds to 5 ft; 3 in. A reasonable height for a human!

31.   **ORGANIZE AND PLAN**   We know one side, S_1, the hypotenuse, and h that the triangle is a right triangle. That is all we need to find everything else we might want to know about this triangle. We first specify the names of the angles and the sides: S_2 will be the unknown side, θ_1 is the angle adjacent to S_2 and θ_2 is the angle adjacent to S_2. We solve for θ_1 using the cosine relation:

$$\cos\theta = \frac{adjacent}{hypotenuse} = \frac{S_1}{h}$$

Isolating for θ_1 yields:

$$\theta_1 = \cos^{-1}\frac{S_1}{h}$$

With two angles in the triangle (θ_1 and 90°) we can find the third by noting that the sum of all interior angles in a triangle must add to give 180°. In other words,

$$\theta_2 = 180° - 90° - \theta_1$$

There are a few ways to find S_2. We can use the Pythagorean theorem directly noting that

$$S_1^2 + S_2^2 = h^2$$

which, upon isolating the unknown S_2 in terms of the known values, yields:

$$S_2 = \sqrt{h^2 - S_1^2}$$

SOLVE   We are given $S_1 = 19.8$ cm and $h = 25.0$ cm. Substituting these values into the expression for θ_1 yields:

$$\theta_1 = \cos^{-1}\frac{19.8\text{ cm}}{25.0\text{ cm}} = 37.6°$$

Using θ_1 and the constraint on the sum of the interior angles yields:

$$\theta_2 = 180° - 90° - 37.6° = 52.4°$$

Finally, the Pythagorean theorem yields:

$$S_2 = \sqrt{(25.0 \text{ cm})^2 - (19.8 \text{ cm})^2} = 31.9 \text{ cm}$$

REFLECT   The theme in all the trig review problems is if you are given two things, two sides, one side and the hypotenuse, an angle and a side then you can find everything else you need. This is useful because it is often easier in practice to obtain directly some measurements and deduce others.

33.   **ORGANIZE AND PLAN**   Traveling h km down a θ down-grade maps to a right triangle that makes an angle of θ with the horizontal and has a hypotenuse of h km. We are asked to find the change in elevation of that stretch (it is a down-grade so our change in elevation will be negative), in other words, the length (y) of the side opposite the known angle in the problem. The trigonometric relation between the known values and the unknown value is the sine function: $\sin \theta = \dfrac{y}{h}$. Isolating for the unknown value yields:

$$y = h \sin \theta$$

SOLVE   Substituting the values in the problem in the above relation yields:

$$y = 5.5 \text{ km} \sin 6° = 575 \text{ m}$$

Therefore, the change in elevation is —575 m

REFLECT   In each problem, the particular trigonometric relationships used arise out of the measurements given. This problem is a little tricky because you have to know you are traveling along the hypotenuse of the triangle.

35.   **ORGANIZE AND PLAN**   Determining the angle a vector makes with the x-axis requires two parts. First, evaluating for the principle value of the angle using equation 3.8b: $\theta = \tan^{-1}\left(\dfrac{y}{x}\right)$. Second, augmenting the principle value: If the vector lies in the second or third quadrant you must add 180° to the principal value to obtain the correct angle. In this problem we have to convert from the cardinal directions to Cartesian coordinates. North and south will be $+\hat{j}$ and $-\hat{j}$ respectively while east and west correspond to $+\hat{i}$ and $-\hat{i}$ respectively.

SOLVE   Converting the English sentence "4.5 km north and 2.3 km west" yields:

$$-2.3 \text{ km}\,\hat{i} + 4.5 \text{ km}\,\hat{j}$$

This vector lies in the second quadrant with a principle angle of

$$\theta_p = \tan^{-1}\frac{-4.5}{2.3} = -62.9°$$

So, for part (a) $\theta = \theta_p + 180° = 117°$.

For part b: 9.9 km west and 3.4 km south translates to

$$-9.9 \text{ km}\,\hat{i} - 3.4 \text{ km}\,\hat{j}$$

This vector lies in the third quadrant with a principle angle of

$$\theta_p = \tan^{-1}\frac{3.4}{9.9} = 19°$$

Since the vector lies in the third quadrant $\theta = 19° + 180°$ or $\theta = 199°$

For part c: 1.2 km east and 4.0 km south translates to

$$1.2 \text{ km}\,\hat{i} - 4.0 \text{ km}\,\hat{j}$$

This vector lies in the fourth quadrant with a principle angle of

$$\theta_p = \tan^{-1}\frac{-4.0}{1.2} = -73.3°$$

The principle angle equals the angle in the first and fourth quadrants so $\theta = -73.3°$ or $\theta = 287°$

REFLECT   It is important to translate the cardinal directions into the proper signed direction in the Cartesian system. Once that is done correctly, the rest is just application of the principle angle expression and remembering the rules for when you have to add 180° and when you don't. It is also helpful to keep a picture of the vectors in your mind so you will know if the angle makes sense. If the magnitude of the y component is larger than the x component, the angle should be between 45° and 135° or between 225° and 315°

37.   **ORGANIZE AND PLAN**   Given vector $\vec{v} = v_x\hat{i} + v_y\hat{j}$ the magnitude is obtained using the Pythagorean Theorem:

$$v = \sqrt{v_x^2 + v_y^2}$$

SOLVE

$$r_1 = \sqrt{r_{1x}^2 + r_{1y}^2} = \sqrt{(2.39 \text{ m})^2 + (-5.07 \text{ m})^2} = 5.61 \text{ m} = r_1$$

$$r_2 = \sqrt{r_{2x}^2 + r_{2y}^2} = \sqrt{(-3.56 \text{ m})^2 + (0.98 \text{ m})^2} = 3.69 \text{ m} = r_2$$

REFLECT   One way to check the answers is to make sure the length of the vector is larger than each component. Also we note that the sign of the component does not have an impact on the length of the whole vector since each component is squared.

39.   **ORGANIZE AND PLAN**   Subtracting vectors is achieved by "combining like terms" where the unit vectors are treated as variables. The difference between vectors $\vec{v}_1$ and $\vec{v}_2$ is a vector given by $\vec{v}_1 - \vec{v}_2 = (v_{1x} - v_{2x})\hat{i} + (v_{1y} - v_{2y})\hat{j}$

SOLVE

$$\vec{r}_1 - \vec{r}_2 = (r_{1x} - r_{2x})\hat{i} + (r_{1y} - r_{2y})\hat{j} = (2.39 + 3.56) \text{ m}\hat{i} + (-5.07 - 0.98) \text{ m}\hat{j} = 5.95 \text{ m}\hat{i} - 6.05 \text{ m}\hat{j} = \vec{r}_1 - \vec{r}_2$$

REFLECT   Sign is important when subtracting components and sometimes can be tricky. You can always check the answer by drawing the vectors tail to tail and obtaining $\vec{r}_1 - \vec{r}_2$ by drawing a vector starting at $\vec{r}_2$ and ending at $\vec{r}_1$. Unlike addition, $\vec{r}_1 - \vec{r}_2 \neq \vec{r}_2 - \vec{r}_1$ so order matters.

41.   **ORGANIZE AND PLAN**   Obtaining the angle between the vectors can be done by taking the difference between the angles the vectors make with the x axis. We know that $\vec{r}_1$ lies in the fourth quadrant and $\vec{r}_2$ lies in the second quadrant.

Recall the principle angle of a vector $\vec{v}$ is given as $\theta = \arctan\dfrac{v_y}{v_x}$ and an angle in the second quadrant is obtained by adding 180° to the principle angle.

SOLVE   Let α_1 and α_2 be the angle that the vectors $\vec{r}_1$ and $\vec{r}_2$ make with the x-axis respectively.

$$\alpha_1 = \arctan\frac{-5.07 \text{ m}}{2.39 \text{ m}} = -65° = 295°$$

$$\alpha_2 = \arctan\frac{-0.98 \text{m}}{-3.56 \text{m}} = -180° = 165°$$

Since $\alpha_1 > \alpha_2$, the angle θ between the two vectors is then

$$\alpha_1 - \alpha_2 = \theta = 130°$$

REFLECT   From the diagram in the previous problem (where vectors were laid tail-to-tail for the graphical subtraction) the angle between the vectors is consistent with the calculated angle.

43.   **ORGANIZE AND PLAN**   Given the magnitude and direction form (or *polar* form) of the vector, the component forms can be extracted out by finding the projection of the vector on the axis using appropriate trigonometric functions. In parts a and b the vectors lie in the first quadrant and have angle measured from the x-axis. In each case, the y component of the vector will be given by $v_y = r\sin\theta$ and the x components will be $v_x = r\cos\theta$ where r is the length of the vector and θ is the angle made with the x-axis.

In part c, the vector is actually given in thinly veiled component form.

SOLVE   (a)

$$v_y = r\sin\theta = 6.4 \text{ m} \sin 80° = 6.30 \text{ m}$$

and

$$v_x = r\cos\theta = 6.4 \text{ m} \cos 80° = 1.11 \text{ m}$$

which yields vector $\vec{A}$ in component form:

$$\vec{A} = 1.11 \text{ m}\hat{i} + 6.30 \text{ m}\hat{j}$$

(b)

$$v_y = r\sin\theta = 13 \text{ m} \sin 30° = 6.5 \text{ m}$$

and
$$v_x = r\cos\theta = 13\,\text{m}\cos 30° = 11.3\,\text{m}$$

which yields vector $\vec{B}$ in component form:
$$\vec{B} = 11.3\,\text{m}\,\hat{i} + 6.5\,\text{m}\,\hat{j}$$

(c) There is no x-component to this vector, it is all -y so vector $\vec{C}$ in component form is
$$\vec{C} = -10\,\text{m}\,\hat{j}$$

REFLECT   Vectors in polar form are easier to imagine in space than the same vector written in Cartesian form. However, in this course, mathematically manipulation of vectors must be done in Cartesian form. Consequently, being able to do these conversions is crucial if you want to be able to manipulate vectors given in polar form.

45.   **ORGANIZE AND PLAN**   Adding the vectors graphically requires first being able to draw a scale drawing (In this problem it is appropriate to let 1 cm = 2 m). Then addition is just moving the tail of one vector to the head of the other and drawing a vector from the tail of one vector to the head of the other.

Adding these vectors in component form first requires a conversion from polar form to Cartesian form, then adding of x components to get x component of the resultant and adding of y components to get y component of resultant. Converting to Cartesian form from polar form $(r, \theta) \rightarrow (r_x, r_y)$ yields:
$$r_x = r\,\cos(\theta)$$
$$r_y = r\,\sin(\theta)$$

Then, adding the vectors in component form looks like this:
$$\vec{S} + \vec{R} = (S_x + R_x)\hat{i} + (S_y + R_y)\hat{j}$$

SOLVE   Graphic solution is shown in figure below

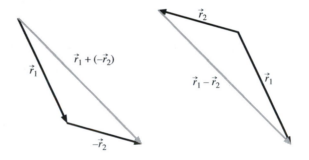

Components of the vectors:
$$S_x = 9.0\,\text{m}\,cos(60°) = 4.5\,\text{m}$$
$$S_y = 9.0\,\text{m}\,sin(60°) = 7.8\,\text{m}$$
$$R_x = 6.0\,\text{m}\,cos(0°) = 6.0\,\text{m}$$
$$R_y = 6.0\,\text{m}\,sin(0°) = 0\,\text{m}$$

With components in hand we can add the vectors:
$$\vec{S} + \vec{R} = (S_x + R_x)\hat{i} + (S_y + R_y)\hat{j}$$
$$= (4.5 + 6.0)\,\text{m}\hat{i} + (7.8)\,\text{m}\hat{j}$$
$$= 10.5\,\text{m}\hat{i} + 7.8\,\text{m}\hat{j}$$

REFLECT   Using the graphical method the resultant vector is 2.2 times the length of $\vec{R}$ or 13.2 m long. The length of the calculated vector is 13.1 m long. Having multiple methods of calculating things is very valuable when the answer is not in the back of the book!

47.   ORGANIZE AND PLAN   This problem is solved just like all the other problems above that involve addition of vectors. However, here we are adding 3 vectors instead of two. The procedure is the same. We write each of the three vectors in component form, we add all the components together and we end up with the resultant vector. In other words:

$$\vec{r}_1 + \vec{r}_2 + \vec{r}_3 = (r_{1x} + r_{2x} + r_{3x})\hat{i} + (r_{1y} + r_{2y} + r_{3y})\hat{j}$$

As always, the components are given by:

$$r_{ix} = r_i \cos(\theta i)$$
$$r_{ix} = r_i \sin(\theta i)$$

SOLVE   The angles to be used in the problem are as follows:

$$\theta_1 = -10° \; \theta_2 = 6° \; \theta_3 = 0$$

The individual x components are then:

$$r_{ix} = 26 \text{ m } \cos(-10°) = 25.6 \text{ m}$$
$$r_{ix} = 15 \text{ m } \cos(6°) = 14.9 \text{ m}$$
$$r_{ix} = 18 \text{ m } \cos(0) = 18 \text{ m}$$

And the individual y components are:

$$r_{ix} = 26 \text{ m } \sin(-10°) = -4.5 \text{ m}$$
$$r_{ix} = 15 \text{ m } \sin(6°) = 1.6 \text{ m}$$
$$r_{ix} = 18 \text{ m } \sin(0) = 0 \text{ m}$$

$$\vec{r}_1 + \vec{r}_2 + \vec{r}_3 = (25.6 \text{ m} + 14.9 \text{ m} + 18 \text{ m})\hat{i} + (-4.5 \text{ m} + 1.6 \text{ m} + 0)\hat{j} = 58.5 \text{ m}\hat{i} - 2.9 \text{ m}\hat{j}$$

So the total displacement is. $58.5 \text{ m}\hat{i} - 2.9 \text{ m}\hat{j}$.

REFLECT   If you can do this for two vectors you can do it for 3 or 4 or ... as many as you want. The biggest source of error as problems like this get more complicated is sign errors and using the right angles.

49.   ORGANIZE AND PLAN   The car's speed at any point is the cars speedometer reading. We are given the time it takes to go around the track once. In the previous problem we calculated the distance travelled over one lap. We obtain the speed by dividing the distance by the time it takes to travel the distance.

The average velocity between two position vectors is the displacement vector divided by the time.

The instantaneous velocity has the magnitude of the speed (we are told its constant around the circle) and the direction tangent to the trajectory.

SOLVE   Obtaining the distance once around the track from last problem:

Car's Speed = 1570 m/55 s = 29 m/s

Average velocity for half lap: We obtain the displacement vector from start to half-lap point from previous problem:

$$\vec{d}_{1/2} = -500 \text{ m}\hat{i}$$

Which yields an average velocity for the half lap of

$$\vec{v}_{1/2} = \frac{-500 \text{ m}}{55 \text{ s}}\hat{i} = -9\frac{\text{m}}{\text{s}}$$

The instantaneous velocity at the end of one lap is directed in the $+\hat{j}$ direction and has the magnitude equal to the speed:

$$\vec{v} = 29 \text{ m}s\hat{j}$$

REFLECT   The moral of the story is to be careful when you read the problem. Words generally have very specific meanings in physics. Get in the habit of using the terms speed, velocity, instantaneous velocity and average velocity in the way physicist use them.

51.   **ORGANIZE AND PLAN**   We obtain average velocity by dividing the displacement vector by the time it takes for the displacement. The displacement vector is given by the end co-ordinates since the snail began at the origin.

The average speed will just be the total distance traveled divided by the time. We will assume the snail has traveled in a straight line.

We shall express everything in units of centimeters per minute.

SOLVE   Displacement vector: $\vec{r} = 5.6 \text{ cm}\hat{i} + 4.3 \text{ cm}\hat{j}$

Average velocity: $\vec{v} = \dfrac{\vec{r}}{\Delta t} = 5.6 \text{ cm/min}\hat{i} + 4.3 \text{ cm/min}\hat{j}$

Average speed: $r/min = \sqrt{5.6^2 + 4.3^2} \text{ cm/min} = 7.1 \text{ cm/min}$

REFLECT   Aside from reinforcing the difference between average velocity and average speed we see in this problem an example of the use of descriptive units. Meters per second would have been technically just as valid but much less transparent.

53.   **ORGANIZE AND PLAN**   Average acceleration is determined by taking the change in velocity over time:

$$\vec{a}_{\text{avg}} = \frac{\vec{v}_f - \vec{v}_i}{\Delta t}$$

We know the initial velocity $\vec{v}_i = 573 \text{ km/hr}\hat{i} + 573 \text{ km/hr}\hat{j}$, the final velocity $\vec{v}_f = -810 \text{ km/hr}\hat{i}$ and the time in seconds.

Because the velocities are in units of km/hr and time is given in seconds we will have to do some unit conversion.

SOLVE

$$\vec{a}_{\text{avg}} = \frac{-810 \text{ km/hr}\hat{j} - (573 \text{ km/hr}\hat{i} + 573 \text{ km/hr}\hat{j})}{45 \text{ s}}$$

$$\vec{a}_{\text{avg}} = \frac{573 \text{ km/hr}\hat{i} - 1383 \text{ km/hr}\hat{j}}{45 \text{ s}}$$

We note 1 $km/hr = 0.278 \ m/s$ so:

$$\vec{a}_{\text{avg}} = \frac{159 \text{ m/s}\hat{i} - 384 \text{ m/s}\hat{j}}{45 \text{ s}}$$

$$\vec{a}_{\text{avg}} = 3.53 \text{ m/s}^2\hat{i} - 8.54 \text{ m/s}^2\hat{j}$$

REFLECT   The magnitude of the acceleration is comparable to the acceleration due to gravity with the largest component in the southerly direction. To be expected from the change in trajectory.

55.   **ORGANIZE AND PLAN**   We are given the time after launch and the velocity at that time. The velocity at 55 s is given in polar coordinates and can be converted to Cartesian as done in earlier problems. The velocity in Cartesian coordinates is given by:

$$\vec{v} = v \cos \theta \hat{i} + v \sin \theta \hat{j}$$

The rocket is launched from rest on the ground. The average acceleration between 0 and 55 s is found by taking the difference between the final velocity and the initial velocity and dividing by the duration.

$$\vec{a}_{\text{avg}} = \frac{\vec{v}_f - \vec{v}_i}{\Delta t} =$$

SOLVE   The velocity is

$$\vec{v} = 950 \text{ m/s} \cos 75°\hat{i} + 950 \text{ m/s} \sin 75°\hat{j} = 246 \text{ m/s}\hat{i} + 918 \text{ m/s}\hat{j}$$

The average acceleration of the time between launch and 55 s is:

$$\vec{a}_{\text{avg}} = \frac{246 \text{ m/s}\hat{i} + 918 \text{ m/s}\hat{j} - 0}{55 \text{ s}} = 4.47 \text{ m/s}^2\hat{i} + 16.7 \text{ m/s}^2\hat{j}$$

REFLECT   What is interesting about this question is what you do not have to know. The path the rocket took is unknown. Under its own power there are many paths possible. However when determining the average acceleration only the end points are required.

57. **ORGANIZE AND PLAN**   We shall define the coordinate system as follows. The trajectory the ball takes to the bat immediately before impact will be the x-axis. We shall denote movement toward the bat as the negative x direction and away from the bat as movement in the positive directions.

The change in the velocity is $\vec{v}_f - \vec{v}_i$. The average acceleration over the contact time of 0.75 ms is obtained by dividing the change in velocity by the contact time.

We will convert milliseconds to seconds in order to get the acceleration in a form we can use for comparison.

SOLVE   Change in velocity is: 36 m/s $\hat{i}$ – 32 m/s($-\hat{i}$) = 68 m/s $\hat{i}$

The average acceleration over the contact time is: $\vec{a}_{avg} = \dfrac{68\ m/s}{0.75\times 10^{-3}\, s} = 91,000\ m/s^2 \hat{i}$

REFLECT   The acceleration is roughly 9300 times the acceleration due to gravity. This large acceleration is due in large part to the very small impact time. If it were less than the acceleration due to gravity we would know we did something wrong!

59. **ORGANIZE AND PLAN**   The acceleration is given in polar form $\vec{a} = (1.15\ \text{m/s}^2, -7.5°)$. The x component is then $a_x = a\times\cos(\theta)$ and the y component is $a_y = a\times\sin(\theta)$

Finding the velocity after 10.0 s undergoing the constant acceleration above is best solved in the polar form. Since the acceleration will always be in the same direction, (the same direction of the velocity), solving this in the polar form reduces the problem to one dimension.

Constant acceleration down the hill allows us to extract the velocity by solving for velocity in the average acceleration relationship. $\Delta\vec{v} = \vec{a}_{avg}\Delta t$

SOLVE   Acceleration written in Cartesian form is:
$$\vec{a} = a_x\hat{i} + a_y\hat{j} = 1.15\ m/s^2 \cos(-7.5°)\hat{i} + 1.15\ m/s^2 \sin(-7.5°) = 1.14\ m/s^2\vec{i} - 0.15\ m/s^2\hat{j}$$

The velocity written in polar form after 10 s assuming zero initial velocity is:
$$\Delta\vec{v} = \vec{v}_f = (1.15\ \text{m/s}^2, -7.5°)10\ s = (11.5\ \text{m/s}, -7.5°)$$

In Cartesian form the final velocity is:
$$\vec{v}_f = 11.4\ \text{m/s}\hat{i} - 1.5\ \text{m/s}\hat{j}$$

The speed is 11.5 m/s

REFLECT   The ability to go back and forth between acceleration and velocity and from velocity to position should start to instill in you a feeling that knowing the acceleration will allow you to find position. We are preparing you to understand the genius of Newton.

61. **ORGANIZE AND PLAN**   Calculating the change in velocity just requires the final velocity minus the initial velocity. The initial and final velocities are given in relation to the side cushion of the billiard table. We first need to convert the velocities in the given form for part a to Cartesian form:
$$\vec{v}_i = 1.80\ \text{m/s}\ \cos 45°(-\hat{i}) + 1.80\ \text{m/s}\ \sin 45°(-\hat{j}) = 1.27\ \text{m/s}(-\hat{i}) + 1.27\ \text{m/s}(-\hat{j}),$$

and the final velocity for part a is given by
$$\vec{v}_f^a = 1.27\ \text{m/s}(-\hat{i}) + 1.27\ \text{m/s}(+\hat{j})$$

For the more realistic case (part b) where the speed is diminished, the final velocity is:
$$\vec{v}_f^b = 1.6\ \text{m/s}\ \cos 45°(-\hat{i}) + 1.60\ \text{m/s}\ \sin 45°(\hat{j}) = 1.13\ \text{m/s}(-\hat{i}) + 1.13\ \text{m/s}(+\hat{j})$$

SOLVE   The change in velocity for part a:
$$\vec{v}_f^a - \vec{v}_i = 2.54\ \text{m/s}\hat{j}$$

The change in velocity for part b:
$$\vec{v}_f^b - \vec{v}_i = 0.14\ \text{m/s}\hat{i} + 2.40\ \text{m/s}\hat{j}$$

REFLECT   The change in velocity is limited to only the direction directly away from the cushion when the speed is unchanged after impact. When there is a speed change after impact, a component along the cushion is present opposite the direction of the initial velocity. If considering average acceleration (which would be in the direction of the change in velocity) interesting questions arise that will be addressed in later chapters.

63.   **ORGANIZE AND PLAN**   Given the initial velocity we can use the relationships derived in question 11 to find the range and the time of flight. The range x_R is given by

$$x_R = \frac{2v_{0y}v_{0x}}{g}.$$

The time of flight was derived to be $T = \frac{2v_{0y}}{g}$.

The maximum height can be determined by considering that half the time time of flight is spent going up and half is spent going down when the launch elevation is the same as the landing elevation. When the projectile is at the apex of the path the component of the velocity in the vertical direction is zero. Using equation 3.19a with zero initial velocity in the direction of acceleration ($y = -\frac{1}{2}gt^2$) and taking the origin of the coordinate system as the apex. The distance of the fall in $t = \frac{T}{2}$ will give us the maximum height.

Known: $\vec{v}_i = 27$ m/s$(\cos 45°\hat{i} + \sin 45°\hat{j})$

SOLVE   The range is:

$$x_R = \frac{2v_{0y}v_{0x}}{g} = \frac{2\times(27)^2 \,\text{m}^2/\text{s}^2}{2\times 9.8 \,\text{m}/\text{s}^2} = 74 \text{ m}$$

The time of flight is:

$$T = \frac{2v_{0y}}{g} = \frac{2\times(27) \,\text{m}/\text{s}}{\sqrt{2}9.8 \,\text{m}/\text{s}^2} = 3.9 \text{ s}$$

The maximum height is:

$$y = \frac{1}{2}gt^2 = \frac{1}{2}9.8 \,\text{m}/\text{s}^2(\frac{3.9 \, s}{2})^2 = 19 \text{ m}$$

REFLECT   The distance between home-plate and the outfield wall is approximately 400 ft. The range found above is consistent with a fly-ball hit into the outfield. The time of flight is also consistent with the experience of watching a ball game and timing the time from bat to glove of a fly-ball.

65.   **ORGANIZE AND PLAN**   **I SUGGEST CHANGING THE DISTANCE STRUCK FROM BASE OF TABLE BE REDUCED FROM 0.65 m TO 0.1 5 m...** Otherwise the height of the table is 24meters... Not really a "table", more like a building.

If we knew the time it took to hit the floor we would be able to use the time, and the initial velocity in the vertical direction (0 in this problem), the acceleration due to gravity and equation 3.19a ($y = v_{0y}t - \frac{1}{2}gt^2$). However, we are not given the time directly, we must extract it from the information given.

As usual with projectile motion problems we will separate the horizontal and vertical components and treat them independently. Since there is no acceleration in the horizontal direction, the initial velocity remains the horizontal component of the velocity throughout the problem. The distance of travel from the base of the table, d will be related to the time of flight t, and the velocity v_{0x} by $v_{0x} \times t = d$. Since both v_{0x} and d are given, the derived value of t is given by $t = \frac{0.15 \text{ m}}{0.30 \text{ m}/\text{s}} = 0.50$ s .

If the ball rolls off the same table at a higher horizontal velocity, the time of flight will remain the same as before since the vertical components of velocity and acceleration are unchanged from part a of the problem. Taking the time of flight and the increased horizontal velocity and using the fact that rate times time equals distance will produce the desired distance.

SOLVE   part a: The distance the ball falls is calculated as

$$y = -\frac{1}{2}gt^2 = -\frac{1}{2}9.8 \,\text{m}/\text{s}^2(0.50 \text{ s})^2 = -1.2 \text{ m}$$

Yielding a height of 1.2 m.

part b: The distance d_b the ball strikes with horizontal velocity of $v_{0x} = 0.60$ m/s is

$$d_b = 0.60 \text{ m}/\text{s} \times 0.5 \text{ s} = 0.30 \text{ m}$$

REFLECT The crucial realization in this problem is that the horizontal and vertical components are completely independent. From that realization the solution of the problem is just a puzzle using the given information to get to the desired information. Beyond the mathematical manipulations, it is beneficial for your understanding of physics as well as your numerical intuition to imagine this occurring in real life (or even try it out for real) to check if the results are consistent with your experience.

67. **ORGANIZE AND PLAN** Again, we treat the horizontal and vertical component separately. This ability allow us to determine the time it takes to hit the ground the same way we would if the problem was one dimensional; namely with equation 3.19a with no initial velocity since the object was launched horizontally and therefore has no initial velocity component in direction of the acceleration.

$$y = -\frac{1}{2}gt^2$$

Isolating for t yields:

$$t = \sqrt{-2y/g}$$

Knowing the time of travel allows us to calculate the horizontal displacement using the $r \times t = d$ relationship. In this case the velocity is the initial velocity and the time is found above.

The final velocity is the vector sum of the horizontal component of the velocity with the vertical velocity. The horizontal velocity is unchanged over the flight. The vertical component of the velocity is obtained via equation 3.19b: $v_y = v_{0y} - gt$. The initial component in the y direction is zero. So the final velocity will be:

$$\vec{v} = v_{0x}\hat{i} - gt\hat{j}$$

SOLVE The elapsed time is:

$$t = \sqrt{2 \times 9.50 \text{ m}/9.8 \text{ m/s}^2} = 1.39 \text{ s}$$

The horizontal distance traveled is:

$$x = 1.39 \text{ s} \times 13.4 \text{ m/s} = 18.7 \text{ m}$$

The final velocity is:

$$\vec{v} = v_{0x}\hat{i} - gt\hat{j} = 13.4 \text{ m/s}\hat{i} - 9.8 \text{ m/s}^2 \times 1.39 \text{ s}\hat{j} = 13.4 \text{ m/s}\hat{i} - 13.6 \text{ m/s}\hat{j}$$

REFLECT This cliff is about the height of a 3 story building. All the results depend on the elapsed time. Aside from checking the units, you can always ask yourself if the answers make sense. A time of 1.39 s is consistent with experience of dropping a ball froma 3 story building.

69. **ORGANIZE AND PLAN** We will find the height above the tip of the hose then add the height above the ground of the hose to obtain the height reached.

As always we will take advantage of the fact that the x and y components are independent. In this case, we only need to consider the vertical components. We will focus our attentions on one water droplet emerging from the hose.

The vertical component of the initial velocity is $v_{0y} = v\sin(75°)$.

Ultimately we want to use equation 3.19a to find the height at the time when the droplet reaches the apex. However, we must determine the time after emergence from the hose to the apex to use 3.19a. We can determine this time by using 3.19b and the fact that the vertical velocity at the apex is zero so the time at apex is $t_m = v_{0y}/g$ With the time in hand, we have what we need to find the max height above the tip of the hose:

$$y_m = v_{0y}t_m - \frac{1}{2}gt_m^2$$

Subbing in t_m yields:

$$y_m = \frac{v_{0y}^2}{2g}$$

Finally, the height above the ground that the stream with reach is $h_m = y_m + y_L$ where y_L is the launch height of 1.5 m.

Preliminary calculation: $v_{0y} = 21.3 \text{ m/s}$.

SOLVE Max height of stream:

$$h_m = \frac{v_{0y}^2}{2g} + 1.5 \text{ m} = \frac{(21.3 \text{ m/s})^2}{2 \times 9.8 \text{ m/s}^2} + 1.5 \text{ m} = 25 \text{ m}$$

REFLECT The firefighters could apply water to approximately the window of the 7th floor.

71. **ORGANIZE AND PLAN** Neglecting air resistance, the maximum range is achieved for a launch angle of $45°$. The relationship for the range in terms of the velocity is $x_R = \frac{2v^2}{g} \cos\theta \sin\theta$. In this problem, we are given the maximum range and asked to find the velocity required to achieve that range. Solving for speed (v) in the range equation with $\theta = 45°$ yields:

$$v = \sqrt{x_r \times g}$$

The shell's flight time is obtained by noting the given range x_r and the horizontal component of the velocity v_x. These quantities yield a time of flight given by

$$t = \frac{x_r}{v_x}$$

Notes on units: Range is given in terms of km and the acceleration due to gravity is in units of m/s. We are going to convert km to meters for unit consistency.

SOLVE The launch speed required to obtain a range of 15 km = 15,000 m is

$$v = \sqrt{15,000 \text{ m} \times 9.8 \text{ m/s}} = 380 \text{ m/s}$$

The horizontal component of the velocity is $v_x = v \times \cos 45° = 270 \text{ m/s}$

The time of flight is

$$t = \frac{15000 \text{ m}}{270 \text{ m/s}} = 56 \text{ s}$$

REFLECT To achieve this range with air resistance the required speed would be greater. The significant digits of this problem would not be sufficient to ensure that the battleship could strike a ship. One would need to have, at least, a precision of tens of meters. So, if you are tasked with actually hitting things you will have to keep more digits!

73. **ORGANIZE AND PLAN** The relationship for the range used above is $x_R = \frac{2v_{0y}v_{0x}}{g}$. Note the proportionality relationship between the range and the gravitational acceleration $x_R \alpha \frac{1}{g}$. If g is doubled the range is halved. If g is halved the range is doubled.

To compare the ranges in the different places we note: $\frac{x_R^M}{x_R^E} = \frac{g^E}{g^M}$, or

$$x_R^M = x_R^E \frac{g^E}{g^M}$$

SOLVE Range on the moon is:

$$x_R^M = x_R^E \frac{g^E}{g^M} = 120 \text{ m} \times \frac{g}{g/6} = 120 \text{ m} \times 6 = 720 \text{ m}$$

REFLECT If you wanted to get in 18 holes on a Sunday on the moon you better have a pretty fast golf cart!

75. **ORGANIZE AND PLAN** We are asked to find the velocity required to clear a fence of known height if the launch angle is given. The vertical component of the motion is all that matters here. Equation 3.19c gives the maximum height when the initial velocity is given:

$$\Delta y = \frac{v_{0y}^2}{2g}$$

It follows that:

$$v_{0y} = v \sin\theta = \sqrt{2\Delta y g}$$

Isolating for the velocity is:

$$v = \frac{\sqrt{2\Delta y g}}{\sin\theta}$$

SOLVE   The minimum velocity required to clear the fence:

$$v = \frac{\sqrt{2 \times 2.1 \text{ m} \times 9.8 \text{ m/s}^2}}{\sin 45°} = 9.1 \text{ m/s}$$

REFLECT   Gazelles can easily reach speeds of 9.1 m/s (about 20 mph) so making this jump (over a tall man for example) would be no problem.

77.   **ORGANIZE AND PLAN**   This problem is similar to problem 76. The relationship derived for the trajectory as a function of x applies generally and can be used for this problem. Written slightly differently from problem 76 we show $y(x)$ as

$$y(x) = x \tan\theta - \frac{1}{2} g \left(\frac{x}{v \times \cos\theta} \right)^2$$

The first complication in this problem is that there are two distinct ranges of angles that can result in a ball in the net. As shown in part a in figure we see for shallow angles the ball will obviously go in the net. As the angle increases it is generally possible that the ball can go over the net. However, further increasing the angle can also result in a goal.

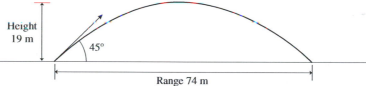

In this problem we are given the launch velocity $v = 19.8 \text{ m/s}$, the distance to the goal $x_f = 11.0 \text{ m}$ and the height of the goal $y_f = 2.44 \text{ m}$ and we are asked to find the angles that obey the following inequality:

$$y_f > y(f_f) = x_f \tan\theta - \frac{1}{2} g \left(\frac{x_f}{v \times \cos\theta} \right)^2$$

The difficulty here is that there is no simple way to isolate θ in the above inequality. When algebraic manipulation fails, one can always resort to numerical methods. In this case we shall plot the function

$$f(\theta) = x_f \tan\theta - \frac{1}{2} g \left(\frac{x_f}{v \times \cos\theta} \right)^2 - y_f$$

When $f(\theta) > 0$ the ball will go over the net. When $f(\theta) < 0$ the ball will either go directly into the net or bounce into the net. The values of θ that result in $f(\theta) = 0$ will be the boundaries of the ranges requested in the problem.

SOLVE   Plotting

$$f(\theta) = 11.0 \text{ m} \tan\theta - \frac{1}{2} 9.8 \text{ m/s}^2 \left(\frac{11.0 \text{ m}}{19.8 \text{ m/s} \times \cos\theta} \right)^2 - 2.44 \text{ m}$$

we obtain part b in figure above. The range of potential goal angles are: $\theta = (0°, 20.6°)$ and $\theta = (81.7°, 90°)$

REFLECT   The actual range on the field is surely $\theta = (0°, 20.6°)$ since the goalie would have a lot of time to locate and act to stop the ball on the high arcing trajectories. However, if the goalie were to pass out from the pressure, both ranges can potentially result in a goal.

79.   **ORGANIZE AND PLAN**   Continuing from the previous problem and using the graphical strategy from problem 77 we will plot two functions: The first will tell us when the ball can go into the net (black plot in figure below)-

$$f(\theta) = 20.0 \text{ m} \tan\theta - \frac{1}{2} 9.80 \text{ m/s}^2 \times \left(\frac{20.0 \text{ m}}{18 \text{ m/s} \times \cos\theta} \right)^2 - 2.44 \text{ m}$$

When $f_1(\theta) > 0$ the ball will go over the net. When $f_1(\theta) < 0$ the ball will either go directly into the net or bounce into the net. The values θ of that result in $f_1(\theta) = 0$ will be set one boundary on the range of potential angles.

The second will tell us when the ball can go over the head of the goalie (red plot in figure below)-

$$f_2(\theta) = -15.0 \text{ m tan}\theta + \frac{1}{2}9.8 \text{ m/s}^2\left(\frac{15.0 \text{ m}}{18 \text{ m/s}\times\cos\theta}\right)^2 + 1.70 \text{ m}$$

When $f_2(\theta) > 0$ the ball will go toward the body of the goalie. When $f_2(\theta) < 0$ the ball will go over the head of the goalie. The values of θ that result in $f_2(\theta) = 0$ will be set one boundary on the range of potential angles.
In order to score a goal the ball must go over the head of the goalie AND not go over the net. In other words, the following two conditions must be met: $f_2(\theta) < 0$ and $f_1(\theta) < 0$. Reading the regions off the plot that obey these conditions will define our range of launch angles that result in a potential goal.

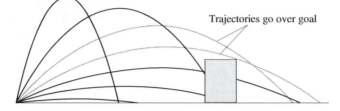

Trajectories go over goal

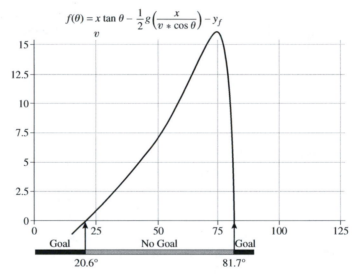

SOLVE   From figure above we identify two regions that satisfy the conditions: $\theta_{goal} = (20.4°, 26.6°)$ and $\theta_{goal} = (70.4°, 76.1°)$

REFLECT   Technically both ranges can result in a goal if the goalie is not paying attention. In practice the more shallow angles are more likely to result in a goal because the time of flight is much smaller, giving the goalie less time to react. Of course, the goal also has width which makes it harder to be a goalie and easier to score a goal.

81.   **ORGANIZE AND PLAN**   The point at a latitude of 38° on the earth is rotating around a circle of radius $r_c = R_E \cos(38°)$ (see figure below). Once the radius is known, we can use the relationship derived above for the centripetal acceleration in terms of the period and the radius $a_r = \frac{4\pi^2 r_c}{T^2}$. Substituting the value r_c for in terms of R_E.

$$a_r = \frac{4\pi^2 R_E \cos(38°)}{T^2} = \cos(38°)\times a_c^{eq}$$

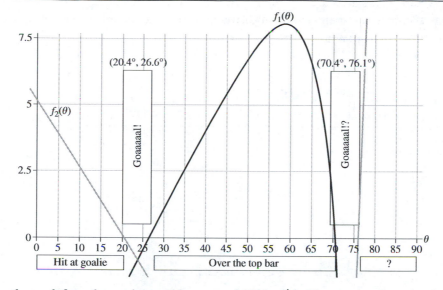

SOLVE   Using the result from the previous problem, $a_c^{eq} = 0.0336$ m/s^2. The centripetal acceleration at 38° latitude (north or south) is:

$$a_r = \cos(38°) \times 0.0336 \text{ m/s}^2 = 0.0265 \text{ cm/s}^2$$

REFLECT   As you move closer to the poles, the centripetal acceleration reduces to zero with a maximum value at the equator.

83.   **ORGANIZE AND PLAN**   The centripetal acceleration a_r is the acceleration required to keep an object moving in a circle. If the acceleration is greater than a_r then the object will move closer to the center of the circle. At the top of the circle, the car is at least experiencing an acceleration due to gravity of 9.8 m/s^2. If this acceleration is more than is required to keep it moving in a circle, the car will move in the direction of the center of the circle. That is bad for the passengers of the car as the car could fall off the track. If this acceleration is less than that required to keep the car moving in a circle, the track itself will provide whatever additional acceleration is required to keep the desired path.

For a given radius of loop R, the minimum speed required to ensure the centripetal acceleration is at least 9.8 m/s^2 is determined by the equation $g \le a_r = \dfrac{v_{\text{safe}}^2}{R}$. Solving for v_{safe} yields:

$$v_{\text{safe}} \ge \sqrt{g^R}$$

The minimum value is when $v_{\min} = \sqrt{g^R}$

SOLVE   Minimum velocity to ensure car does not come off the tracks at the top of the loop is:

$$v_{\min} = \sqrt{9.8 \text{ m/s}^2 \times 7.3 \text{ m}} = 8.5 \text{ m/s}$$

REFLECT   As the loop gets smaller the minimum required velocity decreases. Larger loops require larger velocities. Imagine if the loop were the size of the earth. Solving this problem for a loop that big will give you an idea of the speed of satellites.

85.   **ORGANIZE AND PLAN**   The speed of light is $c = 3 \times 10^5$ km/s. The period of the particles is determined by $T = \dfrac{2\pi R}{v}$. Given the period T and the radius R, the centripetal acceleration is

$$a_r = \frac{4\pi^2 R}{T^2}$$

SOLVE   Assuming the speed is the speed of light to two decimal places the period is

$$T = \frac{2\pi 1.2 \text{ km}}{3 \times 10^5 \text{ km/s}} = 2.5 \times 10^{-5} \text{ s}$$

The centripetal acceleration is

$$a_r = \frac{4\pi^2 1200 \text{ m}}{(2.5 \times 10^{-5} \text{ s})^2} = 7.6 \times 10^{13} \text{ m/s}^2$$

REFLECT This period corresponds to 40,000 trips around the circle per second and the acceleration is roughly 10^{13} times greater than the acceleration due to gravity.

87. **ORGANIZE AND PLAN** The information we are given is not in a form that we can directly use relationships derived in previous problems. We do have the radius $R = 46$ cm $= 0.46$ m but we are given the number of rotations per second, 500 rev/s, and we need the period.

If there are 500 rev/min the period is $T = \dfrac{1}{500\ \text{s}^{-1}} = 0.002$ m $= 0.12$ s.

The centripetal acceleration is $a_r = \dfrac{4\pi^2 R}{T^2}$.

SOLVE Plugging in the numerical values: $a_r = \dfrac{4\pi^2 0.46\ \text{m}}{(0.12\ \text{s})^2} = 1.26 \times 10^3$ m/s

REFLECT This value is roughly 140 times the acceleration due to gravity, hopefully enough to squeeze the water out in the spin cycle.

89. **ORGANIZE AND PLAN** Given radius and rotational period we recall $a_r = \dfrac{4\pi^2 R}{T_2}$.

We want to use standard units so we convert $R = 12$ km $= 1.2 \times 10^4$ m

SOLVE Plugging in numbers:
$$a_r = \frac{4\pi^2 \times 1.2 \times 10^4\ \text{m}}{(1.0\ \text{s})^2} = 4.7 \times 10^5\ \text{m/s}^2$$

Reflect This is a very large centripetal acceleration (about 50,000 g). What keeps the neutron star from ejecting all its mass? What provides the acceleration? Stay tuned!

91. **ORGANIZE AND PLAN** Some things to note about velocity and acceleration. Velocity is always tangent to the trajectory. The change in velocity is the acceleration.

SOLVE Choice (a) is unacceptable because the velocity is not tangent to the trajectory.
Choice (b) is unacceptable because the acceleration away from the center of the circle would result in a velocity in the next time step that was directed away from the center of the circle.
Choice (c) is unacceptable because the acceleration is co-linear with the velocity. This acceleration would result in the bike going off the circle in a straight line with a decreasing speed.
Choice (d) is the correct choice! The acceleration directed toward the center of the circle produces a change in velocity that turns along the circle.
Reflect We have yet to discuss the source of the acceleration but at least we know how accelerations affect velocities. You are ready to understand the brilliance of Newton!

93. **ORGANIZE AND PLAN** We will break up the track into 4 distinct parts. The 2 curved parts and the two straight parts. The length of one curved part is half the circumference of a circle so the length of the two curved parts is just the circumference of a circle $2\pi R$.
The combined length of the two straight parts is easy: $2 \times L$
Finding the displacement of half a lap shown in diagram will require a coordinate system. The runner will have travelled the diameter of the semi-circle down $(\hat{j})$ and the length of the straight side across $(\hat{i})$.

SOLVE Length of a lap:
$$d = 2\pi 38.2\ \text{m} + 2 \times 80\ \text{m} = 400\ \text{m}$$
Displacement of half a lap beginning between straight part and curved part:
$$\vec{r} = 80\ \text{m}\hat{i} - 76.4\ \text{m}\hat{j}$$

REFLECT The distance around the track is approximately 1/4 mi so 4 times around is a mile.

95. **ORGANIZE AND PLAN**   In the case where the ball is launched from a height above where the ball will land the ball will spend a little extra time in the air while it falls the extra distance. There are a number of ways to solve this problem. We will use equation 3.19a:

$$y(t) = v_{0y}t - \frac{1}{2}gt^2$$

This equation assumes an initial position at $y = 0$.

We are given (or are able to directly calculate) the following values:

$$v_{0y} = v \times \sin(32°) = 13.2 \text{ m/s}, y(t_e) = -7.2 \text{ m}$$

where t_e is the time at the end of the flight.

If we use equation 3.19a to find t_e by solving the quadratic equation we can use that time in equation 3.19b to find the vertical component of the velocity at t_e. The time t_e will also be used to find the range $x_R = v_{0y} \times t_e$.

Finding the time:

$$\frac{1}{2}gt_e^2 - v_{0y}t_e + y_f = 0$$

The roots of this equation are found using the solution to quadratic equation:

$$t_e = \frac{v_{0y} \pm \sqrt{v_{0y}^2 - 2gy_f}}{g} = \frac{13.2 \text{ m/s} \pm \sqrt{(13.2 \text{ m/s})^2 + 2 \times 9.8 \text{ m/s}^2 \times 7.2 \text{ m}}}{9.8 \text{ m/s}^2}$$

Yielding $t_e = -0.47$ s and/or $t_e = 3.16$ s. Since the golfer was on his way to striking the ball at $t = -0.47$ s, this choice is not a physical solution to this problem. The time of flight is $t_e = 3.16$ s.

Vertical component of velocity when ball hits the ground:

$$v_y(t_e) = v_{0y} - gt_e = -17.8 \text{ m/s}$$

SOLVE   Velocity at apex (same as last problem):

$$v_{ox}\hat{i} = 25 \text{ m/s } \cos32°\hat{i} = 21 \text{ m/s}\,\hat{i}$$

Velocity when ball reaches ground:

$$\vec{v} = 21 \text{ m/s}\,\hat{i} - 18 \text{ m/s}\,\hat{j}$$

or in polar form:

$$\vec{v} = (28 \text{ m/s}, -40°)$$

Horizontal range:

$$x_R = 21 \text{ m/s} \times 3.16 \text{ s} = 66 \text{ m}$$

REFLECT   Consistent with experience, the ball goes farther when struck at elevation (10 m for this problem) and the angle below the horizon is increased from $-32°$ to $-40°$.

97. **ORGANIZE AND PLAN**   We use the fact that the components of the velocity can be considered independently. The vertical component of your velocity should be the same as the vertical component of the boat you are trying to intercept. Here "vertical" means the velocity of the boat to be intercepted perpendicular to the line connecting the two boats.

We know the velocity of our boat v_1 and the velocity of the boat we are trying to intercept v_2. We are trying to determine the heading we need to set (the angle θ in front of the ship we should point the boat toward). The vertical component of the velocity of our ship $v_{1x} = v_1 \sin\theta = v_2$. Solving for θ yields:

$$\theta = \arcsin\left(\frac{v_2}{v_1}\right)$$

The time (t it will take to reach the ship will be the path-length (d) divided by the velocity: $v_1 : t = \dfrac{d}{v_1}$

The path-length is the hypotenuse of the right triangle formed by the line connecting the boats when the radar located the boat and the path of the other boat. This gives a length of $d = \dfrac{10 \text{ km}}{\cos\theta}$

The vector representing the meeting point (assuming our current position is the origin) will be, in cardinal form $\vec{r} = (d, N(90° - \theta)E)$.

SOLVE   Heading: $\theta = \arcsin\left(\dfrac{25 \text{ km/hr}}{40 \text{ km/hr}}\right) = 39°$, also known as N51° E

Time of interception:

$$t = \frac{d}{v_1} = \frac{10 \text{ km}}{410 \text{ km/hr} \times \cos 39°} = 0.32 \text{ hr} = 19 \text{ min}$$

Meeting point: 13 km N51° E

REFLECT   You can check the answers above by taking useful limits in the situation. Suppose you could increase the speed of your boat. How would that effect the heading? Fast enough and you just need to head straight for the other boat. If you could just go a little faster than the ship you are trying to intercept, in this situation you would have to head north and slowly merge with the ship following almost parallel paths.

99.   **ORGANIZE AND PLAN**   We have discovered that the maximum range for a fixed launch speed is obtained when the launch angle is 45°. We use the equation developed in problem 74 for the range $x_r = \dfrac{v^2 \sin 2\theta}{g}$. Solving for v results in the following relationship:

$$v = \sqrt{\frac{g x_R}{\sin 90°}} = \sqrt{g \times x_R}$$

The time of flight is $t = \dfrac{x_r}{v_x} = \dfrac{x_r}{v \cos(45°)}$

SOLVE   Minimum velocity to achieve $x_R = 12.8$ m : $v_{min} = \sqrt{9.8 \text{ m/s}^2 12.8 \text{ m}} = 11.2$ m/s

The time of flight: $t = \dfrac{12.8 \text{ m}\sqrt{2}}{11.2 \frac{m}{s}} = 1.61$ s

REFLECT   The minimum takeoff speed (25 mph) will achieve the range if the launch angle is optimal. Large or smaller angles will require a faster speed.

101.   **ORGANIZE AND PLAN**   The equation, $R = \dfrac{v_0^2 \sin(2\theta)}{g}$, derived above allows for straight forward answers to these questions.

SOLVE   Maximum range is obtained for a fixed launch speed when $\sin(2\theta)$ is maximum. The function $\sin x$ is maximum when. When $x = 90°$. When $2\theta = 90°$, $\theta = 45°$.

The functions in x is symmetric about $x = 90°$, in other words $\sin(90° - x) = \sin(90° + x)$ as long as $0 < x < 90°$ So, the range of a particle launched at $\theta = 22°$ is equal to the range of a particle launched at $\theta = 68°$ because $90 - x = 2 \times 22°$ yields $x = 46°$ and $90 + 46° = 136° = 2\theta$ yields $\theta = 68°$.

When the launch angle is 90°, $R = 0$ since $\sin 180° = 0$. This jibes with experience because if you throw something straight up in the air it comes right back down. No horizontal component to the velocity results in no horizontal displacement.

REFLECT   Simple when you look at it that way.

103.   **ORGANIZE AND PLAN**   What is a parabola? The standard form for a parabola is
$$y = ax^2 + bx + c$$

Does the equation derived in problem 102 fit this form?

SOLVE   Yes it does. The standard form value for $a = -\left(\dfrac{g}{2v_0^2 \cos^2(\theta)}\right)$, $b = \tan \theta$ and $c = 0$ reflecting the fact that our base equations assumed the trajectory began at the origin.

REFLECT   Now you can impress your friends at parties. In your most stuffy voice, say, "The trajectory of a particle under constant vertical acceleration downward is parabolic. Of course, you must neglect air resistance. Any fool knows that.".

105.   **ORGANIZE AND PLAN**   We are given the period T, and the radius R of the assumed uniform circular motion. With this given information the form of centripetal acceleration that we will use is:

$$a_r = \frac{4\pi^2 R}{T^2}$$

The period is given in terms of days. As we have all chapter, we would like the acceleration in units of m/s^2. Converting

$$T = 27.3 \text{ days} \times 24 \text{ hr/day} \times 3600 \text{ s/hr} = 2.36 \times 10^6 \text{ s}$$

and

$$R = 3.84 \times 10^5 \text{ km} = 3.84 \times 10^8 \text{ m}$$

SOLVE   Centripetal acceleration of the moon required to keep it in orbit:

$$a_r = \frac{4\pi^2 \times 3.84 \times 10^8 \text{ m}}{\left(2.36 \times 10^6 \text{ s}\right)^2} = 2.72 \times 10^{-3} \text{ m/s}^2 = 0.000278 \ g$$

REFLECT   The centripetal acceleration of the moon is about $\dfrac{1}{3600}$ of the gravitational acceleration on the surface of the earth. When you find out where that acceleration originates from (psst... it's gravity) you must conclude that the gravitational pull of the earth depends inversely on distance.

107.   **ORGANIZE AND PLAN**   We are given the number of revolutions per minute and the centripetal acceleration from which we are to determine the radius of the uniform circular motion in units of the gravitational acceleration g. We can convert the revolutions per minute to a rotational period and use the equation:

$$a_r = \frac{4\pi^2 R}{T^2}$$

Isolating for R yields:

$$R = \frac{a_r \times T^2}{4\pi^2}$$

Preliminary calculations:

If there are 3380 revolutions per minute each revolution takes

$$T = \frac{1}{3380} \text{min} \times \frac{60 \text{ s}}{1 \text{ min}} = 1.775 \times 10^{-2} \text{ s}.$$

The centripetal acceleration is $a_r = 1600 \ g = 1600 \times 9.8 \text{ m/s}^2 = 1.568 \times 10^4 \text{ m/s}^2$.

SOLVE   Radius of the test tubes is:

$$R = \frac{1.568 \times 10^4 \text{ m/s}^2 \times \left(1.775 \times 10^{-2} \text{ s}\right)^2}{4\pi^2} = 0.1251 \text{ m} = 12.51 \text{ cm}$$

REFLECT   This result is consistent with the dimensions of a tabletop centrifuge. At these rotational periods the centrifuges must be balanced carefully to prevent damage to the machine.

4

FORCE AND NEWTON'S LAWS OF MOTION

1. **ORGANIZE AND PLAN** In what ways would a more massive golf ball change its flight? We need to think of what forces will act on the golf ball, and how the ball's flight will be affected by these forces.

   SOLVE Once a golf ball is in flight, it experiences forces due to air resistance and wind. Newton's second law of motion ($\vec{F} = m\vec{a}$) says that, for a given force, a more massive golf ball will experience less acceleration than its lighter counterpart. Thus, for example, the drag force due to air resistance will produce less of a deceleration for a heavy golf ball, so the heavier ball will maintain its speed more effectively than the lighter ball. Forces due to spinning of the golf ball ("slice") will also produce less acceleration on the heavy ball, so it will fly straighter.

   REFLECT One might think that for a given golf-club-swing, a heavier golf ball would experience less acceleration than a lighter ball. If only things were so simple! The initial speed of a golf ball depends on a complex relationship between the speed of the golf club swing and the hardness of the ball. Since a heavier ball will likely be harder (more dense), it will react differently than a lighter ball when you hit it, but there is no guarantee that it will experience more acceleration. There are entire books written on the subject of the physics of golf. For those interested, more information is available with a quick internet search.

3. **ORGANIZE AND PLAN** We will consider different forces and how they affect the acceleration and velocity of an object. This involves thinking about Newton's second law ($\vec{F} = m\vec{a}$). Remember that gravity is given by Eq. 4.1 ($\vec{w} = m\vec{g}$), drag forces always act to oppose the velocity of an object, and that the net force on an object is simply the vector sum of all the forces acting on the object.

   SOLVE Just before she opens her parachute,

   1. The force due to gravity is $\vec{w} = m\vec{g}$ (toward the center of the Earth).

   2. The drag force cancels the force due to gravity.

   3. The net force is the sum of gravity and drag, and is zero due to 2.

   4. Because of 3, she is not accelerating (Newton's second law).

   5. Because of 4, her velocity (which is oriented downwards) is constant.

   Just after she opens her parachute:

   1. The force due to gravity is $\vec{w} = m\vec{g}$ (toward the center of the Earth).

   2. The drag force increases dramatically, and is oriented upwards.

   3. Net force is the sum of gravity and drag, and is pointing upwards due to 2.

   4. Because of 3, her acceleration is upwards.

   5. Because of 4, her velocity (which is oriented downwards) decreases.

   As she continues now towards the Earth, the force on her due to gravity will essentially not change. The magnitude of the drag force will decrease as her velocity is reduced, but will always point upward. Because the drag force is changing, her net force will change as well, until the drag force cancels the force due to gravity. As the drag force changes, her acceleration will change (Newton's second law). Since the drag force and hence her acceleration is oriented opposite to her velocity, her velocity will decrease. When she reaches her new terminal velocity (i.e., when drag force cancels gravity so net force is zero), she will no longer accelerate, and her velocity will be constant until she reaches the Earth.

REFLECT   The force due to the Earth's gravity depends on how far you are from the center of the Earth, but for the distances considered in this problem, we can ignore this effect.

5.   **ORGANIZE AND PLAN**   Uniform circular motion is caused a centripetal ("center-seeking") force, which, as its name implies, is oriented toward the center of the circle. Review Figure 4.30 in the text for a clear picture of the forces at work.

SOLVE   For a car on a flat curve (not banked), the normal force exerted by the road on the car is oriented straight up, opposing gravity. On a banked curve, the normal force is still perpendicular to the road, but it is no longer parallel to gravity. The steeper a curve is banked, the larger will be the component of the normal force toward the center of the curve, and this is the direction of force needed to make the car execute circular motion. Thus the normal force from a banked curve aids the car in turning.

If a car is going too fast on an inner section of the track, the friction between the tires and the road will not be able to exert the required centripetal force to execute the turn, and the car will slide outward. Here the track will be more banked, providing an extra force component to assist the tires in maintaining the turn, and (hopefully) the car will regain its circular trajectory.

REFLECT   When running or skiing around a turn, why do these athletes lean their bodies toward the center of the turn?

7.   **ORGANIZE AND PLAN**   Apply Newton's third law to the situation.

SOLVE   Newton's third law says that all forces come in pairs, with each member of the pair being of equal magnitude but opposite direction. Ball B receives a force oriented at 30° above horizontal, so by Newton's third law, ball A must experience a force oriented at $30 - 180 = -150°$ above horizontal, or 150° below horizontal.

REFLECT   If ball A experiences a force at 150° below the horizontal, why is it not moving in that direction? By Newton's second law it is known that ball A was indeed *accelerated* in the direction 150° below the horizontal. However, the acceleration lasted only a brief moment, so was not enough to modify the initial velocity to orient it completely in the new direction.

9.   **ORGANIZE AND PLAN**   The force due to gravity is always oriented toward the center of the Earth, which will be downward in our drawing. The drag force always acts opposite to velocity, so it will act at $25 - 180 = -155°$ above the horizontal, or 155° below the horizontal.

SOLVE

REFLECT   Often baseballs will not fly straight, but will "hook" and go out of play. What force(s) cause this? If a ball has spin on it, the cushion of air will create a pressure difference across the ball, and that will produce a net force that can act transverse to the ball's velocity.

11.   **ORGANIZE AND PLAN**   This problem involves the interaction of two objects: a person (object 1) walking on the Earth (object 2). We will have to think about Newton's third law to resolve this problem. Regarding the difference between static and kinetic friction, remember that static friction requires that the interface between the two objects is static (i.e., not changing) during the time that the static friction force is being applied. For kinetic friction, the interface is dynamic (i.e., changing), as with a skate sliding over ice, or a tire skidding on the pavement.

SOLVE   When you walk you push against the Earth with your feet to propel you forward. From Newton's third law, the Earth exerts a force of equal magnitude but opposite direction to the sole of your shoe. It is the friction between the soles of your shoes and the ground that actually transmits the force to your feet to move you forward. As you step forward, the foot you're pushing with does not move with respect to the ground, so it is static friction that is at work. If you're on a slippery surface and your foot begins to slide as you push off with it, then it is kinetic friction that comes into play.

REFLECT   When your foot slips out from beneath you unexpectedly, what happens to your body? Does it lurch forward? Fall straight down?

13.   **ORGANIZE AND PLAN**   Draw a net force diagram and look at the forces acting in the vertical direction.
SOLVE   If the skydiver is falling with a constant velocity, then she is not accelerating and Newton's second law tells us that there is no net force acting on her. Therefore, the force exerted by her parachute must exactly cancel the force on her due to gravity. This means her parachute will exert a force of magnitude equal to her weight, but in the upward direction.
REFLECT   When the skydiver first opens her parachute, is the force exerted by the parachute greater than, less then, or the same as the force due to gravity? Since the skydiver's velocity decreases, she must be undergoing an acceleration in the upward direction, thus by Newton's second law the parachute exerts an upward force on her — but the magnitude of that force is unknown since the magnitude of her acceleration is unknown.

15.   **ORGANIZE AND PLAN**   Draw a force diagram for the roller coaster at the top of a hill and at the bottom of a valley, and consider in which direction the roller coaster must be accelerating at each point. Now apply Newton's second and third laws.
SOLVE   When the roller coaster is going over a hill, its velocity is changing from an upward direction before the summit of the hill to a downward direction after the summit, so you are experiencing an overall downward acceleration. By Newton's second law, there must be a net downward force on the roller coaster. Since the force due to gravity has not changed, the only other force acting on the system (the normal force from the track) must have been reduced. This is shown in the figure below, where the normal force at the top of the hill is noticeably weaker than the force due to gravity.

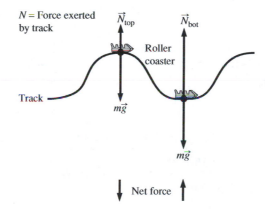

From Newton's third law the track must experience a force equal in magnitude to the normal force, but in the opposite (i.e., downward) direction. Thus, the track going over a summit experiences a force that is less than the weight of the roller coaster.
When the roller coaster is going through a valley, the opposite logic applies. The net force on the roller coaster must be upward. This upward force must be exerted by the track, which is added to the normal force needed to counter the weight. (Note the length of the normal force vector compared to the weight vector in the figure above). Newton's third law states that the track will experience a force of the same magnitude but in the opposite direction. Thus the track going through a valley experiences a force that is greater than the weight of the roller coaster.
REFLECT   At the very top of the summit and at the very bottom of the valley, the track is horizontal. The acceleration, however, is not. It is oriented toward the center of the curve, in other words downward at the summit and upward in the valley.

17.   **ORGANIZE AND PLAN**   We can approach this in two ways. Either we can calculate the acceleration due to both forces individually and then do a vector sum of the accelerations, or we can calculate the net force first and then find the acceleration due to that force. We will take the second approach.
To calculate the net force, do a vector sum of all the forces acting on the box, then use Newton's second law to find the acceleration of the box. Note also that the responses contain vector and scalar accelerations, so we will have to calculate both.

Known: $\vec{F_1} = 105\hat{i} - 87\hat{j}$, $m = 1.4$ kg

SOLVE   To get the acceleration due to each force individually, we use Newton's second law $\vec{F}_{net} = m\vec{a}$. The only force acting on the box is $\vec{F_1}$, so $\vec{F}_{net} = \vec{F_1}$. We thus find an acceleration of [Eq. 1]

$$\vec{a} = \vec{F_1}/m = (105\ \text{N}/1.4\ \text{kg})(\hat{i}) + (-87\ \text{N}/1.4\ \text{kg})(\hat{j}) = 75\ \text{m/s}^2(\hat{i}) - 62\ \text{m/s}^2(\hat{j})$$

This corresponds to response (a).

We will now calculate the magnitude of the net force, and then insert it into Newton's second law to find the magnitude of the acceleration. Inserting the components of $\vec{F_1}$ into the Pythagorean theorem, we find that the magnitude of the net force is [Eq. 2]

$$F_1 = \sqrt{105^2 + (-87)^2} = 136.4\ \text{N}$$

Using Newton's second law to find the magnitude of the acceleration gives [Eq. 3]

$$a = F_1/m = 136.4\ \text{N}/1.4\ \text{kg} = 97.4\ \text{m/s}^2$$

which is response (d).

REFLECT   Which answer gives more information? The first approach tells us the box is accelerated more in the $\hat{i}$ direction than in the $\hat{j}$ direction, but the second approach tells us the overall magnitude of the acceleration without further calculation. If we want to check our math, we can verify that the two approaches give the same magnitude for the acceleration. Using the acceleration components from Eq. (1), we find that magnitude of the acceleration is [Eq. 4]

$$a = \sqrt{(75\ \text{m/s}^2)^2 + (-62\ \text{m/s}^2)^2} = 97.4\ \text{m/s}^2$$

which again is response (d).

19.   **ORGANIZE AND PLAN**   This problem involves adding forces. How many ways can we add two forces together? If we assume the first force defines one axis (let's call it the $\hat{i}$ axis), we can add the other force to it using any angle θ between the two forces. Let's express this mathematically and see what the upper and lower limits are for the magnitude of the resulting net force.

Known: $F_1 = 32$ N ; $F_2 = 51$ N

SOLVE   Assuming $\vec{F_1}$ is along the $\hat{i}$ axis, and $\vec{F_2}$ makes an angle θ with respect to $\vec{F_1}$, then the net force is

$\vec{F}_{net} = [F_1 + F_2\cos(\theta)](\hat{i}) + F_2\sin(\theta)(\hat{j})$ [Eq. 1].

The magnitude of the net force is [Eq. 2]

$$F_{net} = \sqrt{(F_1 + F_2\cos(\theta))^2 + (F_2\sin(\theta))^2} = \sqrt{(32 + 51\cos(\theta))^2 + (51\sin(\theta))^2}\ \text{N}$$

The maximum force will occur when $\theta = 0$ (i.e., the two forces are acting in the same direction), and the minimum force will act when $\theta = 180°$ (i.e., the two forces are acting in the opposite direction). In the former case we find [Eq. 3]

$$F_{net} = \sqrt{(32 + 51)^2} = 83\ \text{N}$$

and in the latter case we find [Eq. 4]

$$F_{net} = \sqrt{(32 - 51)^2} = 19\ \text{N}$$

Thus the net for must fall between 19 and 83 N, or $19N \leq F_{net} \leq 83$ N [Eq. 5] which is response (c).

REFLECT   Notice that the maximum force attainable is greater than either of the two forces alone, which is reasonable. In this case, the minimum force attainable is less than either of the two forces alone, but is this always so?

REFLECT   We find that the force required to accelerate the 5.0 kg mass is five times the force required to impart the same acceleration to the 1.0 kg mass. This is consistent with Newton's second law.

21.   ORGANIZE AND PLAN   A drawing of the situation will clarify things.

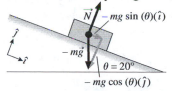

Note that the coordinate system chosen is aligned to the incline. To find the acceleration of the block of ice we need to find the net force, and then we can use Newton's second law to find the acceleration.

Known: angle of incline $\theta = 20°$

SOLVE   The normal force exerted on the block must cancel the $\hat{j}$ component of the force due to gravity, otherwise the block would move through or away from the inclined surface. Thus the net force is the just the $\hat{i}$ component of the force due to gravity, or [Eq. 1] $\vec{F}_{net} = mg\sin(\theta)(\hat{i})$.

From Newton's second law, we can find the acceleration due to this net force [Eq. 2],

$$\vec{F}_{net} = mg\sin(\theta)(\hat{i}) = m\,\vec{a}$$

$$\vec{a} = g\sin(20°)(\hat{i}) = 3.4 \text{ m/s}^2(\hat{i}).$$

which is response (c).

REFLECT   The acceleration must be less than that due to gravity, since the net force exerted on the block is less than the force due to gravity. We have also found more information than asked for, since we know the direction of the acceleration as well as its magnitude.

23.   ORGANIZE AND PLAN   For this problem we can use Eq. 4.9, which gives the force required for an object to execute uniform circular motion, $F_r = mv^2/R$. This force is directed toward the center of the circle and will be provided by the frictional force, $f = \mu_s n = \mu_s mg$, so we can equate these two expressions and solve for the coefficient of static friction.

Known: $R = 275$ m; $v = 21$ m/s

SOLVE   Equating the frictional force to the force needed to execute centripetal motion gives [Eq. 1]:

$$\mu_s mg = mv^2/R$$

$$\mu_s = v^2/(Rg) = \frac{(21 \text{ m/s})^2}{(275 \text{ m})(9.8 \text{ m/s}^2)} = 0.16$$

which is response (a).

REFLECT   We label the coefficient of friction as static because the part of the tire that is in contact with the road is stationary while it touches the road. Is this a reasonable static coefficient of friction for a tire on a concrete road? Looking at Table 4.1, we see that a tire on a dry concrete road has a static coefficient of friction of 1.0, which is reduced to 0.3 on wet concrete. Thus the static coefficient of friction of 0.16 needed to execute this turn is well within the normal range, and we should have no problem executing this turn, even on wet concrete.

25.   ORGANIZE AND PLAN   Find the horizontal forces acting on the puck, and then use Newton's second law to find the horizontal acceleration of the puck. Use this acceleration in the kinematic equations from Chapter 2 to find the initial speed needed to travel 61 m. Choose a coordinate system where the initial velocity of the puck is in the $\hat{i}$ direction. Also, note that the puck's final velocity is 0 *m/s*.

Known: $\mu_k = 0.015$; distance $d = 61$ m; final puck speed $v = 0$ m/s.

SOLVE   The horizontal force acting on the puck is due to kinetic friction, since the puck is moving with respect to the ice. The magnitude of the kinetic friction force is given by Eq. 4.6, $f_k = \mu_k n$, where n is the magnitude of the normal force. The direction of the kinetic friction force will be opposite to the velocity of the puck, or in the $-\hat{i}$ direction. The normal force n must have the same magnitude as the force due to gravity, otherwise the puck would experience a vertical acceleration, so $n = mg$. Using Newton's second law to find the acceleration due to kinetic friction, we have $\vec{F}_{net} = \vec{f}_k = m\vec{a}$ [Eq. 1].

Using Eq. 4.6 to express the kinetic friction force in terms of the normal force, and recalling that the kinetic friction force acts in the $-\hat{i}$ direction, we have [Eq. 2]

$$-\mu_k n\left(\hat{i}\right) = m\vec{a},$$

$$-\mu_k mg\left(\hat{i}\right) = m\vec{a},$$

$$\vec{a} = -\mu_k g\left(\hat{i}\right)$$

To find the initial speed needed for the puck to travel 61 meters, use Eq. 2.9 (reproduced here as Eq. 3)

$$x = x_0 + v_0 t + 1/2 at^2$$

Since the distance traveled $d = x - x_0$, we have [Eq. 4]

$$d = v_0 t + 1/2 at^2$$

This equation has two unknowns, v_0 and t, so another equation is needed relating these two quantities. Use Eq. 2.8, $v = v_0 + at$ [Eq. 5], where v is the final puck speed (i.e., zero). Using Eq. 5 to eliminate t in Eq. 4, we find [Eq. 6]

$$d = v_0\left(\frac{-v_0}{a}\right) + 1/2 a\left(\frac{-v_0}{a}\right)^2$$

or

$$v_0 = \sqrt{-2ad}$$

Using Eq. 2 for the acceleration, we find that [Eq. 7]

$$v_0 = \sqrt{-2\left(-\mu_k g\right)d} = \sqrt{(-2)(\text{-}0.015 \cdot 9.8 \text{ m/s}^2)(61 \text{ m})} = 4.2 \text{ m/s}$$

which is response (c).

REFLECT   Looking at the units under the radical in the last expression for v_0, we see that they are m²/s², so that when we take the square root we are left with m/s, which are the correct units for a velocity. Also notice that we had to be careful about the directions of the velocity and the acceleration, otherwise we may have found a negative value under the radical in Eq. 7.

27.   **ORGANIZE AND PLAN**   We can take the ratio of the Eqs.1 and 3 from the previous problem to answer this question.

     SOLVE   Taking the ratio of centripetal forces of the SUV to the compact car, we have [Eq. 1]

$$\frac{F_v^{\text{SUV}}}{F_v^{\text{cc}}} = \frac{m_{\text{SUV}} v_{\text{SUV}}^2 / R}{m_{\text{cc}} v_{\text{cc}}^2 / R} = \frac{(2430 \text{ kg})(12.5 \text{ m/s})^2}{(810 \text{ kg})(25 \text{ m/s})^2} = \frac{3}{4}$$

which is response (c). Thus the centripetal force acting on the SUV is only 75% of that acting on the compact car.

     REFLECT   Notice that we did not need to know the radius of the turn, since we are only interested in the ratio of the forces. Does the answer make sense? Compared with the compact car, the SUV has three times the mass (increasing the centripetal force by a factor of three), but only ¼ the speed-squared (reducing the centripetal force by a factor of 4). Thus it is reasonable that we get an overall factor of ¾ for the ratio of the centripetal force applied to the SUV to that applied to the compact car.

29.   **ORGANIZE AND PLAN**   To find the net force acting on an object, we perform a *vector addition* of all the forces acting on the object. In this case there are only two forces, and they are anti-parallel (i.e., parallel but in opposite directions), so the vector addition is quite simple. We chose a coordinate system where the $\hat{i}$ direction is toward the right.

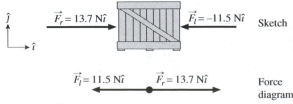

Known: $\vec{F}_r = 13.7 \text{ N}\left(\hat{i}\right); \quad \vec{F}_l = -11.5 \text{ N}\left(\hat{i}\right)$

SOLVE The net force (or total force) acting on the crate is simply the vector sum of all the forces acting on it, or $\vec{F}_{net} = \vec{F}_r + \vec{F}_l = 13.7 \text{ N}(\hat{i}) + 11.5 \text{ N}(-\hat{i}) = 2.2 \text{ N}(\hat{i})$ [Eq. 1], which is 2.2 N toward the right.

REFLECT Although for this quite simple problem we did not necessarily need to make a drawing showing the forces, it is often helpful to do so to avoid making mistakes or omissions.

31. **ORGANIZE AND PLAN** We can use Newton's second law to find the acceleration of the rocket. The net force on the rocket will be the vector sum of all the forces acting on it, which are the force due to gravity (straight down) and the force from the rocket engine (straight up). We will use a coordinate system where $\hat{j}$ represents the upward vertical direction.

Known: $m = 3.5 \text{ kg}$; $\vec{F}_{engine} = 95.3 \text{ N}(\hat{j})$

SOLVE The net force acting on the rocket is [Eq. 1]

$$\vec{F}_{net} = \vec{F}_{engine} + m\vec{g} = 95.3 \text{ N}(\hat{j}) + (3.5 \text{ kg})(9.8 \text{ m/s}^2)(-\hat{j}) = 61 \text{ N}(\hat{j}).$$

From Newton's second law, the acceleration of the toy rocket is [Eq. 2]

$$\vec{a} = \vec{F}_{net}/m = 61 \text{ N}(\hat{j})/3.5 \text{ kg} = 17.4 \text{ m/s}^2(\hat{j}).$$

REFLECT The rocket is accelerating upward at almost twice the acceleration due to gravity. This is a significant acceleration, compared to what we experience on a daily basis. To appreciate the acceleration due to gravity, try to catch a stick that falls from a stationary position, with your hand initially positioned above the stick. It can be done, but it's not easy. Imagine now the rocket accelerating at almost twice that rate!

33. **ORGANIZE AND PLAN** If the net force acting on the air track glider is constant, then by Newton's second law the acceleration it generates must be constant as well. This means the acceleration generated is just the average acceleration, so use Eq. 2.6 for average acceleration (reproduced here as Eq.1),

$$\vec{a}_{ave} = (\vec{v}_f - \vec{v}_i)/(t_f - t_i)$$

Knowing the average acceleration, use Newton's second law to find the force that acted on the air track glider to produce this acceleration. Choose a coordinate system where $\hat{i}$ indicates right ($-\hat{i}$ indicates left).

Known: $t_i = 0 \text{ s}; \vec{v}_i = 2.0 \text{ m/s}(\hat{i}); t_f = 4.0 \text{ s}, \vec{v}_f = -1.2 \text{ m/s}(\hat{i}); m = 0.230 \text{ kg}$

SOLVE Inserting the known values into Eq. 1 gives [Eq. 2]

$$\vec{a}_{ave} = \frac{(-1.2 \text{ m/s} - 2.0 \text{ m/s})(\hat{i})}{(4.0 \text{ s} - 0 \text{ s})} = -0.8 \text{ m/s}^2(\hat{i})$$

We now use Newton's second law to find the force needed to produce this acceleration [Eq. 3].

$$\vec{F}_{net} = m\vec{a}_{ave} = (0.23 \text{ kg})(0.8 \text{ m/s}^2)(-\hat{i}) = 0.18 \text{ N}(-\hat{i})$$

Since the force is acting in the $-\hat{i}$ direction, it is acting to the left.

REFLECT It is reasonable that the force is acting to the left, since we needed to reverse the puck's velocity from rightward to leftward. How much force is this in pounds? Since 1 pound is approximately 4.5 N, $(0.18 \text{ N})(1 \text{ lb}/4.5 \text{ N}) = 0.04 \text{ lb}$, which is not very much.

35. **ORGANIZE AND PLAN** This problem involves the application of Newton's second law in vector form, $\vec{F} = m\vec{a}$.

Known: $m = 100 \text{ g} = 0.1 \text{ kg}$; $\vec{a} = 0.255 \text{ m/s}^2(-\hat{i}) + 0.650 \text{ m/s}^2(\hat{j})$

SOLVE Inserting the known quantities (with the correct units!) into Newton's second law, we have [Eq. 1]

$$\vec{F} = m\vec{a} = (0.1 \text{ kg})\left(0.255 \text{ m/s}^2(-\hat{i}) + 0.650 \text{ m/s}^2(\hat{j})\right) = 0.0255 \text{ N}(-\hat{i}) + 0.0650 \text{ N}(\hat{j})$$

This is the force needed to generate the desired acceleration. The direction of this force is [Eq. 2]

$$\theta = \text{atan}\left(\frac{-0.0255}{0.0650}\right) = -68.6°, 111.4°.$$

From Eq. (1) we know the force vector is in the second quadrant, so we the direction of the force must be 111.4° above the horizontal. The magnitude of the force is [Eq. 3]

$$F = \sqrt{\left(-0.255 \text{ N}\right)^2 + \left(0.0650 \text{ N}\right)^2} = 0.0698 \text{ N}$$

REFLECT   Note that we had to convert the mass from grams to kilograms so that the result of our calculation would be in SI units (i.e., Newtons).

37.   **ORGANIZE AND PLAN**   Let's draw a schematic representation of the situation and label all the points of interest. To avoid confusion, we can make a separate drawing for the upward and downward trip.
SOLVE

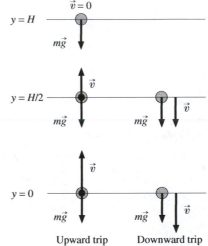

Upward trip   Downward trip

REFLECT   Note that the force acting on the ball during its entire trajectory does not change; it is simply the force due to gravity. This force generates an acceleration on the ball, so the velocity of the ball changes, in both magnitude and direction (compare the upward and downward trips). Note also that we do not know the relationship between the velocity and the force due to gravity, since we do not know the mass of the ball, so beware of assigning undue meaning to the length of the velocity vector compared with that of the force vector. However, since there is no air resistance, we do know that the magnitude of the velocity is the same at the equal heights on the upward and downward trips (note that the length of the velocity vectors are the same at $y = 0$ and $y = H/2$).

39.   **ORGANIZE AND PLAN**   This problem involves a twist on the usual problem of finding the force on an object due to Earth's gravitational field. But, as the hint suggests, if we first calculate the force on the object we can calculate the force on the Earth by applying Newton's third law $\vec{F}_{AB} = -\vec{F}_{BA}$. We can then use Newton's second law $\vec{F} = m\vec{a}$ to find the acceleration of the Earth.
We will define our coordinate system so that $\hat{j}$ is pointing radially away from the center of the Earth. The section on Newton's third law indicates that the mass of the Earth is 6×10^{24} kg.
Known: $m_{\text{object}} = 1000$ kg; $m_{\text{earth}} = 6 \times 10^{24}$ kg; $\vec{g} = 9.8 \text{ m/s}^2 \left(-\hat{j}\right)$

SOLVE   Solving first for the force on the object gives [Eq. 1]

$$\vec{F}_{\text{earth-object}} = m\vec{g} = \left(1000 \text{ kg}\right)\left(9.8 \text{ m/s}^2\right)\left(-\hat{j}\right)$$

$$\vec{F}_{\text{earth-object}} = 9800N\left(-\hat{j}\right).$$

In view of the coordinate system, this force is pulling the object toward the center of the Earth. Newton's third law indicates that [Eq. 2]

$$F_{\text{object-Earth}} = -\vec{F}_{\text{Earth-object}}$$

$$\vec{F}_{\text{object-Earth}} = 9400 \text{ N}\left(\hat{j}\right)$$

We have found that the force exerted on the Earth by the 1000 kg object is 9800 N radially away from the center of the Earth (i.e., up). To find the Earth's acceleration, we apply Newton's second law [Eq. 3],

$$\vec{F}_{\text{object-earth}} = m\vec{a}_{\text{earth}}$$

$$\vec{a}_{\text{earth}} = (9800 \ N\hat{j})/(6\times10^{24} \ \text{kg}) = 9.8/(6\times10^{21} \ \text{m/s}^2)(\hat{j}) \approx 1.6\times10^{-21}\text{m/s}^2(\hat{j}).$$

REFLECT Note that we simplified our calculation in the last step by using the rules for dividing quantities with exponents. Had we simply thrown the numbers into our calculator, we might have had trouble deciphering the results due to all the leading zeros.

This acceleration seems very small. Let's see how long it would take for the Earth to move 1 mm (= 10^{-3} m) if it accelerated at this rate from a stationary start [Eq. 4].

$$x = x_0 + v_0 t + 1/2 at^2$$

$$x = 0 + 0 + 1/2(1.6\times10^{-21} \ \text{m/s}^2)t^2$$

$$t = \sqrt{2x/a} = \sqrt{(2\times10^{-3} \ \text{m})/(1.6\times10^{-21} \ \text{m/s}^2)}$$

$$t = \sqrt{3.2} \times10^9 \text{s} \approx 31 \ \text{years!}$$

41. **ORGANIZE AND PLAN** This problem is the same as the previous one, with the astronaut replaced by the cannon ball and the spacecraft replaced by the cannon, and with the caveat that we are asked for speed instead of velocity. We will use the same approach to solve this problem.

   Known: $m_{\text{cb}} = 3.5$ kg; $m_{\text{cannon}} = 920$ kg; $v_f^{\text{cb}} = 95$ m/s

   SOLVE In view of the similarity between this problem and the previous one, we can simply map the variables of Problem 40 to their counterparts in this problem [Eq. 1]:

$$v_f^{\text{sc}} \rightarrow v_f^{\text{cannon}}$$

$$v_f^{\text{ast}} \rightarrow v_f^{\text{cb}}$$

$$m^{\text{sc}} \rightarrow m^{\text{cannon}}$$

$$m^{\text{ast}} \rightarrow m^{\text{cb}}$$

   Using this mapping and Eq. 8 from Problem 40, we find [Eq. 2]

$$v_f^{\text{cannon}} = -v_f^{\text{cb}} \frac{m^{\text{cb}}}{m^{\text{cannon}}} = -(95 \ \text{m/s})(3.5 \ \text{kg})/(920 \ \text{kg}) = -0.36 \ \text{m/s}$$

   Since we are only interested in the speed of the recoiling cannon, we take the magnitude of the velocity, which is 0.36 m/s.

   REFLECT Note that we reported the result to the same number of significant figures (2) as the lowest-precision datum used in calculating the result.

43. **ORGANIZE AND PLAN** The mass of the plane and the magnitude of the desired acceleration is known, so use Newton's second law to find the magnitude of the force needed to attain that acceleration.

   Known: $m = 247,000$ kg; $a = 3.2$ m/s^2

   SOLVE Inserting the known quantities into Newton's second law gives [Eq. 1]

$$F_{\text{net}} = ma = (247,000 \ \text{kg})(3.2 \ \text{m/s}^2) = 790,400 \ \text{N}$$

   REFLECT What is this in pounds? Using the conversion factor 1 N ~ 0.225 lbs, we find that this is a force of 177,840 lbs.

45. **ORGANIZE AND PLAN** We are given the magnitude of the net force on the object, and its mass, so we can use Newton's second law to calculate its acceleration. We chose a coordinate system in which $\hat{i}$ is aligned with the net force. With the object's acceleration known, use the kinematic equations to find its speed after 2.5 seconds.

   Known: $m = 5.0$ kg; $\vec{F}_{\text{net}} = 150 \ \text{N}(\hat{i})$

SOLVE   Using Newton's second law to find the acceleration gives [Eq. 1]

$$\vec{F}_{net} = m\vec{a},$$

$$\vec{a} = \vec{F}_{net} / m = \frac{(150 \text{ N})}{(5.0 \text{ kg})}(\hat{i}) = 30 \text{ m/s}^2 (\hat{i})$$

Inserting this into the equation of motion for velocity, and recalling that $\vec{v}_0 = 0$, we have [Eq. 2]

$$\vec{v} = \vec{v}_0 + \vec{a}t = (30 \text{ m/s}^2)(2.5 \text{ s})(\hat{i}) = 75 \text{ m/s}(\hat{i}).$$

REFLECT   If we could not remember the formula for velocity as a function of acceleration, we could have let the units guide us. We know that velocity has units of m/s, and acceleration has units of m/s², so we could deduce that we have to multiply acceleration by time to get velocity.

47.   **ORGANIZE AND PLAN**   The change in the balls velocity per unit time is given, so use kinematic equation 2.8 to find the acceleration. Explicitly, $\vec{v}_f = \vec{v}_0 + \vec{a}t$ [Eq. 1].

With the acceleration known, use Newton's second law to find the force.

Known: $\vec{v}_0 = 3.25 \text{ m/s}(-\hat{i})$; $\vec{v}_f = 4.56 \text{ m/s}(\hat{i})$; $\vec{F}_{net} = 35.2 \text{ N}(\hat{i})$

SOLVE   Inserting the known quantities into Eq. 1 gives [Eq. 2]

$$(4.56 \text{ m/s})(-\hat{i}) = (3.25 \text{ m/s})(\hat{i}) + \vec{a}(3.50 \text{ s})$$

$$\vec{a} = \frac{(3.25 \text{ m/s})(\hat{i}) + (4.56 \text{ m/s})(\hat{i})}{(3.50 \text{ s})}$$

$$\vec{a} = 2.23 \text{ m/s}^2 (\hat{i})$$

Using Newton's second law, we find the mass of the ball is [Eq. 3]

$$F_{net} = ma$$
$$m = 35.2 \text{ N} /(2.23 \text{ m/s}^2)$$
$$m = 15.8 \text{ kg}$$

REFLECT   The acceleration we found in Eq. 1 is in the same direction as the force, which must always be the case (Newton's second law). Had this not been the case, we could have been sure there was a typo in the problem. Also, note that we reported the mass to the same number of significant figures as the data used to calculate it.

49.   **ORGANIZE AND PLAN**   We can use Newton's second law to solve this problem. Since this problem involves forces and an acceleration with multiple components, we must be careful when treating the components.

Known: $\vec{F}_1 = 32 \text{ N}(\hat{i}) + 48 \text{ N}(-\hat{j})$; $\vec{a} = 5.17 \text{ m/s}^2 (-\hat{i}) + 2.5 \text{ m/s}^2 (\hat{j})$; $m = 24 \text{ kg}$

SOLVE   Newton's second law gives [Eq. 1]

$$\vec{F}_{net} = \vec{F}_1 + \vec{F}_2 = m\vec{a}$$
$$\vec{F}_2 = m\vec{a} - \vec{F}_1 = (24 \text{ kg})\left[5.17 \text{ m/s}^2 (-\hat{i}) + 2.5 \text{ m/s}^2 (\hat{j}) \right] - \left[32 \text{ N}(\hat{i}) + 48 \text{ N}(-\hat{j}) \right] = 156 \text{ N}(-\hat{i}) + 108 \text{ N}(\hat{j})$$

REFLECT   Since there were 2 components for the force and the acceleration, care was needed to treat each component individually.

51.   **ORGANIZE AND PLAN**   Draw a diagram of the situation that includes all the forces involved. Compute the net force and use Newton's law to calculate the acceleration. From this we can find the final speed.

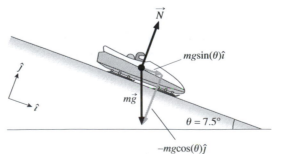

Known: $\theta = 7.5°$; $t = 25$ s; $g = 9.8$ m/s^2

SOLVE   The net force is [Eq. 1] $\vec{F}_{net} = mg\sin(\theta)\left(\hat{i}\right)$. From Newton's second law, this gives an acceleration of [Eq. 2]

$$\vec{a} = \vec{F}_{net}/m = g\sin(\theta)\left(\hat{i}\right)$$
$$\vec{a} = \left(9.8 \text{ m/s}^2\right)\sin\left(7.5°\right)\left(\hat{i}\right) = 1.3 \text{ m/s}^2\left(\hat{i}\right)$$

After 25 s, the final speed will be [Eq. 3] $\vec{v}_f = \vec{v}_0 + \vec{a}t = 0 + \left(1.3 \text{ m/s}^2\right)\left(25 \text{ s}\right)\left(\hat{i}\right) = 32 \text{ m/s}\left(\hat{i}\right)$.

REFLECT   Comparing to Problem 50, we see that the bobsled is traveling at about 3 times the speed of your average dog.

53.   **ORGANIZE AND PLAN**   Calculate the mass of the various body parts [Eq. 1].

$$m_B = 65 \text{ kg}$$
$$m_H = (0.06)(65 \text{ kg}) = 3.9 \text{ kg}$$
$$m_L = (0.345)(65 \text{ kg}) = 22.4 \text{ kg}$$

The subscripts B, H, and L refer to the body, head, and legs, respectively. Make a force diagram and include all the forces acting on *each* relevant body part.

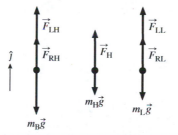

We can assume that each hand and each hip joint supports the same force so [Eq. 2]

$$\vec{F}_{RH} = \vec{F}_{LH}$$
$$\vec{F}_{RL} = \vec{F}_{LL}$$

where the subscripts RH and LH refer to the right and left hand, respectively, and RL and LL refer to the right and left leg, respectively. Sum the forces in each situation to find the net force acting on the various body parts, and then use Newton's second law to find the unknown forces.

Known: masses of body parts (see Eq. 1), $\vec{g} = 9.8 \text{ m/s}^2\left(-\hat{j}\right)$, $\vec{a} = 0.50g = 4.9 \text{ m/s}^2\left(\hat{j}\right)$

SOLVE   (a) Sum the forces in the first force diagram to find the net force. Use Newton's second law and Eqs. 1 and 2 to find the force exerted by the rope on a *single* hand [Eq. 3]:

$$\vec{F}_{net} = \vec{F}_{RH} + \vec{F}_{LH} + m_B\vec{g} = m_B\vec{a}$$
$$2\vec{F}_{RH} = m_B\vec{a} - m_B\vec{g}$$
$$\vec{F}_{RH} = \frac{m_B\vec{a} - m_B\vec{g}}{2} = \frac{\left(65 \text{ kg}\right)\left[4.9 \text{ m/s}^2 - \left(-9.8 \text{ m/s}^2\right)\right]\left(\hat{j}\right)}{2} = 479 \text{ N}\left(\hat{j}\right)$$

Thus, the force acting on each hand is $479 \text{ N}\left(\hat{j}\right)$.

(b) Repeat the same procedure, but for the middle force diagram [Eq. 4]:

$$\vec{F}_{net} = \vec{F}_H + m_H\vec{g} = m_H\vec{a}$$
$$\vec{F}_H = m_H\vec{a} - m_H\vec{g} = \left(3.9 \text{ kg}\right)\left[4.9 \text{ m/s}^2 - \left(-9.8 \text{ m/s}^2\right)\right]\left(\hat{j}\right) = 57 \text{ N}\left(\hat{j}\right)$$

Thus, the rope exerts $57 \text{ N}\left(\hat{j}\right)$ on her head.

(c) Repeat the same procedure for the final force diagram [Eq. 5]:

$$\vec{F}_{net} = \vec{F}_{RL} + \vec{F}_{LL} + m_L \vec{g} = m_L \vec{a}$$

$$2\vec{F}_{RL} = m_L \vec{a} - m_L \vec{g}$$

$$\vec{F}_{RL} = \frac{m_L \vec{a} - m_L \vec{g}}{2} = \frac{(22.4 \text{ kg})\left[4.9 \text{ m/s}^2 - (-9.8 \text{ m/s}^2)\right](\hat{j})}{2} = 165 \text{ N}(\hat{j})$$

Thus the rope exerts $165 \text{ N}(\hat{j})$ on each leg.

REFLECT When there are two structural members sustaining a given force, each member sustains a fraction of the total force. This is the case with her hands, each of which sustained half the total force exerted by the rope on her body.

55.   **ORGANIZE AND PLAN** Draw a force diagram of the situation (see the figure below). Note that we have set up a coordinate system for each mass, and that the $\hat{i}$ direction for the glider corresponds to the $\hat{i}$ direction for the hanging mass. In other words, if you pull on the hanging mass in its $\hat{i}$ direction, the cable will pull on the glider in the glider's $\hat{i}$ direction. Doing this makes it more apparent that the tension forces are opposite in sign. Furthermore, since the pulley is frictionless (and the cable weightless), the magnitude of the tension on each element is the same. Thus we can write [Eq. 1] $\vec{T}_1 = -\vec{T}_2 = \vec{T}$. In addition, since the cable length does not change, the magnitude of the acceleration will be the same for the block and the hanging mass, so [Eq. 2] $a_1 = a_2 \equiv a$. Knowing all the forces acting on each object, we can use Newton's second law to calculate their acceleration. We can omit the vector notation, since the directions of the forces are clear from the diagram.

Known: $m_1 = 0.230$ kg ; $m_2 = 0.100$ kg

SOLVE (a) See the force diagram below.

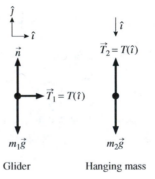

Glider   Hanging mass

(b) The net force acting on the block is [Eq. 3]

$$F_{net,1} = T = m_1 a$$

On the hanging mass, the net force is [Eq. 4]

$$F_{net,2} = m_2 g - T = m_2 a$$

Summing Eqs. (3) and (4) gives [Eq. 5]

$$m_2 g = (m_1 + m_2) a$$

$$a = g \frac{m_2}{m_1 + m_2} = (9.8 \text{ m/s}^2)\left(\frac{0.100 \text{ kg}}{0.230 \text{ kg} + 0.100 \text{ kg}}\right) = 2.97 \text{ m/s}^2$$

Therefore the magnitude of the acceleration of each block is 2.97 m/s², in the $\hat{i}$ directions indicated for each element in the figure above.

REFLECT If the pulley were not frictionless, we would not have been able to equate the tension forces on each object. However, we could have still solved the problem, but with a different relationship between the tension forces $\vec{T}_1$ and $\vec{T}_2$. Also note that the weight and the normal force on the glider must cancel each other, since the block does not move in the vertical direction.

57.   **ORGANIZE AND PLAN**   Use a coordinate system where $\hat{j}$ indicates the vertical upward direction. Using the kinematic equations 2.8 and 2.9, calculate the velocity gained from falling 1.60 m from a stationary start. Begin by calculating the time elapsed during the fall [Eq. 1]:

$$\vec{x} = \vec{x}_0 + \vec{v}_0 + (1/2)\vec{g}t^2$$

$$x - x_0 = -(1/2)gt^2$$

$$t = \sqrt{\frac{2(x_0 - x)}{g}} = \sqrt{\frac{2(1.60 \text{ m})}{9.8 \text{ m/s}^2}}$$

$$t = 0.571 \text{ s}$$

Now calculate the final velocity of the fall [Eq. 2]:

$$\vec{v}_f = \vec{v}_0 + \vec{g}t = 0 + (gt)(-\hat{j}) = (9.8 \text{ m/s}^2)(0.571 \text{ s})(-\hat{j}) = 5.60 \text{ m/s}(-\hat{j})$$

Knowing his final velocity, calculate the acceleration he experiences upon hitting the floor. Following this use Newton's second law to find the force exerted upon him by the floor.

Known: $m = 78$ kg; $t_{stop} = 0.750$ s

SOLVE   (a) Begin by calculating the acceleration he experiences upon hitting the floor. At this point, $\vec{v}_f = 0$ m/s because he is stationary after the acceleration period, and, from Eq. 2, $\vec{v}_0 = 5.60$ m/s$(-\hat{j})$ [Eq. 3]:

$$\vec{v}_f = \vec{v}_0 + \vec{a}t_{stop}$$

$$\vec{a} = -(\vec{v}_0/t_{stop}) = \frac{-5.60 \text{ m/s}}{0.750 \text{ s}}(-\hat{j}) = 7.47 \text{ m/s}^2(\hat{j})$$

To generate this acceleration, the floor must exert a net force on him of [Eq. 4]

$$\vec{F}_{net} = m\vec{a} = (78 \text{ kg})(7.47 \text{ m/s}^2)(\hat{j}) = 582 \text{ N}(\hat{j}).$$

Since the floor exerts this net force on his two feet, each foot receives half this net force, or $291 \text{ N}(\hat{j})$, or 65 lbs $\hat{j}$.

(b) If he comes to a stop in 0.100 s, then we recalculate his acceleration from Eq. 3 with $t_{stop} = 0.100$ s. This gives [Eq. 5]

$$\vec{v}_f = \vec{v}_0 + \vec{a}t_{stop}$$

$$\vec{a} = -(\vec{v}_0/t_{stop}) = \frac{-5.60 \text{ m/s}}{0.100 \text{ s}}(-\hat{j}) = 56.0 \text{ m/s}^2(\hat{j})$$

and the net force on his two legs is [Eq. 6]: $\vec{F}_{net} = m\vec{a} = (78 \text{ kg})(56.0 \text{ m/s}^2)(\hat{j}) = 4368 \text{ N}(\hat{j}).$

Thus the force on each leg is $2184 \text{ N}(\hat{j})$, or 491 lbs $\hat{j}$.

(c) He is more likely to sustain an injury in case (b) because each foot must sustain a much larger force.
REFLECT   By absorbing his fall somewhat with his legs, he can reduce the force from 491 lbs to about 65 lbs; a significant gain. This shows the importance of "knowing how to fall."

59.   **ORGANIZE AND PLAN**   See the figure below for a force diagram of the situation. Sum the forces on the sled to find the net force, and then use Newton's second law to find the acceleration due to that force. Chose the incline angle appropriately so that this acceleration gives the desired velocity.

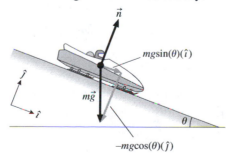

Known: $\vec{v}_0 = 0$ ms; $\vec{v}_f = 40$ m/s$(\hat{i})$; $t = 30$ s

SOLVE   Since the sled presumably does not accelerate off the surface of the ice, the forces in the $\hat{j}$ direction must sum to zero (Newton's second law). Therefore the net force on the sled is [Eq. 1] $\vec{F}_{net} = mg\sin(\theta)(\hat{i}) = m\vec{a}$. So [Eq. 2] $\vec{a} = g\sin(\theta)(\hat{i})$.

Inserting this into the equation of motion gives [Eq. 3]

$$\vec{v}_f = \vec{v}_0 + \vec{a}t$$

$$g\sin(\theta)(\hat{i}) = \vec{v}_f / t$$

$$\theta = a\sin\left(\frac{v_f}{gt}\right) = a\sin\left(\frac{40 \text{ m/s}}{(9.8 \text{ m/s}^2)(30 \text{ s})}\right) = 7.8°$$

REFLECT   Notice that the argument of the a sin function is dimensionless, as it should be. Also, if $\vec{v}_f = 0$, then $\theta = 0$; i.e., there would be no incline.

61.   **ORGANIZE AND PLAN**   Make a force diagram of the situation, including both the neck brace and the cable bifurcation.

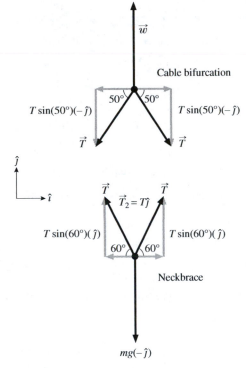

Note that the tension T in the cable is the same in the neck brace force diagram and the cable bifurcation force diagram. Use Newton's second law to find T in terms of w from the cable bifurcation force diagram, and then use the result for T to find the total upward force on the neck brace.
Known: $w = 100$ N

SOLVE   Applying Newton's second law to the cable bifurcation gives [Eq. 1]

$$\vec{F}_{net}^{vert} = (T\sin(50°) + T\sin(50°))(-\hat{j}) + w(\hat{j})$$

$$T = \frac{w}{2\sin(50°)}$$

where the last step is justified because the net force is zero (no acceleration). The total upward force acting on the neck brace is [Eq. 2] $\vec{F}_{up} = 2T\sin(60°)$.

Inserting the result for T from Eq. (1) into Eq. (2) gives (Eq. 3)

$$\vec{F}_{up} = \frac{2w\sin(60°)}{2\sin(50°)}(\hat{j}) = 113 \text{ N}(\hat{j})$$

Thus the neck brace experiences an upward force of 113 N.

REFLECT This value is greater than the total weight of the hanging mass because the cables attached to the neck brace are under the same tension as the cables attached to the bifurcation, but at the neck brace the tension has a larger vertical component. Notice that in the limit that the bifurcation angle tends to zero or 90°, the upward force tends to w (i.e., 100 N).

63. **ORGANIZE AND PLAN** Since the blocks cannot separate from each other due to the application of this force, the acceleration of each block will be the same. Therefore, consider the 3-block set initially as a single large block. To calculate the net forces on each individual block, apply Newton's second law using the mass of the block in question. Use a coordinate system in which $\hat{i}$ indicates the direction in which the force is applied.

Known: $m_1 = 1.5$ kg; $m_2 = 1.0$ kg; $m_3 = 0.5$ kg; $\vec{F} = 36$ N $(\hat{i})$

SOLVE (a) Using Newton's second law to find the acceleration of the 3-block set gives [Eq. 1]

$$\vec{F} = (m_1 + m_2 + m_3)\vec{a}$$

$$\vec{a} = \frac{36 \text{ N}}{1.5 \text{ kg} + 1.0 \text{ kg} + 0.5 \text{ kg}}(\hat{i})$$

$$= 12 \ m/s^2 (\hat{i}).$$

(b) Since the vertical net force must be zero, we only need find the net force in the $\hat{i}$ direction. The net force on block 1 is [Eq. 2] $\vec{F}_1^{net} = m_1\vec{a} = (1.5 \text{ kg})(12 \text{ m/s}^2)(\hat{i}) = 18$ N $(\hat{i})$.

The net force on block 2 is [Eq. 3] $\vec{F}_2^{net} = m_2\vec{a} = (1.0 \text{ kg})(12 \text{ m/s}^2)(\hat{i}) = 12$ N $(\hat{i})$.

The net force on block 3 is [Eq. 4] $\vec{F}_3^{net} = m_3\vec{a} = (0.5 \text{ kg})(12 \text{ m/s}^2)(\hat{i}) = 6$ N $(\hat{i})$.

(c) Block 2 must push on block 3 with the force needed to accelerate block 3 at 12 m/s², which is just $\vec{F}_3^{net} = 6N(\hat{i})$. Therefore, we can say that [Eq. 5] $\vec{F}_{2-3} = \vec{F}_3^{net} = 6$ N $(\hat{i})$.

Block 1 must push on block 2 with enough force to accelerate block 2 and 3 at 12 m/s². Therefore [Eq. 6] $\vec{F}_{1-2} = \vec{F}_2^{net} + \vec{F}_3^{net} = 18N(\hat{i})$.

REFLECT Notice that the sum of the net forces on the individual blocks equals the force applied to the 3-block set.

65. **ORGANIZE AND PLAN** Knowing that the cannon ball's acceleration in free-fall is just that due to gravity, find the velocity of the cannon ball the moment it impacts the Earth. From this, calculate the acceleration the ball must experience to come to a stop within 0.130 m. Finally, use Newton's second law to find the force supplied by the Earth to produce that acceleration.

We choose a coordinate system in which $\hat{j}$ indicates the vertical upward direction.

Known: $m = 2.5$ kg, $\vec{x}_f^t - \vec{x}_0^t = 58.4$ m $(-\hat{j})$, $\vec{x}_f^g - \vec{x}_0^g = 0.130$ m $(\hat{j})$, $\vec{v}_0^t = 0$, $\vec{v}_f^g = 0$, $\vec{v}_f^t = \vec{v}_0^g$, $\vec{g} = 9.8$ m/s² $(-\hat{j})$

SOLVE To find the velocity of the ball the moment it impacts the ground, we use the kinematic equations [Eq. 1]

$$\vec{v}_f^t = \vec{v}_0^t + \vec{g}t$$

$$\vec{v}_f^t = \vec{g}t$$

and [Eq. 2]

$$\vec{x}_f^t = \vec{x}_0^t + \vec{v}_0^t t + 1/2\vec{g}t^2$$

$$t = \pm\sqrt{\frac{2(x_f^t - x_0^t)}{g}}$$

We choose the positive square root since the negative square root is unphysical. Inserting Eq. 2 into Eq. 1 gives [Eq. 3]

$$\vec{v}_f^t = \vec{g}\sqrt{\frac{2(x_f^t - x_0^t)}{g}} = -\sqrt{2g(x_f^t - x_0^t)}$$

The acceleration needed to reduce this velocity to zero within 0.130 m is found using Eqs. 4 and 5:

$$\vec{v}_f^g = \vec{v}_0^g + \vec{a}t$$

$$\vec{v}_0^g = -\vec{a}t$$

$$\vec{x}_f^g = \vec{x}_0^g + \vec{v}_0^g t + 1/2\vec{a}t^2$$

$$t = \pm \sqrt{\frac{2\left(x_f^g - x_0^g\right)}{a}}$$

Combining Eqs. 3, 4, and 5, and using $\vec{v}_f^t = \vec{v}_0^g$, gives

$$\vec{v}_0^g = \sqrt{2a\left(x_f^g - x_0^g\right)}\left(-\hat{j}\right)$$

$$\vec{a} = \frac{v_0^{g2}}{2\left(x_f^g - x_0^g\right)}\left(-\hat{j}\right) = \frac{g\left(x_f^t - x_0^t\right)}{x_f^g - x_0^g}\left(\hat{j}\right) = 4402 \text{ m/s}^2\left(\hat{j}\right)$$

From Newton's second law, the force needed to produce this acceleration is

$$\vec{F}_{net} = m\vec{a} = (2.5 \text{ kg})(4402 \text{ m/s}^2)\hat{j} = 11{,}006 \text{ N}(\hat{j}).$$

REFLECT   Eq. 6, line 3 shows that the ratio of the braking acceleration to the acceleration due to gravity is equal to the ratio of the stopping distance to the height of the tower.

67.   **ORGANIZE AND PLAN**   Draw a force diagram for each situation. Assume that each crutch carries an equal amount of weight, so [Eq.1] $\vec{F}_R = \vec{F}_L$.

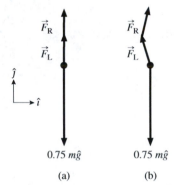

(a)   (b)

Note that in the force diagram, the weight the crutches must carry is 75% of the person's total weight.

Known: m = 78 kg; $\theta = 15°$

SOLVE   (a) The force applied by each crutch is found by calculating the net vertical force and setting it equal to zero, since there is no acceleration in the vertical direction. Using Eq. 1, this gives [Eq. 2]

$$\vec{F}_{net} = 0$$

$$0 = \vec{F}_L + \vec{F}_R + 0.75 \, m\vec{g} = 2\vec{F}_L - 0.75(78 \text{ kg})(9.8 \text{ m/s}^2)(\hat{j})$$

$$\vec{F}_L = 287 \text{ N}(\hat{j})$$

Thus, by Eq. 1, each crutch applies a force of $287 N(\hat{j})$ to the person.

(b) With the crutches inclined at 15° from vertical [Eq. 3], $\vec{F}_{R,L} \rightarrow F_{R,L}\cos(\theta)(\hat{j})$.

Using this mapping in Eq. 2 gives [Eq. 4]

$$\vec{F}_{net} = 0$$

$$0 = 2F_L\cos(\theta)(\hat{j}) - 0.75(78 \text{ kg})(9.8 \text{ m/s}^2)(\hat{j})$$

$$\vec{F}_L = 297 \text{ N}(\hat{j})$$

Thus, each crutch applies 297 N to the person.

REFLECT   If we are not sure whether to use $\sin(\theta)$ or $\cos(\theta)$ in Eq. 3, we can just look at a limiting situation. When $\theta \rightarrow 0$ Eq. 4 must reduce to Eq. 2, which is the case only for $\cos(\theta)$.

69.   **ORGANIZE AND PLAN**   Use Newton's second law to find the acceleration of the block as a function of the cable tension, then set this tension equal to the maximum value allowed to find the maximum acceleration possible for the block.

Known: $m = 350$ kg , $T_{max} = 4,200$ N

SOLVE   Newton's second law gives us [Eq. 1]

$$\vec{F}_{net} = m\vec{g} + T_{max}\left(\hat{j}\right) = m\vec{a}_{max}$$

$$\vec{a}_{max} = \frac{m\vec{g} + T_{max}\left(\hat{j}\right)}{m} = \frac{(350 \text{ kg})(9.8 \text{ m/s}^2)\left(-\hat{j}\right) + (4,200 \text{ N})\left(\hat{j}\right)}{350 \text{ kg}} = 2.2 \text{ m/s}^2\left(\hat{j}\right)$$

REFLECT   This acceleration is about 25% that of the acceleration due to the earth's gravity.

71.   **ORGANIZE AND PLAN**   Using Eq. 4.7, the magnitude of the rolling friction force is [Eq. 1] $\vec{f}_r = \mu_r n$. The friction force will act to oppose the velocity, which we assume is in the $\hat{i}$ direction. Therefore [Eq. 2], $\vec{f}_r = \mu_r mg\left(-\hat{i}\right)$.

From this and Newton's second law we can find the acceleration, and then the distance traveled.

Known: $\mu_r = 0.045$; $\vec{v}_0 = 2.45 \text{ m/s}\left(\hat{i}\right)$; $\vec{v}_f = 0$ m/s

SOLVE   (a) Using Newton's second law and Eq. 2, the acceleration of the golf ball is [Eq. 3]

$$\vec{F}_{net} = \vec{f}_r = m\vec{a}$$

$$\vec{a} = \frac{\mu_r mg}{m}\left(-\hat{i}\right) = 0.045(9.8 \text{ m/s}^2)\left(-\hat{i}\right) = 0.44 \text{ m/s}^2\left(-\hat{i}\right)$$

(b) To find the distance traveled before stopping, we use [Eq. 4] $\vec{v}_f = \vec{v}_0 + \vec{a}t$ and [Eq. 5] $\vec{x}_f = \vec{x}_0 + \vec{v}_0 t + 1/2\vec{a}t^2$ to find [Eq. 6]

$$\vec{x}_f - \vec{x}_0 = \vec{v}_0\frac{v_0}{a} + 1/2\vec{a}\left(\frac{v_0}{a}\right)^2 = \frac{(2.45 \text{ m/s})^2}{0.44 \text{ m/s}^2}(\hat{i}) + \frac{(2.45 \text{ m/s})^2}{2(0.44 \text{ m/s}^2)}\left(-\hat{i}\right) = 6.8 \text{ m}\left(\hat{i}\right)$$

Thus the ball travels 6.8 m in the direction of the initial velocity before it stops.

REFLECT   Notice that we have to be careful with the direction of the vector quantities in Eq. 6 to ensure the correct result.

73.   **ORGANIZE AND PLAN**   Use the kinematic equations to find the time it takes the stone to stop, and its acceleration. Knowing the acceleration, use Newton's second law to find the coefficient of kinetic friction. Use a coordinate system where the initial velocity is in the $\hat{i}$ direction.

Known: $m = 19$ kg; $\vec{v}_0 = 1.50 \text{ m/s}\left(\hat{i}\right)$; $\vec{v}_f = 0$ m/s; $\vec{x} - \vec{x}_0 = 28.4 \text{ m}\left(\hat{i}\right)$

SOLVE   (a) The force diagram is shown in the figure below.

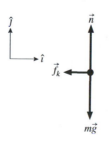

From this diagram, the acceleration is [Eq. 1]

$$\vec{F}_{net} = \vec{f}_k = m\vec{a}$$

$$\vec{a} = \mu_k g\left(-\hat{i}\right).$$

The result for μ_k, calculated in part (c), is 0.04. Inserting this into Eq. (1) gives [Eq. 2]

$a = (0.04)(9.8 \text{ m/s}^2) = 0.039 \text{ m/s}^2$.

(b) The time it takes for the stone to stop may be found from the kinematic equations [Eq. 3] $\vec{v}_f = \vec{v}_0 + \vec{a}t$ and [Eq. 4] $\vec{x}_f = \vec{x}_0 + \vec{v}_0 t + 1/2\vec{a}t^2$, which gives [Eq. 5]

$$t = \frac{v_0}{a} = \frac{2(x-x_0)}{v_0} = \frac{2(28.4 \text{ m})}{1.50 \text{ m/s}} = 37.9 \text{ s}$$

(c) Inserting the result of Eq. 5 into Eq. 3 gives [Eq. 6]

$$\vec{a} = -\vec{v}_0/t = -\frac{(1.50 \text{ m/s})}{(37.7 \text{ s})}(\hat{i}) = 0.0396 \text{ m/s}^2(-\hat{i})$$

REFLECT   The coefficient of kinetic friction is dimensionless, as expected.

75.   **ORGANIZE AND PLAN**   Use Newton's second law to find the force due to kinetic friction, and then find the corresponding coefficient of kinetic friction.

Known: $\theta = 28°$; $\vec{a} = 3.85 \text{ m/s}^2(\hat{i})$

SOLVE   (a) The force diagram is shown in the figure below.

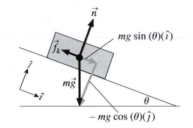

(b) Using Newton's second law and Eq. 4.6, we get [Eq. 1]

$$\vec{F}_{net} = \vec{f}_k + mg\sin(\theta)(\hat{i}) = m\vec{a}$$

$$\mu_k mg\cos(\theta)(-\hat{i}) + mg\sin(\theta)(\hat{i}) = ma(\hat{i})$$

$$\mu_k = \frac{g\sin(\theta) - a}{g\cos(\theta)} = \frac{(9.8 \text{ m/s}^2)\sin(28°) - 3.85 m/s^2}{(9.8 \text{ m/s}^2)\cos(28°)}$$

$$\mu_k = 0.087$$

REFLECT   If $\theta \to 0$ then the expression for μ_k reduces to that found for the case of an object sliding on a horizontal surface (e.g., Eq. 6, Problem 73), as expected. If $\theta \to 90°$, then $\mu_k \to \infty$. Is this reasonable?

77.   **ORGANIZE AND PLAN**   The force diagram is the same as for Problem 75, except that the force due to kinetic friction is replaced by the force due to static friction since the book is at rest. Use Newton's second law to find the magnitude of each force.

Known: $\theta = 15°$; $\vec{a} = 0 \text{ m/s}^2$; $m = 1.75 \text{ kg}$

SOLVE   (a) See the figure below for the force diagram.

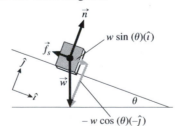

(b) The force on the book due to gravity (i.e., the weight) is [Eq. 1]

$$\vec{w} = m\vec{g} = mg\left[\sin(\theta)(\hat{i}) + \cos(\theta)(-\hat{j})\right] = (1.75 \text{ kg})(9.8 \text{ m/s}^2)\left[\sin(15°)(\hat{i}) + \cos(15°)(-\hat{j})\right] = 4.44 \text{ N}(\hat{i}) + 16.6 \text{ N}(-\hat{j})$$

The magnitude of the weight is [Eq. 2] $w = mg = (1.75 \text{ kg})(9.8 \text{ m/s}^2) = 17.2 \text{ N}$.

To find the normal force, we can use Newton's second law in the $\hat{j}$ direction. This gives [Eq. 3]

$$\vec{n} + mg\cos(\theta)(-\hat{j}) = 0$$

$$\vec{n} = mg\cos(\theta)(\hat{j}) = 16.6 \text{ N}(\hat{j})$$

Finally, by applying Newton's second law in the $\hat{i}$ direction, the force due to friction is [Eq. 4]

$$\vec{f}_s + mg\sin(\theta)(\hat{i}) = 0$$

$$\vec{f}_s = mg\sin(\theta)(-\hat{i})$$

$$\vec{f}_s = 4.44 \text{ N}(-\hat{i})$$

To summarize, the forces are [Eq. 5]:

$$w = 17.2 \text{ N}$$

$$n = 16.6 \text{ N}$$

$$f_s = 4.44 \text{ N}$$

REFLECT   We can find the coefficient of static friction from the ratio of the force due to static friction to the normal force (see Eq. 4.8). This gives [Eq. 6]

$$\mu_s \leq \frac{n}{f_s} = \frac{16.66 \text{ N}}{4.44 \text{ N}}$$

$$\mu_s \leq 3.73$$

Notice that this (upper limit) coefficient is much greater than the typical kinetic coefficients we found in the previous several problems.

79. **ORGANIZE AND PLAN**   Knowing the coefficient of rolling friction, we can find the acceleration as a function of gravity due to the rolling frictional force. With the acceleration, we can calculate the distance traveled before the car stops. We choose a coordinate system in which the initial velocity of the car is in the $\hat{i}$ direction. Be careful to convert the velocity to the appropriate units for the calculation.

Known: $\vec{v}_0 = 50 \text{ km/h}(\hat{i}); \quad \vec{v}_f = 0 \text{ km/h}; \quad \mu_r = 0.023$

SOLVE   Using Newton's second law and the expression for the rolling frictional force (Eq. 4.7), we have {Eq. 1]

$$\vec{F}_{net} = \vec{f}_r = m\vec{a}$$

$$\mu_r n(-\hat{i}) = m\vec{a}$$

$$\mu_r mg(-\hat{i}) = m\vec{a}$$

$$\vec{a} = \mu_r g(-\hat{i}) = (0.023)(9.8 \text{ m/s}^2)(-\hat{i})$$

$$\vec{a} = 0.225 \text{ m/s}^2(-\hat{i})$$

To find the distance traveled before stopping we use the kinematic equations [Eq. 2] $\vec{v}_f = \vec{v}_0 + \vec{a}t$ and $\vec{x}_f = \vec{x}_0 + \vec{v}_0 t + 1/2\vec{a}t^2$.

Solving for $\vec{x}_f - \vec{x}_0$, we find [Eq. 4]

$$\vec{x}_f - \vec{x}_0 = \frac{v_0^2}{2a}(\hat{i}) = \frac{(50 \text{ km/h})^2}{2(0.225 \text{ m/s}^2)}\left(\frac{1000 \text{ m}}{1 \text{ km}}\right)^2\left(\frac{1 \text{ h}}{60 \text{ min}}\right)^2\left(\frac{1 \text{ min}}{60 \text{ s}}\right)^2(\hat{i}) = 429 \text{ m}(\hat{i})$$

REFLECT   Notice how all the units cancel in Eq. 4 to leave us with meters.

81.   **ORGANIZE AND PLAN** This problem is essentially the same as Problem 79, where we found that the acceleration due to a frictional force for an object moving on a horizontal surface is [Eq. 1] $\vec{a} = \mu_k g\left(-\hat{i}\right)$, where $\hat{i}$ indicates the direction of the object's initial velocity.

Known: $m_1 = 1000\,\text{kg}$; $m_2 = 2000\,\text{kg}$; $\mu_k = 0.25$; $\vec{v}_0 = 50\,\text{km/h}\left(\hat{i}\right)$

SOLVE (a) Solving Eq. 1 to find the acceleration of both the car and the truck, we find [Eq. 2]

$$\vec{a} = (0.25)(9.8\,\text{m/s}^2)\left(-\hat{i}\right) = 2.45\,\text{m/s}^2\left(-\hat{i}\right)$$

To find the distance traveled by both the car and the truck before stopping we use the kinematic equations [Eq. 3] $\vec{v}_f = \vec{v}_0 + \vec{a}t$ and [Eq. 4] $\vec{x}_f = \vec{x}_0 + \vec{v}_0 t + 1/2\vec{a}t^2$.

Solving for $\vec{x}_f - \vec{x}_0$, we find [Eq. 5]

$$\vec{x}_f - \vec{x}_0 = \frac{v_0^2}{2a}\left(\hat{i}\right) = \frac{(50\,\text{km/h})^2}{2(2.45\,\text{m/s}^2)}\left(\frac{1000\,\text{m}}{1\,\text{km}}\right)^2\left(\frac{1\,\text{h}}{60\,\text{min}}\right)^2\left(\frac{1\,\text{min}}{60\,\text{s}}\right)^2\left(\hat{i}\right) = 39.4\,\text{m}\left(\hat{i}\right)$$

independent of the mass of the vehicle.

(b) The stopping distance of two 1000 kg cars, one with an initial velocity of $\vec{v}_{0,1} = 50\,\text{km/h}\left(\hat{i}\right)$ and the other with an initial velocity of $\vec{v}_{0,2} = 100\,\text{km/h}\left(\hat{i}\right)$, is given by Eq. 5. We have already found the stopping distance for the first car (see Eq. 5). For the second car, we find [Eq. 6]

$$\vec{x}_f - \vec{x}_0 = \frac{v_0^2}{2a}\left(\hat{i}\right) = \frac{(100k\,\text{m/h})^2}{2(2.45\,\text{m/s}^2)}\left(\frac{1000\,\text{m}}{1\,\text{km}}\right)^2\left(\frac{1\,\text{h}}{60\,\text{min}}\right)^2\left(\frac{1\,\text{min}}{60\,\text{s}}\right)^2\left(\hat{i}\right) = 157\,\text{m}\left(\hat{i}\right)$$

REFLECT Why do the car and the truck both travel the same distance before stopping? The frictional force on the truck is twice that on the car, but its mass is twice the car's mass, so the two cancel and the acceleration of the car and the truck is the same. Notice as well that the stopping distance is proportional to the initial velocity *squared*, so that if the initial velocity is doubled, the stopping distance is quadrupled. Keep this in mind when you speed.

83.   **ORGANIZE AND PLAN** Draw a force diagram of the situation and use Newton's second law to find the force exerted by the student.

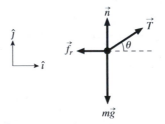

Note that the acceleration of the suitcase is zero since it is moving with constant velocity.

Known: $m = 22\,\text{kg}$; $\theta = 50°$; $\vec{f}_r = 75N\left(-\hat{i}\right)$; $\vec{a} = 0$

SOLVE (a) Apply Newton's law in the $\hat{i}$ direction [Eq. 1].

$$f_r\left(-\hat{i}\right) + T\cos(\theta)\left(\hat{i}\right) = m\vec{a} = 0$$

$$T = \frac{f_r}{\cos(\theta)} = 117\,\text{N}$$

Thus the magnitude of the force exerted by the student is 117 N. The direction of the force is at 50° above the horizontal.

(b) The force due to friction is given by [Eq. 2] $\vec{f}_r = \mu_r n\left(-\hat{i}\right)$,

and the normal force can be found by applying Newton's second law in the $\hat{j}$ direction [Eq. 3]:

$$\vec{n} + mg\left(-\hat{j}\right) + T\sin(\theta)\left(\hat{j}\right) = 0$$

$$\vec{n} = \left[mg - T\sin(\theta)\right]\left(\hat{j}\right)$$

Thus the coefficient of rolling friction is [Eq. 4]:

$$\mu_r = \frac{f_r}{n} = \frac{f_r}{mg - T\sin(\theta)} = \frac{75\ \text{N}}{(22\ \text{kg})(9.8\ \text{m/s}^2) - (117\ \text{N})\sin(50°)} = 0.595$$

REFLECT   The upward pull of the student reduces the normal force exerted by the floor, and with it the force due to friction.

85.   **ORGANIZE AND PLAN**   Draw a force diagram for this problem (see the figure below). Use the same strategy as for Problem 84, but referring to the force diagram for this problem.

Known: hanging mass $m_h = 0.100$ kg, mass of wooden block $m_w = 0.300$ kg, $\theta = 15°$

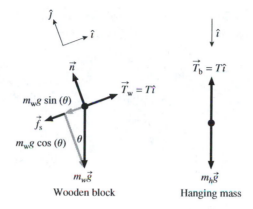

Wooden block   Hanging mass

SOLVE   (a) To find the static friction coefficient, apply Newton's second law in the $\hat{i}$ direction to the wooden block. This gives [Eq. 1]

$$m_w a = -f_s - m_w g\sin(\theta) + T$$
$$0 = -\mu_s m_h g\cos(\theta) - m_w g\sin(\theta) + m_h g$$
$$\mu_s = \frac{m_h - m_w\sin(\theta)}{m_w\cos(\theta)} = 0.077$$

(b) Assume the wooden block slides up the incline. In this case the static friction force will be oriented as drawn in the figure above (i.e., opposing the motion of the wooden block). Applying Newton's second law in the $\hat{i}$ directions of both the wooden block and the hanging mass gives [Eq. 2]

$$m_w a = -f_s - m_w g\sin(\theta) + T$$
$$-m_h a = m_h g - T$$

where a is the *magnitude* of the acceleration (its direction has been accounted for by the signs). Solving for the acceleration a gives [Eq. 3]

$$a = g\left(\frac{m_h - m_w\left[\mu_k\cos(\theta) + \sin(\theta)\right]}{m_w + m_h}\right)$$

$$= (9.8\ \text{m/s}^2)\frac{0.100\ \text{kg} - (0.300\ \text{kg})\left[(0.150)\cos(15°) + \sin(15°)\right]}{0.100\ \text{kg} + 0.300\ \text{kg}}$$

$$= -0.053\ \text{m/s}^2$$

which is not physical since the magnitude of the acceleration must be greater than or equal to zero. Therefore, the hanging mass does not supply enough tension to pull the wooden block up the incline.

Assume now that the wooden block slides down the incline. Since the friction force always acts to oppose the motion, it will act in the $\hat{i}$ direction of the wooden block in this case. Again applying Newton's second law to both blocks gives [Eq. 4]

$$-m_w a = f_k - m_w g\sin(\theta) + T$$
$$-m_h a = m_h g - T$$

where a is again the magnitude of the acceleration. Solving for the acceleration a gives [Eq. 5]

$$a = g\left(\frac{m_w\left[\sin(\theta) - \mu_k \cos(\theta)\right] - m_h}{m_w + m_h}\right)$$

$$= (9.8 \text{ m/s}^2)\left(\frac{(0.300 \text{ kg})\left[\sin(15°) - (0.150)\cos(15°)\right] - 0.100 \text{ kg}}{0.300 \text{ kg} + 0.100 \text{ kg}}\right)$$

$$= -1.61 \text{ m/s}^2$$

which is again unphysical because the magnitude of the acceleration cannot be negative. Thus the block does not move.

REFLECT   Since the block does not move, we should have used the static friction coefficient in part (b), but since $\mu_s \geq \mu_k$, this would only make the results more negative, so our solution remains valid.

87.   **ORGANIZE AND PLAN**   Use Newton's second law applied to each case (incline and horizontal). Note that the final speed for case (b) is the initial speed for case (c).

Known: $\theta = 7.5°$; $\vec{x}_b - \vec{x}_{0,b} = 40 \text{ m}\left(\hat{i}_b\right)$; $\vec{v}_{0,b} = 0 \text{ m/s}$; $m = 35 \text{ kg}$; $\mu_k = 0.060$

SOLVE   (a) See the force diagram below.

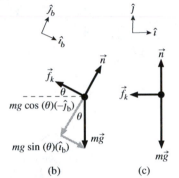

(b)   (c)

(b) To find the sled's speed at the bottom of the incline, we fist need to find the acceleration of the sled. Applying Newton's second law in the $\hat{j}_b$ direction, we find [Eq. 1]

$$\vec{n} + mg\cos(\theta)\left(-\hat{j}_b\right) = m\vec{a} = 0$$

$$\vec{n} = mg\cos(\theta)\left(-\hat{j}_b\right)$$

Applying Newton's law in the $\hat{i}_b$ direction, we find [Eq. 2]

$$\vec{f}_k + mg\sin(\theta)\left(\hat{i}_b\right) = m\vec{a}_b$$

$$m\vec{a}_b = \mu_k n\left(-\hat{i}_b\right) + mg\sin(\theta)\left(\hat{i}_b\right) = \mu_k mg\cos(\theta)\left(-\hat{i}_b\right) + mg\sin(\theta)\left(\hat{i}_b\right)$$

$$\vec{a}_b = \mu_k g\cos(\theta)\left(-\hat{i}_b\right) + g\sin(\theta)\left(\hat{i}_b\right) = (9.8 \text{ m/s}^2)\left[-(0.060)\cos(\theta) + \sin(\theta)\right]\left(\hat{i}_b\right) = 0.70 \text{ m/s}^2\left(\hat{i}_b\right)$$

Using the acceleration found in Eq. 2, we find the velocity at the bottom of the incline using the kinematic equations [Eq. 3] $\vec{v}_b = \vec{v}_{0,b} + \vec{a}_b t$ and [Eq. 4] $\vec{x}_b = \vec{x}_{0,b} + \vec{v}_{0,b}t + 1/2\vec{a}_b t^2$. Solving for $\vec{v}_b$, we find [Eq. 5]

$$\vec{v}_b = \sqrt{2\left(\vec{x}_b - \vec{x}_{0,b}\right)a_b}\left(\hat{i}_b\right) = \sqrt{2(40 \text{ m})(0.70 \text{ m/s}^2)}\left(\hat{i}_b\right) = 7.5 \text{ m/s}\left(\hat{i}_b\right)$$

(c) To find the distance traveled on the flat, we repeat the calculations from case (b), but using the force diagram (c) in the figure above, and recalling that the initial speed is 7.5 m/s from Eq. 5. Applying Newton's second law in the $\hat{j}$ direction, we find [Eq. 6]

$$\vec{n} + mg\left(-\hat{j}\right) = m\vec{a} = 0$$

$$\vec{n} = mg\left(\hat{j}\right)$$

Applying Newton's law to the $\hat{i}$ direction, we find [Eq. 7]

$$\vec{f}_k = m\vec{a}_c$$

$$m\vec{a}_c = \mu_k n\left(-\hat{i}\right) = \mu_k mg\left(-\hat{i}\right)$$

$$\vec{a}_c = \mu_k g\left(-\hat{i}\right) = \left(9.8 \text{ m/s}^2\right)(0.060)\left(-\hat{i}\right) = 0.59 \text{ m/s}^2\left(-\hat{i}\right)$$

Using the acceleration found in Eq. 7, and the kinematic equations (Eqs. 3 and 4), we solve for $\vec{x}_c - \vec{x}_{0,c}$ and find [Eq. 8]

$$\vec{x}_c - \vec{x}_{0,c} = \frac{v_{0,c}^2}{2a_c}\left(\hat{i}\right) = \frac{(7.5 \text{ m/s})^2}{2(0.59 \text{ m/s}^2)}\left(\hat{i}\right) = 47.7 \text{ m}\left(\hat{i}\right)$$

(d) The total time for the ride is given by summing the solutions for t in Eq. 3 for each case (b) and (c). For case (b), this gives [Eq. 9]

$$t_b = \frac{v_b}{a_b} = \frac{7.5 \text{ m/s}}{0.70 \text{ m/s}^2} = 10.7 \text{ s}$$

And for case (c) we get [Eq. 10]

$$t_c = \frac{v_c}{a_c} = \frac{7.5 \text{ m/s}}{0.59 \text{ m/s}^2} = 12.7 \text{ s}$$

The sum is [Eq. 11] $t_b + t_c = 23.4$ s, which is the total time for the trip.

REFLECT   To find the distance traveled in case (c) we could have used the equality between the ratio of the accelerations and the ratio of the distances between cases (b) and (c). Explicitly, this gives [Eq. 12]

$$\frac{0.70 \text{ m/s}^2}{0.59 \text{ m/s}^2} = \frac{40 \text{ m}}{x_c - x_{0,c}}$$

$$x_c - x_{0,c} = \frac{0.70 \text{ m/s}^2}{0.59 \text{ m/s}^2} 40 \text{ m} = 47.5 \text{ m}$$

where the difference with the result of Eq. 8 is due to rounding errors.

89.   **ORGANIZE AND PLAN**   This problem is similar to Problem 83, and the force diagram for that problem will be useful for solving this problem. We choose a coordinate system in which $\hat{i}$ indicates the direction of the force we apply to the crate, and $\hat{j}$ indicates the upward vertical direction.

Known: $m = 173$ kg; $\mu_s = 0.57$; $\vec{T} = 900 \text{ N}\left(\hat{i}\right)$

SOLVE   (a) The maximum force due to static friction (see Eq. 4.8) is [Eq. 1]

$$f_s = \mu_s n = \mu_s mg = (0.57)(173 \text{ kg})(9.8 \text{ m/s}^2) = 966 \text{ N}$$

This is greater than the 900 N we can apply, so we will not be able to move the crate.

(b) To calculate the angle at which we must pull to move the crate, assume the maximum force due to static friction is applied, and calculate the angle at which the net horizontal force on the crate is zero. Using the figure in Problem 83 to guide us, we get [Eq. 2]

$$f_s\left(-\hat{i}\right) + T\cos(\theta)\left(\hat{i}\right) = m\vec{a} = 0$$

$$T\cos(\theta) = \mu_s n$$

The normal force is given by [Eq. 3]

$$\vec{n} + mg\left(-\hat{j}\right) + T\sin(\theta)\left(\hat{j}\right) = 0$$

$$\vec{n} = \left[mg - T\sin(\theta)\right]\left(\hat{j}\right)$$

Inserting this normal force into Eq. 2, we get [Eq. 4]

$$T\cos(\theta) = \mu_s\left[mg - T\sin(\theta)\right]$$

$$\theta = 8.57°$$

REFLECT   If you continue increasing the angle at which you pull, at what angle will you no longer be able to move the crate?

91.   **ORGANIZE AND PLAN**   To make the box accelerate at the same rate as the truck, the static friction force must be strong enough to prevent the box of apples from sliding in the flat bed of the truck. Draw a force diagram for the box of apples with the truck driving on level ground and for the truck driving up Snoqualmie Pass. Use Newton's second law to find the acceleration that corresponds to the static friction force.

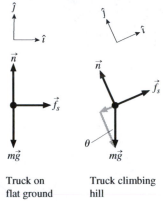

Truck on
flat ground

Truck climbing
hill

Known: $m = 3.0$ kg; $\mu_s = 0.38$; $\theta = 4.5°$

SOLVE   (a) With the truck driving on level ground, Newton's second law applied to the box of apples gives [Eq. 1]

$$\vec{f_s} = m\vec{a}$$

$$\mu_s n(\vec{i}) = m\vec{a}$$

$$\mu_s mg(\vec{i}) = m\vec{a}$$

$$\vec{a} = \mu_s g(\vec{i}) = (0.38)(9.8 \text{ m/s}^2)(\vec{i}) = 3.7 \text{ m/s}^2(\vec{i})$$

(b) Applying Newton's second law to the box of apples when the truck is going up Snoqualmie Pass gives [Eq. 2]

$$\vec{f_s} + mg\sin(\theta)(-\hat{i}) = m\vec{a}$$

$$\mu_s n(\vec{i}) + mg\sin(\theta)(-\hat{i}) = m\vec{a}$$

$$\mu_s mg\cos(\theta)(\vec{i}) + mg\sin(\theta)(-\hat{i}) = m\vec{a}$$

$$\vec{a} = \mu_s g\cos(\theta)(\vec{i}) + g\sin(\theta)(-\hat{i})$$

$$= (9.8 \text{ m/s}^2)[(0.38)\cos(4.5°) - \sin(4.5°)](\vec{i}) = 2.9 \text{ m/s}^2(\vec{i})$$

REFLECT   When the truck is going uphill, the box of apples can tolerate less acceleration by the truck since gravity is already contributing a non-zero acceleration component.

93.   **ORGANIZE AND PLAN**   The problem states that the strings have negligible weight, and we furthermore assume that they have no elasticity so that the three blocks accelerate at the same rate. We can thus treat the three blocks as one block whose weight is the sum of the three individual weights.

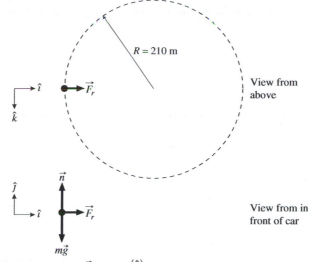

Known: $m_1 = 4$ kg; $m_2 = 6$ kg, $m_3 = 10$ kg; $\vec{F} = 10$ N$(\hat{i})$; $\mu_k = 0.10$

SOLVE   (a) The acceleration of the entire system is given by Newton's second law, using as the mass of the entire system. Using the appropriate force diagram, we find [Eq. 1]

$$f_k^T = \mu_k n = \mu_k (m_1 + m_2 + m_3) g (0.10)(20 \text{ kg})(9.8 \text{ m/s}^2) = 19.6 \text{ N}$$

This is the total *kinetic* frictional force. The total *static* frictional force will be greater still. Since the total kinetic frictional force of 19.6 N is already greater than the force applied of 10 N, the blocks will not move.

(b) To calculate the tension in the string between blocks 2 and 3, we first need to calculate the tension in the string between blocks 1 and 2. The reason for this is that the friction between the ground and block 1 will reduce the tension on the string between the blocks that follow. Applying Newton's second law to block 1 (see force diagram above), we find [Eq. 2]

$$\vec{F} + \vec{F}_{2-1} + \vec{f}_{k,1} = m_1 \vec{a} = 0$$

$$\vec{F}_{2-1} = -\vec{F} - \vec{f}_{k,1} = -10N\left(\hat{i}\right) - \mu_k m_1 g\left(-\hat{i}\right)$$

$$= -10 \text{ N} + (0.10)(4 \text{ kg})(9.8 \text{ m/s}^2)\left(\hat{i}\right) = 6.1 \text{ N}\left(-\hat{i}\right)$$

This is the response to part (c). By Newton's third law [Eq. 3] $\vec{F}_{1-2} = -\vec{F}_{2-1}$. Using this and Newton's second law applied to block 2 gives [Eq. 4]

$$\vec{F}_{1-2} + \vec{F}_{3-2} + \vec{f}_{k,2} = m_2 \vec{a} = 0$$

$$\vec{F}_{3-2} = -\vec{F}_{1-2} - \vec{f}_{k,1} = -\left(-6.1 \text{ N}\right)\left(-\hat{i}\right) - \mu_k m_2 g\left(-\hat{i}\right)$$

$$= -6.1 \text{ N} + (0.10)(6 \text{ kg})(9.8 \text{ m/s}^2)\left(\hat{i}\right) = 0.2 \text{ N}\left(-\hat{i}\right)$$

Thus the tension on the string between the 10 kg and the 6 kg blocks is 0.2 N.

(c) From the solution to part (b), the tension on the string between the 6 kg and the 4 kg blocks is 6.1 N.

REFLECT   Notice that friction only acts to counter any motion generated by other forces. It cannot generate motion of its own. Also notice that we should have used the coefficient of static friction for parts (b) and (c), but since we were provided only with the coefficient of kinetic friction, we used that.

95.   **ORGANIZE AND PLAN**   Use Eq. 4.9 to find the centripetal force holding the Moon and the Earth in orbit. Convert values from Appendix E to SI units.

Known: $m_{\text{earth}} = 5.97 \times 10^{24}$ kg; $m_{\text{moon}} = 7.35 \times 10^{22}$ kg; $R_{\text{earth}} = 150 \times 10^9$ m; $R_{\text{moon}} = 3.85 \times 10^8$ m; $v_{\text{earth}} = 3 \times 10^4$ m/s; $v_{\text{moon}} = 10^3$ m/s

SOLVE   (a) The force holding the Moon in orbit around the Earth is a centripetal force, and is given by [Eq. 1]

$$F_r^{\text{moon}} = \frac{m_{\text{moon}} v_{\text{moon}}^2}{R_{\text{moon}}} = \frac{\left(7.35 \times 10^{22} \text{ kg}\right)\left(10^3 \text{ m/s}\right)^2}{3.85 \times 10^8 \text{ m}} = 1.91 \times 10^{20} \text{ N}$$

(b) The force holding the Earth in orbit around the sun is a centripetal force, and is given by [Eq. 2]

$$F_r^{earth} = \frac{m_{earth} v_{earth}^2}{R_{earth}} = \frac{(5.97 \times 10^{24} \text{ kg})(3 \times 10^4 \text{ m/s})^2}{150 \times 10^9 \text{ m}} = 3.58 \times 10^{22} \text{ N}$$

which is about 200 times greater than the centripetal force holding the Moon in orbit around the earth.

REFLECT   We had to be careful to convert all values from Appendix E to SI units before doing the calculation.

97.   **ORGANIZE AND PLAN**   Draw a force diagram and apply Newton's second law in the vertical direction, with zero acceleration. In the horizontal direction apply Newton's second law to uniform circular motion (Eq. 4.9).

Known: $R = 225$ m; $\mu_k = 0.65$

SOLVE   (a) See the force diagram below.

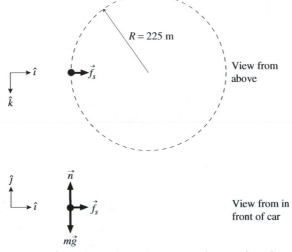

Note that the force due to static friction is the force that causes the centripetal acceleration.

(b) Applying Newton's law in the $\hat{j}$ direction, we find [Eq. 1]

$$\vec{n} + m\vec{g} = m\vec{a} = 0$$

$$\vec{n} = mg\left(\hat{j}\right)$$

Using Eq. 2 in Eq. 4.9, we find the maximum speed possible for the car before slipping [Eq. 2].

$$f_s = \frac{mv^2}{R}$$

$$v = \sqrt{\frac{f_s R}{m}} = \sqrt{\frac{\mu_s mgR}{m}} = \sqrt{\mu_s gR} = \sqrt{(0.65)(9.8 \text{ m/s}^2)(225 \text{ m})} = 37.9 \text{ m/s}$$

REFLECT   Notice that the centripetal force is not *generated* by turning, but is a force that is necessary to execute a turning motion. If must be supplied by an outside object, in this case the road.

99.   **ORGANIZE AND PLAN**   Use Eq. 4.9, which is Newton's second law applied to uniform circular motion, to find the orbital speed of Venus, and then calculate the time it takes to execute one orbit.

Known: $R = 1.08 \times 10^{11}$; $F_r = 5.56 \times 10^{22}$ N

(a) The orbital speed is [Eq. 1]

$$F_r = \frac{mv^2}{R}$$

$$v = \sqrt{\frac{F_r R}{m}}$$

Using Eq. 1, we find the orbital period to be [Eq. 2]

$$T = \frac{2\pi R}{v} = \frac{2\pi R}{\sqrt{F_r R/m}} = \frac{2\pi(1.08 \times 10^{11} \text{ m})}{\sqrt{(5.56 \times 10^{22} \text{ N})(1.08 \times 10^{11} \text{ m})/(4.9 \times 10^{24} \text{ kg})}} = 1.9 \times 10^7 \text{ s}$$

or about 224 days.

REFLECT   Taking into account only the gravity from the sun, our calculated orbital period is within about 0.5% of the observed value. This indicates that the sun's gravity is the dominant force acting on Venus.

101.   **ORGANIZE AND PLAN**   A weightless object may be one that experiences no gravitational force, or one that is accelerating without additional forces in a gravitational field. This latter case is the situation of the pilot at the top of the circle. Therefore, the only force acting on him is the force due to gravity, and the pilot's acceleration at that point is $\vec{g}$ [Eq. 1] $\vec{a}_a = \vec{g}$.

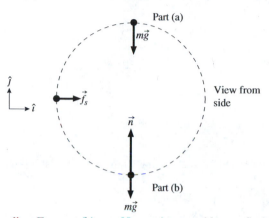

Use Eq. 4.9 to find the circle's radius. For part (b), use Newton's second law to find the weight of the pilot at the low point of the circle.

Known: $v = 90$ m/s

SOLVE   (a) See the force diagram above part (a). Using Eq. 4.9 and Eq. 1 to find the circle's radius, we have [Eq. 2]

$$F_{net} = ma = \frac{mv^2}{R}$$

$$mg = \frac{mv^2}{R}$$

$$R = \frac{v^2}{g} = \frac{(90 \text{ m/s})^2}{9.8 \text{ m/s}^2} = 826.5 \text{ m}$$

(b) See the force diagram for part (b) in the figure above. Applying Newton's second law at the bottom of the circle and using Eq. 2 gives [Eq. 3]

$$F_{net} = m\vec{a}$$

$$-mg + n = \frac{mv^2}{R}$$

$$n = \frac{mv^2}{R} + mg = 2(mg) \text{ N}$$

Thus the pilot's apparent weight at the bottom of the circle is twice his normal weight.

REFLECT   Note that the result for part (b) is independent of the radius of the circle.

103.   **ORGANIZE AND PLAN**   Make a force diagram for the pilot at the top and bottom of the circle.

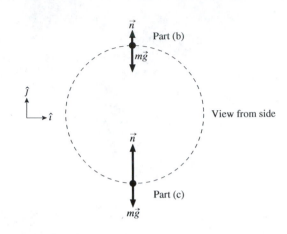

At both positions the pilot must be accelerating in uniform circular motion toward the center of the circle ($a = v^2/R$).

Known: $m = 63$ kg; $R = 1850$ m; $v = 235$ m/s

SOLVE   (a) When the plane flies horizontally with a constant velocity, the scale reads normally [Eq. 1],
$w = mg = (63 \text{ kg})(9.8 \text{ m/s}^2) = 617.4$ N.

(b) Applying Newton's second law at the bottom of the loop (see force diagram above) gives [Eq. 2]

$$\vec{n} + mg(-\hat{j}) = \frac{mv^2}{R}(\hat{j})$$

$$\vec{n} = \frac{mv^2}{R}(\hat{j}) - mg(-\hat{j}) = \left[\frac{(63 \text{ kg})(235 \text{ m/s})^2}{1850 \text{ m}} + (63 \text{ kg})(9.8 \text{ m/s}^2) \right](\hat{j}) = 2{,}498 \text{ N}(\hat{j})$$

Thus, the scale pushes up with 2498 N, so by Newton's third law the pilot pushes down on the scale with the same magnitude of force, so the scale will read 2498 N.

(c) Applying Newton's second law at the top of the loop (see force diagram above) gives [Eq. 3]

$$\vec{n} + mg(-\hat{j}) = \frac{mv^2}{R}(-\hat{j})$$

$$\vec{n} = \frac{mv^2}{R}(-\hat{j}) - mg(-\hat{j}) = \left[\frac{(63 \text{ kg})(235 \text{ m/s})^2}{1850 \text{ m}} - (63 \text{ kg})(9.8 \text{ m/s}^2) \right](-\hat{j}) = 1263 \text{ N}(-\hat{j})$$

so the scale must push down with (i.e., read) 1263 N of force.

REFLECT   How is it possible for the scale to push downward on the pilot at the top of the loop? The pilot is upsidedown at the top of the loop and the scale is therefore above him, so it can push down. Note as well that the normal force as drawn in the force diagram is oriented in the wrong direction, but the calculation worked nonetheless since we kept track of the vector directions.

105.   **ORGANIZE AND PLAN**   Apply Newton's second law to the center of mass of the person to calculate the velocity. Assume that, for each step, the person's center of mass executes uniform circular motion about the point of contact with the ground, and that gravity provides the centripetal force.

Known: mass $= M$; length of leg $= L$

SOLVE   (a) At the top of the arc, the horizontal velocity of the center of mass can be obtained by applying Newton's second law to the center of mass of the person [Eq. 1].

$$F_r = Ma$$

$$Mg = \frac{Mv^2}{L}$$

$$v = \sqrt{Lg}$$

If the velocity exceeds $\sqrt{Lg}$, then the force due to gravity will be insufficient to provide the centripetal acceleration needed to maintain circular motion, and the center of mass will depart from circular motion about the foot. Assuming this occurs at the top of the arc, the person will lose contact with the ground.

(b) The maximum walking speed for an adult male is about 2.5 m/s. Using Eq. 1, this corresponds to a leg length of

$$L = \frac{v^2}{g} = \frac{\left(2.5 \ \text{m/s}^2\right)}{9.8 \ \text{m/s}^2} = 0.64 \ \text{m}$$

or 64 cm.

REFLECT   Eq. 1 shows why people with longer legs have the capacity to walk faster than those with shorter legs.

107.   **ORGANIZE AND PLAN**   Draw a force diagram as seen from above and from the side. Include all forces acting on the mass.

Note that the radius of the circle about which the ball moves is $L \sin \theta$.

Known: $R = L \sin \theta$

SOLVE   (a) See the force diagram above.

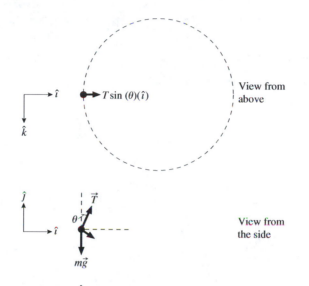

(b) Applying Newton's second law in the $\hat{j}$ direction gives [Eq. 1]

$$T \cos\left(\theta\right)\left(\hat{j}\right) + mg\left(-\hat{j}\right) = 0$$

$$T = \frac{mg}{\cos\left(\theta\right)}$$

For the ball to execute circular motion, the horizontal component of the tension (T) must supply the force necessary to generate the centripetal acceleration. Thus, using Eq. 4.9 and Eq. 1, we find the velocity of the ball to be [Eq. 2]

$$F_r = T \sin\left(\theta\right) = \frac{mv^2}{R}$$

$$v = \sqrt{\frac{gL \sin^2\left(\theta\right)}{\cos\left(\theta\right)}}$$

Using this velocity to calculate the period of the circular motion gives [Eq. 3]

$$P = \frac{2\pi R}{v} = 2\pi L \sin\left(\theta\right)\sqrt{\frac{\cos\left(\theta\right)}{gL \sin^2\left(\theta\right)}} = 2\pi\sqrt{\frac{L \cos\left(\theta\right)}{g}}$$

(c) As $\theta \to 0$, the period $P \to 2\pi\sqrt{\dfrac{L}{g}}$.

REFLECT   What happens if $\theta \to 90°$? Is this physically possible?

109.   **ORGANIZE AND PLAN**   Use the kinematic equations (see, e.g., Problem 42, Eq. 3) to calculate the acceleration. Use Newton's second law to calculate the force due to kinetic friction from the acceleration.

Known: $m = 2.90$ kg; $x - x_0 = 3.25$ m; $v_0 = 2.10$ m/s

SOLVE   From, e.g., Problem 42, Eq. 3, we calculate the magnitude of the acceleration [Eq. 1].

$$a = \frac{v_0^2}{2(x - x_0)} = \frac{(2.10 \text{ m/s})^2}{2(3.25 \text{ m})} = 0.678 \text{ m/s}^2$$

Using Newton's second law and Eq. 4.6, we can find the coefficient of kinetic friction [Eq. 2].

$$F = ma$$

$$\mu_k n = ma$$

$$\mu_k mg = ma$$

$$\mu_k = a / g = \frac{0.678 \text{ m/s}^2}{9.8 \text{ m/s}^2} = 0.069$$

REFLECT   The attentive reader might have noticed a sign difference between Eq. 3, Problem 42, and Eq. 1 of this problem. The difference is that the former is a vector equation, while the later relationship is scalar. Since the acceleration in this problem is due to the kinetic friction force, the direction of the acceleration vector must be antiparallel to the velocity vector.

111.   **ORGANIZE AND PLAN**   Use the kinematic equations (e.g., Eq. 1 from Problem 109) to calculate the subject's acceleration, and then use Newton's second law to calculate the net force experienced by the subject.

Known: $\vec{v}_0 = 286 \ m/s\left(-\hat{i}\right)$; $\vec{x} - \vec{x}_0 = 112.7 \ m\left(-\hat{i}\right)$; $m = 77.0$ kg; $v = 0$ m/s

SOLVE   (a) The force diagram for the subject while he is slowing down is shown in the figure below.

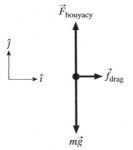

The upward force is the buoyancy force due to water, and the friction force is due to drag in the water.

(b) Using Eq. 1 of Problem 109 to find the acceleration gives [Eq. 1]

$$a = \frac{v_0^2}{2(x - x_0)} = \frac{(286 \text{ m/s})^2}{2(112.7 \text{ m})} = 363 \text{ m/s}^2$$

in the $\hat{i}$ direction (since the drag force acts counter to the velocity). The net force acting on the subject is given by Newton's second law [Eq. 2].

$$\vec{F}_{net} = m\vec{a} = (77 \text{ kg})(363 \text{ m/s}^2)\left(\hat{i}\right) = 27{,}943 \text{ N}\left(\hat{i}\right) = 37w\left(\hat{i}\right)$$

The last line of Eq. 2 gives the force as a multiple of the subject's weight, and is obtained by dividing the acceleration he experienced by the acceleration due to gravity.

(c) The force acted for a time [Eq. 3]

$$\vec{v} = \vec{v}_0 + \vec{a}t$$

$$t = \frac{v_0}{a} = \frac{286 \text{ m/s}}{363 \text{ m/s}^2} = 0.79 \text{ s}$$

REFLECT   Do you think the astronaut survived this test?

113.   **ORGANIZE AND PLAN**   Draw a force diagram of the situation. Be careful to convert the known quantities to SI units.

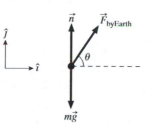

Known: $m = 12.3$ mg; $v_0 = 0$ m/s; $\theta = 58°$; $t = 1.0$ m/s

SOLVE   (a) See the force diagram in the figure above.
(b) The magnitude of the acceleration is given by [Eq.1]

$$v = v_0 + at$$

$$a = v/t = \frac{2.8 \text{ m/s}}{1 \text{ ms}} \left(\frac{10^3 \text{ m/s}}{1 \text{ s}} \right) = 2800 \text{ m/s}^2$$

Referring to the force diagram, we can decompose the acceleration into its horizontal and vertical components. This gives [Eq. 2]

$$\vec{a} = a\cos(\theta)(\hat{i}) + a\sin(\theta)(\hat{j}) = (2800 \text{ m/s}^2) \left[\cos(58°)(\hat{i}) + \sin(58°)(\hat{j}) \right] = (1484 \text{ m/s}^2)(\hat{i}) + (2375 \text{ m/s}^2)(\hat{j}).$$

(c) Use Newton's second law to calculate the force exerted by the ground on the froghopper [Eq. 3].

$$\vec{F} = m\vec{a} = (12.3 \text{ mg}) \left(\frac{1 \text{ kg}}{10^6 \text{ mg}} \right) (1484 \text{ m/s}^2)(\hat{i}) + (12.3 \text{ mg}) \left(\frac{1 \text{ kg}}{10^6 \text{ mg}} \right) (2375 \text{ m/s}^2)(\hat{j})$$

$$= 1.83 \times 10^{-2} \text{ N}(\hat{i}) + 2.92 \times 10^{-2} \text{ N}(\hat{j}) = 151.4w(\hat{i}) + 242.2w(\hat{j})$$

where the last line gives the force as a multiple of the insect's weight *w*, and is calculated by dividing the insect's acceleration by the acceleration due to gravity.

REFLECT   Compare this result with that of Problem 111 (astronaut training). The froghopper can apparently sustain a much larger force relative to their weight than a human. Perhaps froghoppers would make good astronauts.

115.   **ORGANIZE AND PLAN**   To calculate the time it takes the glider to stop, first use Newton's second law to find the acceleration of the glider.
Known: $v_0 = 1.25$ m/s; $v = 0$ m/s; $\theta = 5°$

SOLVE   (a) Since the track is frictionless, there is no force due to friction, so the force diagram is the same for the glider going up and going down the track.

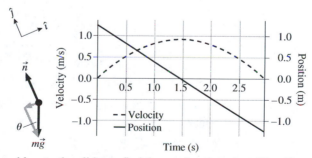

(b) Applying Newton's second law to the glider to find the acceleration gives [Eq. 1]

$$mg\sin(\theta)(-\hat{i}) = m\vec{a}$$

$$\vec{a} = g\sin(\theta)(-\hat{i})$$

Using the kinematic equations to find the distance traveled before the glider stops gives [Eq. 2]

$$x - x_0 = \frac{v_0^2}{2a} = \frac{(1.25 \text{ m/s})^2}{2(9.8 \text{ m/s})^2 \sin(5°)} = 0.915 \text{ m}$$

(c) The time it takes the glider to stop is [Eq. 3]

$$v_0 = v + at$$

$$t = \frac{v_0}{a} = \frac{1.25 \text{ m/s}}{(9.8 \text{ m/s})^2 \sin(5°)} = 1.46 \text{ s}$$

(d) Since there is no friction, the acceleration on the downward trip will be the same as on the upward trip. Thus, the speed of the glider when it returns to the starting point is the same as its initial speed, 1.25 m/s (but the velocity is antiparallel to the initial velocity).

(e) See graphs of velocity and position versus time in the figure above.

REFLECT   We could have used conservation of energy to find the speed of the glider when it returns to its launching point. Since there is no friction, there is no energy lost during the round trip, so energy conservation tells us that the initial speed and the final speed must be the same.

117.   **ORGANIZE AND PLAN**   Make a force diagram of both situations (i.e., with and without the velocity-dependent drag force).

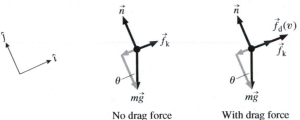

No drag force   With drag force

Apply Newton's second law to find the coefficient of kinetic friction.

Known: $\theta = 3.4°$; $m = 365$ kg; $a = 0.51$ m/s^2

SOLVE   (a) Applying Newton's second law gives [Eq. 1]

$$f_k(\hat{i}) + mg\sin(\theta)(-\hat{i}) = ma(-\hat{i})$$

$$\mu_k mg\cos(\theta)(\hat{i}) + mg\sin(\theta)(-\hat{i}) = ma(-\hat{i})$$

$$\mu_k = \frac{-a + g\sin(\theta)}{g\cos(\theta)}$$

$$= -\frac{-(0.51 \text{ m/s}^2) + (9.8 \text{ m/s}^2)\sin(3.4°)}{(9.8 \text{ m/s}^2)\cos(3.4°)} = 0.007$$

(b) Use Newton's second law to the force diagram that includes the drag force and with an acceleration of zero to find the magnitude of the drag force [Eq. 2].

$$\vec{f}_k + \vec{f}_d + mg\sin(\theta)(-\hat{i}) = m\vec{a} = 0$$

$$f_d = mg\sin(\theta) - f_k = mg\sin(\theta) - \mu_k mg\cos(\theta)$$

$$= (365 \text{ kg})(9.8 \text{ m/s}^2)\left[\sin(3.4°) - 0.007\cos(3.4°)\right] = 186 \text{ N}(\hat{i})$$

REFLECT   Notice that the coefficient of kinetic friction is quite low — the force due to kinetic friction is only [Eq. 3] $f_k = \mu_k mg\cos(\theta) = 26$ N, meaning that the force due to air drag is the dominant force slowing the bobsled. You can see why they try to make the bobsleds as aerodynamic as possible.

119. **ORGANIZE AND PLAN** Draw force diagrams for the person at the bottom and top of the Ferris wheel.

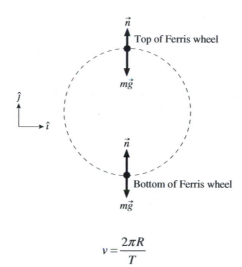

Use the relation [Eq. 1]

$$v = \frac{2\pi R}{T}$$

to convert the period T to speed.

Known: $T = 36$ s; $R = 9.6$ m

SOLVE (a) The Ferris wheel supplies a centripetal force of constant magnitude directed toward the center of the wheel. At the top of the Ferris wheel, this force is oriented opposite to the normal force supplied by the scale, while at the bottom of the Ferris wheel the opposite is true (see force diagram in figure above). Thus the normal force at the top of the wheel is smaller than at the bottom of the wheel.

The scale reads the magnitude of force applied to it, which by Newton's third law is equal to the magnitude of the normal force applied by the scale. Since this normal force is reduced (augmented) by the centripetal force at the top (bottom) of the Ferris wheel, the scale reading at each location will be different.

(b) Apply Newton's law to the situation at the top and bottom of the Ferris wheel, with the acceleration being the centripetal acceleration in both cases. Use Eq. 1 to express the speed in terms of the period. At the bottom of the Ferris wheel, this gives [Eq. 2]

$$\vec{n} + mg\left(-\hat{j}\right) = ma_r\left(\hat{j}\right)$$

$$n = \left[m\frac{v^2}{R} + mg\right]\left(\hat{j}\right) = \left[m\frac{4\pi^2 R}{T^2} + mg\right]\left(\hat{j}\right)$$

$$= \left(60 \text{ kg}\right)\left(\frac{4\pi^2\left(9.6 \text{ m}\right)}{\left(36 \text{ s}\right)^2} + 9.8 \text{ m/s}^2\right)\left(\hat{j}\right) = 605.5 \text{ N}\left(\hat{j}\right)$$

At the top of the Ferris wheel, this gives [Eq. 3]

$$\vec{n} + mg\left(-\hat{j}\right) = ma_r\left(-\hat{j}\right) = \left(60 \text{ kg}\right)\left(-\frac{4\pi^2\left(9.6 \text{ m}\right)}{\left(36 \text{ s}\right)^2} + 9.8 \text{ m/s}^2\right)\left(\hat{j}\right) = 570.5 \text{ N}\left(\hat{j}\right)$$

REFLECT Notice that the only difference between Eqs. (2) and (3) is the sign of the centripetal acceleration.

WORK AND ENERGY

CONCEPTUAL QUESTIONS

1. **SOLVE**   The work done by gravity is always the force of gravity $F_g = -mg$ multiplied by the vertical displacement Δy. Here we have made our usual definition of the positive y-direction as upwards, so the force of gravity is negative. When you climb the mountain, gravity does negative work on you equal to $W_g = F_g \Delta y = -mg\Delta y$, where the elevation change Δy is a positive number. To climb the mountain, you have to do positive work opposing the negative work by gravity. This explains why it's so tiring to walk uphill. When you descend to you starting point, gravity does positive work equal to equal to $W_g = F_g \Delta y = -mg\Delta y$, where the elevation change Δy now is a negative number. However, your body does not have a way of converting this work by gravity into stored energy in your body. Your muscles convert chemical energy into mechanical energy, but they can't convert mechanical energy back into chemical energy, so your body does not gain any energy back during the descent. Instead, the positive work by gravity is dissipated as heat.

    REFLECT   Even when walking on level ground you eventually feel tired. Why is that? When you walk, your body moves up and down a small distance with each step, using chemical energy from the muscles which is converted into mechanical energy. For these very small elevation changes, some of the mechanical energy can be stored as potential energy in your tendons, which act like springs. However, the tendons have a limited storage capacity and there are also some internal drag and frictional forces (energy losses), so eventually all the mechanical energy is dissipated as heat, similar to the climbing example above.

3. **SOLVE**   In casual conversation, "work" often refers to our jobs. "I work at an office." "I work in a coal mine." Other uses of the word "work" exist as well: "Does the clock work?" "Beethoven's Ninth Symphony is a great work of music." In physics, work is precisely defined as the product of force and the displacement in the direction of the force.

    REFLECT   Sometimes we refer to an activity as "hard work." For example, "working in a coal mine is hard work." Often (but not always!), an activity referred to as "hard work" is one where a relatively large amount of work as defined by physics is produced. Certainly, a coal mine worker produces more work as defined by physics than an office worker.

5. **SOLVE**   The net work on an object equals its change in kinetic energy. The net work is also equal to the net force multiplied by the displacement in the direction of the net force. If the kinetic energy is constant, the net work must be zero. If the net work is zero, there is either no net force or there is no displacement in the direction of the net force. Since we were told there is a nonzero net force, the displacement must be perpendicular to the net force, i.e., the object is moving perpendicular to the force, which is the same as to say the velocity is perpendicular to the force.

    REFLECT   This is why the centripetal force does not change the speed or the kinetic energy of an object.

7. **SOLVE**   The work by kinetic friction is the frictional force multiplied by the displacement. The frictional force is proportional to the normal force. The normal force from the floor to the box does not change depending on how fast we go, or if we are accelerating or moving at a constant velocity, so the frictional force is the same in both cases. The displacement is also the same (2 m) in both cases. Consequently, the work done by kinetic friction is the same in both cases.

REFLECT   The analysis above holds true when we drag the force with a pulling force which is parallel to the floor. If the pulling force is at an angle to the floor (for example by pulling the box with a rope at an angle to the floor) the normal force will be different in the two cases.

9.   **SOLVE**   If you roll an object on the floor or the ground, it eventually comes to a stop due to rolling friction. Because the rolling friction has reduced the total mechanical energy of the object, rolling friction is a nonconservative force.

REFLECT   All frictional and drag forces are nonconservative.

11.   **SOLVE**   (a) The kinetic energy of an object at rest is zero. If the speed of the object increases, so does the kinetic energy. This means that the kinetic energy is always positive (or zero). Mathematically, the kinetic energy equals one half times the mass times the velocity squared. Mass is always positive and the velocity is squared is also always positive. Consequently, kinetic energy can never be negative.

(b) Gravitational potential energy is always defined relative to a zero elevation where the potential energy is zero. We are free to define the zero elevation anywhere we want (sea level, ground level, bottom of the sea, on top of Mt. Everest, etc). If an object is below our zero level, the gravitational potential energy is negative. It is important to understand that the absolute number of a gravitational potential energy is not important, only changes in a gravitational potential energy is important.

(c) The potential energy of a spring increases if the spring is either stretched or compressed from its equilibrium length. This means that the potential energy of a spring is always positive (or zero). Mathematically, the potential energy of a spring equals one half times the spring constant times the stretch squared. The spring constant is always positive and the stretch squared is also always positive. Consequently, the potential energy of a spring can never be negative.

(d) The total mechanical energy of an object is a sum of its kinetic and potential energies. Because the potential energy can be negative (for example for a gravitational potential energy) the total mechanical energy can be negative.

(e) The work done by air on a projectile is a drag force. Drag forces and frictional forces always point in the opposite direction of the motion. Work is the product of a force and the displacement in the direction of the force, so the work of any drag force (such as air resistance) or frictional force is always negative.

REFLECT   Could the mass of an object ever be negative? The mass of all known matter and all known particles in physics is positive (or zero). This includes anti-matter and anti-particles. There is no experimental result, no observation, and no accepted theory that predicts the existence of negative mass particles.

Could the spring constant of an object ever be negative? For a regular spring, the answer is no, so the potential energy of a spring is always positive. However, it is possible to construct a system (for example using force-feedback sensors and computer-controlled motors) that responds like a spring but with a negative spring constant, at least over a limited range. For such a system one could talk of a negative potential energy.

13.   **SOLVE**   Here's a simplified description of the energy transformations throughout a pole vault. First, the athlete converts food energy to mechanical energy, more precisely kinetic energy, as he accelerates from rest to his final running speed. After he has planted his pole, the kinetic energy is quickly converted into stored potential energy in the pole as it bends. The stored potential energy in the pole then lifts the athlete up towards the bar, converting the potential energy in the pole to both kinetic energy and gravitational potential energy. At some elevation above the ground the upward kinetic energy has reached a maximum, and from this point on both the remaining potential energy in the pole and the kinetic energy is converted to gravitational potential energy. When the athlete reaches his maximum height, all of the potential energy in the pole and all of the upward kinetic energy has been converted to gravitational potential energy. After crossing over the bar and having let go of the pole, the athlete falls toward the cushion, converting the gravitational potential energy to kinetic energy. Hitting the cushion, the kinetic energy is dissipated as heat and the athlete comes to rest.

REFLECT   This is a broad-strokes description of the pole vault, omitting some smaller details. For example, not all of the kinetic energy is converted into stored potential energy in the pole. A small amount of kinetic energy remains to carry the athlete forward at a slow speed. Also, the athlete pushes off from the pole using his arms to gain a small additional amount of elevation.

MULTIPLE-CHOICE PROBLEMS

15.   SOLVE   The work is the force multiplied by the displacement:

$$W = F\Delta y = -mg(y_1 - y_0) = -(0.50\text{ kg})(9.80\text{ m/s}^2)((1.5\text{ m}) - (12.5\text{ m})) = 54\text{ J}$$

Here we have used the common mathematical notation of indicating starting quantities with the subscript 0, so the starting elevation is y_0. Subsequent quantities get incrementally higher subscripts, so the next elevation is y_1. The difference in elevation is $\Delta y = y_1 - y_0$ and is a negative number in this example.

REFLECT   Gravity does positive work on objects moving downward.

17.   SOLVE   The work equals the change in the gravitational potential energy:

$$W = \Delta U = mg\Delta y = (185\text{ kg})(9.80\text{ m/s}^2)(0.550\text{ m}) = 997\text{ J}$$

REFLECT   Gravity did work on the barbell equal to $W_g = -997$ J. The work done by gravity was negative because the barbell moved upwards. The change in potential energy was positive.

19.   SOLVE   The work is the difference in potential energy stored in the spring (see Equation 5.14):

$$W = \Delta U = \tfrac{1}{2}kx_1^2 - \tfrac{1}{2}kx_0^2 = \tfrac{1}{2}(250\text{ N/m})(0.40\text{ m})^2 - \tfrac{1}{2}(250\text{ N/m})(0.30\text{ m})^2 = 8.8\text{ J}$$

REFLECT   The potential energy stored in a spring equals the work done extending the spring from its equilibrium position.

21.   SOLVE   The direction of motion is not important when calculating kinetic energy, only the speed. Since the speed doubled, the kinetic energy must have quadrupled, because kinetic energy is proportional to the square of the speed. The correct answer is (c) $4\ K$.

REFLECT   The kinetic energy can never be negative.

23.   SOLVE   The change in gravitational potential energy is given by Equation 5.13:

$$\Delta U = mg\Delta y = (70\text{ kg})(9.80\text{ m/s}^2)(8850\text{ m}) = 6.1\times10^6\text{ J}$$

REFLECT   This is only about 1,450 food calories, less than the required food consumption of most adults.

25.   SOLVE   When the projectile rises it is gaining gravitational potential energy and losing kinetic energy at equal rates, keeping its total mechanical energy constant. The change in kinetic energy is:

$$\Delta K = -\Delta U = -mg\Delta y = -(1.25\text{ kg})(9.80\text{ m/s}^2)(12.8\text{ m}) = -157\text{ J}$$

REFLECT   Can you calculate the projectile's change in speed with the information given to you?

27.   SOLVE   The acceleration is in the direction of the net force, so because the net force is in the opposite direction of the velocity, the box is slowing and (a) is true.

The net work equals the net force times the displacement in the direction of the net force. Because the net force and the displacement are in opposite directions, the net work must be negative and (b) is true.

The work done by gravity equals the gravitational force times the displacement in the direction of the gravitational force. Because the motion the gravitational force is vertical and there is no vertical displacement (the motion is horizontal) gravity does zero work on the box and (c) is false.

Because the net force is acting to slow the box down, eventually the speed of the box will be zero, i.e., the box will not be moving. What happens after the box has stopped is not clear; if the net force is a drag force, the box will remain at rest. If the net force has a finite value regardless of the speed of the box, the box will begin moving to the left a moment after it has stopped. Either way the box is not moving for at least an instant, so (d) is true. The final answer is (c).

REFLECT   The work-energy theorem states that the net work equals the change in kinetic energy. This also tells us that (b) is true.

PROBLEMS

29.   **ORGANIZE AND PLAN**   The force you apply on the box is in the same direction as the displacement, so F_x and Δx must have the same sign. Let's define this direction to be our positive x-direction.
Known: $F_x = 540$ N; $\Delta x = 3.5$ m.

SOLVE   We compute the work you do by using Equation 5.1:

$$W = F_x \Delta x = (540 \text{ N})(3.5 \text{ m}) = 1.89 \text{ kJ}$$

REFLECT   You do positive work, against the negative work done by friction.

31.   **ORGANIZE AND PLAN**   Constant braking and drag forces mean a constant acceleration. If we first find this acceleration, we can find the force from Newton's second law and then calculate the work by multiplying the force by the displacement. The acceleration can be found from the speed of the car at two different locations – we can call the starting location x_0 and the final location x_1 – and the distance between the locations $\Delta x = x_1 - x_0$.
Known: $m = 1320$ kg; $v_0 = 21.5$ m/s; $v_1 = 0$ m/s; $\Delta x = 145$ m.

SOLVE   The speed of the car is $v = v_0 + at$. The acceleration a is negative because the car is slowing down, and we can express this acceleration in terms of the time t_1 when the car comes to a stop: $a = -v_0/t_1$. At this time the car has traveled a distance:

$$\Delta x = \frac{at_1^2}{2} + v_0 t_1 = \frac{(-v_0/t_1)t_1^2}{2} + v_0 t_1 = -\frac{v_0 t_1}{2} + v_0 t_1 = \frac{v_0 t_1}{2}$$

Solve for the time to stop the car:

$$t_1 = 2\Delta x/v_0 = 2(145 \text{ m})/(21.5 \text{ m/s}) = 13.5 \text{ s}$$

Now we can calculate the acceleration

$$a = -\frac{v_0}{t_1} = -\frac{(21.5 \text{ m/s})}{(13.5 \text{ s})} = -1.59 \text{ m/s}^2$$

the force from Newton's Second Law

$$F_x = ma = (1320 \text{ kg})(-1.59 \text{ m/s}^2) = -2.10 \text{ kJ}$$

and the work from Equation 5.1

$$W = F_x \Delta x = (-2.10 \text{ kN})(145 \text{ m}) = -305 \text{ kJ}$$

REFLECT   The work done by the braking forces can be calculated in a much easier way by using the methods of Section 5.3. The work equals the initial kinetic energy but with an opposite sign.

33.   **ORGANIZE AND PLAN**   Friction acts in a direction opposite to the motion, doing negative work. The frictional force is the product of the normal force and the coefficient of friction. Work is done only when the book is moving, so it is appropriate to use the coefficient of kinetic friction.
Known: $m = 1.52$ kg; $\mu_k = 0.140$; $\Delta x = 1.24$ m.

SOLVE   The normal force is:

$$n = mg = (1.52 \text{ kg})(9.80 \text{ m/s}^2) = 14.9 \text{ N}.$$

If the motion is in the positive x-direction, then the frictional force is in the negative x-direction:

$$f_k = -\mu_k n = -(0.140)(14.9 \text{ N}) = -2.09 \text{ N}.$$

The work done by friction is:

$$W_f = f_k \Delta x = (-2.09 \text{ N})(1.24 \text{ m}) = -2.59 \text{ J}.$$

REFLECT   The book can slide in a curved path with a total length of 1.24 m and the result would be the same.

35. **ORGANIZE AND PLAN**   The work is the force times the displacement in the direction of the force.
Known: $F_x = 2.34$ N; $F_y = 1.06$ N; $\Delta x_a = \Delta x_c = \Delta y_c = 2.50$ m; $\Delta x_b = -2.50$ m; $\Delta y_a = \Delta y_b = 0$.

SOLVE   The work in part (a) is:
$$W_a = F_x \Delta x_a + F_y \Delta y_a = (2.34 \text{ N})(2.50 \text{ m}) + (1.06 \text{ N})(0) = 5.85 \text{ J}$$

The work in part (b) is:
$$W_b = F_x \Delta x_b + F_y \Delta y_b = (2.34 \text{ N})(-2.50 \text{ m}) + (1.06 \text{ N})(0) = -5.85 \text{ J}$$

The work in part (c) is:
$$W_c = F_x \Delta x_c + F_y \Delta y_c = (2.34 \text{ N})(2.50 \text{ m}) + (1.06 \text{ N})(2.50 \text{ m}) = 8.5 \text{ J}$$

REFLECT   When the force is perpendicular to the displacement it does no work.

37. **ORGANIZE AND PLAN**   The rocket accelerates upward because the engine force is larger than gravity. We expect the work done by the engine force to be larger than the work done by gravity. All work in this problem can be calculated as the product of a force times the displacement.
Known: $m = 1.85$ kg; $F_e = 46.2$ N; $\Delta y = 100$ m.

SOLVE   We repeatedly use Equation 5.1 to calculate work done by the different forces.
(a) The work done by the engine force is:
$$W_e = F_e \Delta y = (46.2 \text{ N})(100 \text{ m}) = 4.62 \text{ kJ}$$

(b) The work done by gravitational force $F_g = -mg$ is:
$$W_g = -mg\Delta y = -(1.86 \text{ kg})(9.80 \text{ m/s}^2)(100 \text{ m}) = -1.82 \text{ kJ}$$

(c) The net force is $F_{net} = F_e + F_g = F_e - mg$, which we can use to calculate the net work with $W_{net} = F_{net}\Delta y$. Or an easier solution is to simply add up the work by the engine and gravitational forces to obtain the net work:
$$W_{net} = W_e + W_g = (4.62 \text{ kJ}) + (-1.82 \text{ kJ}) = 2.80 \text{ kJ}$$

REFLECT   What has the work by the engine force accomplished? It has increased the potential energy of the rocket (by 1.82 kJ) and it has increased the kinetic energy (by 2.80 kJ). Kinetic energy is discussed in Section 5.3 and potential energy in Section 5.4.

39. **ORGANIZE AND PLAN**   Because the block is moving at constant speed, Newton's first law tells us that there is no net force on the block. The work is force times displacement in the direction of the force. The net work is the sum of all work done on the block by all forces.
Known: $m = 1.25$ kg; $\theta = 15°$; $d = 0.60$ m.

SOLVE   (a) There are three forces acting on the block: gravity, the normal force from the frictionless incline, and the pulling force $\vec{F}$. As shown in the figure below, gravity is downwards, the normal force is normal to the incline, and the pulling force is as stated in the problem text directed up the incline. Newton's first law tells us that the pulling force must oppose the component of the gravitational force which is directed along the incline.

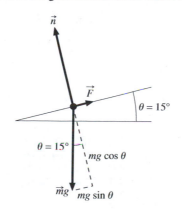

The magnitude of $\vec{F}$ is:

$$F = mg\sin\theta = (1.25\text{ kg})(9.80\text{ m/s}^2)\sin15° = 3.17\text{ N}$$

The horizontal and vertical components of $\vec{F}$ are:

$$F_x = F\cos\theta = (3.17\text{ N})\cos15° = 3.06\text{ N}$$
$$F_y = F\sin\theta = (3.17\text{ N})\sin15° = 0.821\text{ N}$$

so in vector form the pulling force $\vec{F}$ is:

$$\vec{F} = F_x\hat{i} + F_x\hat{j} = (3.06\text{ N})\hat{i} + (0.821\text{ N})\hat{j}$$

(b) The work done by $\vec{F}$ in moving the block up the incline is the force multiplied by the displacement in the direction of the force. In this case, the force is parallel to the displacement, so the work is:

$$W_F = Fd = (3.17\text{ N})(0.60\text{ m}) = 1.9\text{ J}$$

(c) The work done by gravity is the component of gravity along the incline multiplied by the displacement along the incline:

$$W_g = -mg\sin\theta d = -(1.25\text{ kg})(9.80\text{ m/s}^2)\sin(15°)(0.60\text{ m}) = -1.9\text{ J}$$

(d) The normal force and the component of gravity perpendicular to the incline are both perpendicular to the displacement and do no work. The net force is the sum of all work and is zero.

REFLECT   When the velocity of an object remains constant, no net work is done on the object.

41.   **ORGANIZE AND PLAN**   The forces acting on the metal block are gravity and the tension in the string. Both of these forces are vertical. The forces acting on the glider are gravity, the normal force from the air track, and the string tension. The vertical forces acting on the glider (gravity and the normal force) must be balanced. The string tension pulls the glider horizontally.

Known: $m_1 = 0.15\text{ kg}$; $m_2 = 0.10\text{ kg}$; $\Delta x_1 = -\Delta y_2 = 0.50\text{ m}$.

SOLVE   (a) The acceleration is as usual found by applying Newton's second law. The glider is accelerated by a tension T in the string:

$$T = m_1 a_1$$

The metal block is accelerated downward by the net force acting on it, and that net force is gravity minus the string tension:

$$F_{2,\text{net}} = m_2 g - T = m_2 a_2$$

which we can rewrite as an expression for the string tension:

$$T = m_2(g - a_2)$$

The objects are accelerating together, $a = a_1 = a_2$, so if we combine the two expressions for the string tension above, we can solve for the acceleration:

$$m_1 a = m_2(g - a)$$
$$(m_1 + m_2)a = m_2 g$$
$$a = \frac{m_2 g}{m_1 + m_2}$$

This expression makes a lot of sense and we could probably have written it down without prelude. It tells us that the gravity on the metal block has to move the mass of both objects. If we insert our known values we can calculate the acceleration:

$$a = \frac{m_2 g}{m_1 + m_2} = \frac{(0.10\text{ kg})(9.80\text{ m/s}^2)}{(0.15\text{ kg}) + (0.10\text{ kg})} = 3.92\text{ m/s}^2$$

(b) The net work done on each object is the net force multiplied with the displacement. For the glider the net force is the string tension:

$$F_{1,\text{net}} = T = m_1 a = (0.15\text{ kg})(3.92\text{ m/s}^2) = 0.588\text{ N}$$

The net work done on the glider is:

$$W_{1,\text{net}} = F_{1,\text{net}}\Delta x_1 = (0.588 \text{ N})(0.50 \text{ m}) = 0.294 \text{ J}$$

The net force on the metal block is:

$$F_{2,\text{net}} = m_2 g - T = (0.10 \text{ kg})(9.80 \text{ m/s}^2) - (0.588 \text{ N}) = 0.392 \text{ N}$$

The net force is directed downward, so with our usual convention of the positive y-direction being upward, the net work done on the metal block is:

$$W_{2,\text{net}} = -F_{2,\text{net}}\Delta y_2 = -(0.392 \text{ N})(-0.50 \text{ m}) = 0.196 \text{ J}$$

(c) We have already calculated the work done by the string on the glider; it is 0.294 J. The work done by the string on the metal block is:

$$W_{2,T} = T\Delta y_2 = (0.588 \text{ N})(-0.50 \text{ m}) = -0.294 \text{ J}$$

(d) The work done by gravity on the metal block is:

$$W_{2,g} = -m_2 g \Delta y_2 = -(0.10 \text{ kg})(9.80 \text{ m/s}^2)(-0.50 \text{ m}) = 0.490 \text{ J}$$

REFLECT   It makes sense that the total work done by the string on both objects is zero, because the string does not have some kind of internal energy source with which it can do work. Note that in this problem we have assumed not only that the air track is frictionless, but also that there is no friction in the pulley. In such a case, the string tension remains the same across the pulley, something you can convince yourself of with a simple experiment.

43.   **ORGANIZE AND PLAN**   Hooke's law (Equation 5.7) says that the spring constant is the ratio of the force applied to the spring and the displacement of the spring's end.

Known: $m_a = 0.150$ kg; $m_b = 1.00$ kg; $x_a = 0.125$ m.

SOLVE   For part (a), calculate the spring constant as the ratio of force and displacement:

$$k = \frac{F_a}{x_a} = \frac{m_a g}{x_a} = \frac{(0.150 \text{ kg})(9.80 \text{ m/s}^2)}{(0.125 \text{ m})} = 11.8 \text{ N/m}$$

For part (b), we again use Hooke's law to find the total stretch of the spring as the force divided by the spring constant:

$$x_b = \frac{F_b}{k} = \frac{m_b g}{k} = \frac{(1.00 \text{ kg})(9.80 \text{ m/s}^2)}{(11.8 \text{ N/m})} = 0.833 \text{ m}$$

REFLECT   The stretch is proportional to the force, i.e., proportional to the mass we hang on the spring. Since we increased the mass from part (a) to part (b) by a ratio $\frac{1.00}{0.15} = \frac{20}{3}$, the stretch increases by the same ratio, giving a new stretch, which is: $\frac{20}{3} \times 0.125$ m $= 0.833$ m.

45.   **ORGANIZE AND PLAN**   Equation 5.8 gives the work done stretching (or compressing) a spring. We can use this equation to find the spring constant if we already know the work.

Known: $W = 13.4$ J; $x = 2.37$ cm.

SOLVE   Rewrite Equation 5.8 to solve for the spring constant:

$$k = \frac{2W}{x^2} = \frac{2(13.4 \text{ J})}{(2.37 \text{ cm})^2} = 47.7 \text{ kN/m}$$

REFLECT   The work $13.4 \text{ J} = 3.2 \times 10^{-3}$ food calories doesn't sound like a lot, but many people would have a hard time keeping this spring compressed by 2.37 cm for very long – what is the required force?

47.   **ORGANIZE AND PLAN**   Equation 5.8 gives the work done stretching (or compressing) a spring. We will use this equation twice, first calculating the work to extend the spring from no extension to 0.30 m, then subtracting the work to extend the spring from no extension to 0.10 m.

Known: $k = 150$ N/m; $x_0 = 0.10$ m; $x_1 = 0.30$ m.

SOLVE   The work required to extend the spring from no extension to 0.30 m is:

$$W_1 = \frac{kx_1^2}{2} = \frac{(150 \text{ N/m})(0.30 \text{ m})^2}{2} = 6.8 \text{ J}$$

The work required to extend the spring from no extension to 0.10 m is:

$$W_0 = \frac{kx_0^2}{2} = \frac{(150 \text{ N/m})(0.10 \text{ m})^2}{2} = 0.75 \text{ J}$$

Consequently, the work required to extend the spring from 0.10 m to 0.30 m is:

$$\Delta W = W_1 - W_0 = (6.8 \text{ J}) - (0.75 \text{ J}) = 6.0 \text{ J}$$

REFLECT   It's important to realize that this work is not the same as extending the spring from no extension to 0.20 m!

49.   **ORGANIZE AND PLAN**   In a force-versus-position graph, the work is the area under the curve. This curve consists of two parts. First, the portion between 0 and 10 cm where the force increases linearly to reach 35 N at 10 cm. In the second portion, above 10 cm, the force is constant at 35 N.
Known: Force-versus-position.

SOLVE   For part (a), the work is the triangular area under the curve from 0 cm to 10 cm:

$$W_a = \frac{(35 \text{ N})(10 \text{ cm})}{2} = 1.75 \text{ J}$$

For part (b), the work is the answer from (a) minus the triangular area under the curve from 0 cm to 5 cm:

$$W_b = (1.75 \text{ J}) - \frac{(17.5 \text{ N})(5 \text{ cm})}{2} = 1.31 \text{ J}$$

For part (c), the work is the answer from (a) plus the rectangular area under the curve from 10 cm to 15 cm:

$$W_c = (1.75 \text{ J}) + (35 \text{ N})(5 \text{ cm}) = 3.50 \text{ J}$$

$$k = \frac{F}{x} = \frac{mg}{x} = \frac{(125 \text{ kg})(9.80 \text{ m/s}^2)}{(2.66 \text{ mm})} = 461 \text{ kN/m}$$

The force does negative work when the displacement is in a direction opposite the force. Consequently, the answer to part (d) is the same as part (a) but with the opposite sign: –1.75 J.

REFLECT   For the portion from 0 cm to 10 cm the force is linearly proportional to the position, i.e., Hooke's law applies and we could calculate a spring constant for this portion.

51.   **ORGANIZE AND PLAN**   In a force-versus-position graph, the work is the area under the curve. When the curve is below the x-axis, the force does negative work equal to the area above the curve. This curve consists of three parts. First, the portion between 0 and 2 m where the force increases linearly from –40 N to 20 N. In the second portion, between 2 and 3 m, the force is constant at 20 N. In the third portion, from 3 m to 5 m, the force increases linearly from 20 N to 60 N.
Known: Force-versus-position.

SOLVE   For part (a), we must first calculate where the graph crosses the x-axis. The force in this portion of the graph is linear and can be written similar to Hooke's law, except with an added constant offset:

$$F = kx + F_0$$

where $F_0 = -40$ N is the offset force for zero displacement and $k = 30$ N/m so that the force reaches 20 N by $x = 2$ m.

Solve this expression for zero force:

$$x = -\frac{F_0}{k} = -\frac{(-40 \text{ N})}{(30 \text{ N/m})} = 1.33 \text{ m}$$

This is where the graph crosses the x-axis. Between 0 and 1.33 m the force does negative work equal to the triangular area above the curve. Between 1.33 and 2 m the force does positive work equal to the area below the curve. Add these two together for the total work from 0 to 2 m:

$$W_a = -\frac{(40\,\text{N})(1.33\,\text{m})}{2} + \frac{(20\,\text{N})(0.67\,\text{m})}{2} = -20\,\text{J}$$

For part (b), the work is the rectangular area under the curve from 2 m to 3 m:

$$W_b = (20\,\text{N})(1\,\text{m}) = 20\,\text{J}$$

For part (c), the work is the area under the curve from 3 m to 5 m. This area consists of both a triangular and a rectangular part:

$$W_c = \frac{(40\,\text{N})(2\,\text{m})}{2} + (20\,\text{N})(2\,\text{m}) = 80\,\text{J}$$

For part (d), the work is the sum of the work for parts (a), (b), and (c), so $W_d = 80\,\text{J}$. For part (e), the force does the same work as in part (a) but with the opposite sign, so $W_e = 20\,\text{J}$.

REFLECT In all portions of the graph, we can write the force in the form $F = kx + F_0$, but with different values on the spring constants k and the offset forces F_0.

53. **ORGANIZE AND PLAN** We will use Hooke's law (Equation 5.7) to find additional displacement. We can either make the calculation with four springs holding the combined four passengers or with each spring holding one passenger.

    Known: $k = 63.4\,\text{kN/m}$; $m = 90\,\text{kg}$.

    SOLVE The extra displacement from Hooke's law:

    $$x = \frac{F}{k} = \frac{mg}{k} = \frac{(90\,\text{kg})(9.80\,\text{m/s}^2)}{(63.4\,\text{kN/m})} = 1.4\,\text{cm}$$

    REFLECT Obviously, multiplying both m and k with 4 produces the same result.

55. **ORGANIZE AND PLAN** Part (a) is a simple application of Hooke's law, calculating the displacement for the gravitational force from the 0.100-kg mass. For parts (b) and (c) we will have to consider the energy stored up in the spring, i.e., the work done by the gravitational force on the spring. The stored energy is equal to the work done by the gravitational force.

    Known: $k = 25.0\,\text{N/m}$; $m = 0.100\,\text{kg}$; $h = 10.0\,\text{cm}$.

    SOLVE For part (a), the displacement from Hooke's law is:

    $$x = \frac{F}{k} = \frac{mg}{k} = \frac{(0.100\,\text{kg})(9.80\,\text{m/s}^2)}{(25.0\,\text{N/m})} = 3.92\,\text{cm}$$

    For part (b), the work by gravity must equal the stored energy in the spring. This gives us an equation that we can solve for the displacement x:

    $$mgx = \tfrac{1}{2}kx^2$$
    $$mg = \tfrac{1}{2}kx$$
    $$x = \frac{2mg}{k} = \frac{2(0.100\,\text{kg})(9.80\,\text{m/s}^2)}{(25.0\,\text{N/m})} = 7.84\,\text{cm}$$

    Part (c) is the same as part (b), but with additional work by gravity from the 10.0 cm drop. This gives us a quadratic equation for x:

    $$mg(x+h) = \tfrac{1}{2}kx^2$$
    $$\tfrac{1}{2}kx^2 - mgx - mgh = 0$$
    $$x = \frac{mg}{k} \pm \sqrt{\left(\frac{mg}{k}\right)^2 + \frac{2mgh}{k}}$$

If we set $h=0$ in this equation we obtain the answer for part (b). With the extra work by gravity corresponding to a 10.0 cm drop, the answer for part (c) is:

$$x = \frac{(0.100 \text{ kg})(9.80 \text{ m/s}^2)}{(25.0 \text{ N/m})} + \sqrt{\left(\frac{(0.100 \text{ kg})(9.80 \text{ m/s}^2)}{(25.0 \text{ N/m})}\right)^2 + \frac{2(0.100 \text{ kg})(9.80 \text{ m/s}^2)(10.0 \text{ cm})}{(25.0 \text{ N/m})}} = 13.6 \text{ cm}$$

REFLECT   Why is the answer for part (b) twice that of part (a)?

57.   **ORGANIZE AND PLAN**   The quantities in this problem are related through Equation 5.10:

$$K = \tfrac{1}{2}mv^2$$

Known: $v_a = 12.4$ m/s; $K_a = 305$ J; $v_b = 2v_a = 24.8$ m/s; $v_c = \tfrac{1}{2}v_a = 6.20$ m/s.

SOLVE   (a) Rewrite Equation 5.10 to express mass as a function of kinetic energy and speed, and calculate:

$$m = \frac{2K}{v^2} = \frac{2(305 \text{ J})}{(12.4 \text{ m/s})^2} = 3.97 \text{ kg}$$

(b) With its speed doubled, the kinetic energy is:

$$K = \tfrac{1}{2}(3.97 \text{ kg})(24.8 \text{ m/s})^2 = 1.22 \text{ kJ}$$

(c) With its speed halved, the kinetic energy is:

$$K = \tfrac{1}{2}(3.97 \text{ kg})(6.20 \text{ m/s})^2 = 76.3 \text{ J}$$

REFLECT   When the speed doubles, the kinetic energy quadruples. When the speed is halved, the kinetic energy is reduced to a quarter of its original value.

59.   **ORGANIZE AND PLAN**   The work required is the kinetic energy of the airliner going at 250 km/h. This work equals force times displacement, so we can calculate the force once the work is known.
Known: $m = 68{,}000$ kg; $v = 250$ km/h; Δx=1.20 km.

SOLVE   (a) The work required equals the kinetic energy and is calculated from Equation 5.10:

$$K = \tfrac{1}{2}mv^2 = \tfrac{1}{2}(68{,}000 \text{ kg})(250 \text{ km/h})^2 = 164 \text{ MJ}$$

(b) The minimum force to achieve takeoff in a distance Δx is calculated from Equation 5.1:

$$F_x = \frac{W}{\Delta x} = \frac{K}{\Delta x} = \frac{(164 \text{ MJ})}{(1.20 \text{ km})} = 137 \text{ kN}$$

The answer to part (c) is yes, because the combined force from the two engines is 234 kN, which is larger than 137 kN.

REFLECT   In addition to considering drag forces, aircraft engineers have to design the engines to produce enough force to safely operate the aircraft should one engine fail. This includes getting the plane in the air with one engine should the other engine fail late in the takeoff run, or being able to stop the aircraft (with brakes and reverse thrust from one engine) before the end of the runway should the other engine fail early in the takeoff run.

61.   **ORGANIZE AND PLAN**   The work done by the drag force is the difference in kinetic energy between pitch and reaching the home plate. This work equals force times displacement, so we can calculate the force once the work is known.
Known: $m = 0.145$ kg; $v_0 = 39.0$ m/s; $v_1 = 36.2$ m/s; $\Delta x = 18.4$ m.

SOLVE   (a) The kinetic energy is calculated from Equation 5.10. The ball is pitched with a kinetic energy:

$$K_0 = \tfrac{1}{2}mv_0^2 = \tfrac{1}{2}(0.145 \text{ kg})(39.0 \text{ m/s})^2 = 110 \text{ J}$$

The ball reaches the home plate with a kinetic energy:

$$K_1 = \tfrac{1}{2}mv_1^2 = \tfrac{1}{2}(0.145 \text{ kg})(36.2 \text{ m/s})^2 = 95.0 \text{ J}$$

So the work done by the drag force is:

$$W = K_1 - K_0 = (95.0 \text{ J}) - (110 \text{ J}) = -15.3 \text{ J}$$

(b) Because the force is assumed constant, we can calculate it from Equation 5.1:

$$F_x = \frac{W}{\Delta x} = \frac{(-15.3 \text{ J})}{(18.4 \text{ m})} = -0.830 \text{ N}$$

REFLECT   The work done by drag forces is always negative, because the drag force acts in a direction opposite that of the displacement. In reality, the drag forces must be slightly larger than what we calculated, because the pitcher stands on a mound which is higher than the home plate. This means the ball gains some kinetic energy from the work of gravity. The height of this mound was lowered from 15 inches to 10 inches in 1969 to reduce the dominance of the pitchers. How much did this change affect the velocity (vertical and horizontal speed) of a baseball as it reaches the home plate?

63.   **ORGANIZE AND PLAN**   The kinetic energy is one half times mass times velocity squared. When the ball reaches its maximum height it is no longer moving, and gravity has done negative work equal to the original kinetic energy. When the ball is falling back down to the ground, gravity does positive work on the ball. As the ball falls past its starting point, gravity has done zero work total on the ball. Then gravity does a small amount of additional positive work as the ball falls the final 1.20 m to the ground.

Known: $m = 0.145 \text{ kg}$; $v_0 = 21.8 \text{ m/s}$; $y_0 = 1.20 \text{ m}$.

SOLVE   For part (a), the kinetic energy of the ball when it leaves the bat is calculated from Equation 5.10:

$$K_0 = \tfrac{1}{2}mv_0^2 = \tfrac{1}{2}(0.145 \text{ kg})(21.8 \text{ m/s})^2 = 34.5 \text{ J}$$

For part (b), gravity has done work $W_{up} = -K_0 = -34.5 \text{ J}$ on the ball once the ball reaches its maximum height. For part (c), the maximum height can be found from the answer in part (b) by using Equation 5.5:

$$W_{up} = -mg\Delta y$$

$$\Delta y = -\frac{W_{up}}{mg} = -\frac{(-34.5 \text{ J})}{(0.145 \text{ kg})(9.80 \text{ m/s}^2)} = 24.2 \text{ m}$$

The ball is thus 24.2 m above its starting location, or $y_1 = y_0 + \Delta y = (1.2 \text{ m}) + (24.2 \text{ m}) = 25.4 \text{ m}$ above ground. For part (d), gravity does positive work on the ball as it falls to the ground:

$$W_{down} = mgy_1 = (0.145 \text{ kg})(9.80 \text{ m/s}^2)(25.4 \text{ m}) = 36.2 \text{ J}$$

to which we have to add the negative work of part (b), so the total work gravity does on the ball is:

$$W = W_{down} + W_{up} = (36.2 \text{ J}) + (-34.5 \text{ J}) = 1.71 \text{ J}$$

For part (e), the work gravity has done on the ball falling down equals the kinetic energy of the ball at the ground. We can rewrite equation 5.10 to calculate the ball's speed at the ground:

$$v_{ground} = \sqrt{\frac{2K_{ground}}{m}} = \sqrt{\frac{2W_{down}}{m}} = \sqrt{\frac{2(36.2 \text{ J})}{(0.145 \text{ kg})}} = 22.3 \text{ m/s}$$

REFLECT   The answer for part (d) is equal to the potential energy of the ball 1.20 m above ground. Potential energy is discussed in Section 5.4.

65.   **ORGANIZE AND PLAN**   The kinetic energy of the vertical downward speed when the rock hits the ground equals the work done by gravity.

Known: $h = 10 \text{ m}$.

SOLVE   For part (a), setting the kinetic energy of the rock when it hits the ground equal to the work done by gravity we can

SOLVE for the vertical speed:

$$\tfrac{1}{2}mv_y^2 = mgh$$

$$\tfrac{1}{2}v_y^2 = gh$$

$$v_y = -\sqrt{2gh} = -\sqrt{2(9.80 \text{ m/s}^2)(10 \text{ m})} = -14 \text{ m/s}$$

As usual, the positive y-direction is up.

For part (b), using the same equality we can SOLVE for the height when the speed is half that in part (a):

$$\tfrac{1}{2}mv_y^2 = mgh$$

$$\tfrac{1}{2}v_y^2 = gh$$

$$h = \frac{v_y^2}{2g} = \frac{(-7.0 \text{ m/s})^2}{2(9.80 \text{ m/s}^2)} = 2.5 \text{ m}$$

REFLECT We could have solved this problem without considering kinetic energy and work by gravity by simply using the equations of motion for a constantly accelerating object (see Equation 2.12). That would make the answer in part (b) obvious: a constantly accelerating object moves four times longer in double the time, and the speed doubles when the time doubles.

67. **ORGANIZE AND PLAN**   The average force does work equal to the force times the displacement. This work must equal the original kinetic energy of the bullet but with the opposite sign. If we first find the kinetic energy, we can easily calculate the average force.
Known: $m = 25$ g; $v = 310$ m/s; $\Delta x = 15$ cm.

SOLVE   We can calculate the kinetic energy of the bullet using Equation 5.10:

$$K = \tfrac{1}{2}mv^2 = \tfrac{1}{2}(25 \text{ g})(310 \text{ m/s})^2 = 1.2 \text{ kJ}$$

The force does work $W_f = -K = -1.2$ kJ on the bullet. We can calculate the average force from Equation 5.1:

$$F_x = \frac{W_f}{\Delta x} = \frac{(-1.2 \text{ kJ})}{(15 \text{ cm})} = -8.0 \text{ kN}$$

REFLECT   The force is a drag force and all drag forces are negative, i.e., in the opposite direction of the displacement.

69. **ORGANIZE AND PLAN**   When the crane lifts at constant speed, it exerts a force on the girder equal to the force of gravity (but in the opposite direction). The work done by this force is the force times the displacement. When the girder is accelerating by 1.20 m/s^2 we can calculate what velocity when it has traveled 8.85 m. The crane has then done work equal to kinetic energy of the girder plus the work done by gravity in part (a).
Known: $m = 750$ kg; $h = 8.85$ m; $a = 1.20$ m/s^2.

SOLVE   For part (a), the crane has done work equal to the weight times the displacement:

$$W_a = mgh = (750 \text{ kg})(9.80 \text{ m/s}^2)(8.85 \text{ m}) = 65.0 \text{ kJ}$$

For part (b), we first calculate the time it takes to accelerate the girder to a height of 8.85 m:

$$h = \tfrac{1}{2}at^2$$

$$t = \sqrt{\frac{2h}{a}} = \sqrt{\frac{2(8.85 \text{ m})}{(1.20 \text{ m/s}^2)}} = 3.84 \text{ s}$$

When the girder accelerates at a constant rate for this period of time it reaches a velocity:

$$v = at = (1.20 \text{ m/s}^2)(3.84 \text{ s}) = 4.61 \text{ m/s}$$

This means that the girder has a kinetic energy:

$$K = \tfrac{1}{2}mv^2 = \tfrac{1}{2}(750 \text{ kg})(4.61 \text{ m/s})^2 = 7.97 \text{ kJ}$$

The total work done by the crane in the accelerating case is the sum of the kinetic energy and the work equal to the weight times the displacement:

$$W_b = K + W_a = (7.97 \text{ kJ}) + (65.0 \text{ kJ}) = 73.0 \text{ kJ}$$

REFLECT When it's time to stop the girder at a certain elevation, the work done to build up kinetic energy is returned in the sense that the girder will travel the last distance before reaching its final elevation with no further work required by the crane.

71. ORGANIZE AND PLAN We will treat this problem the same as the previous problem, except that the box starts with some kinetic energy associated with the initial 1 m/s speed. The force then adds the same amounts of kinetic energy as in the previous problem to this initial amount.
Known: Force-versus-position. $m = 1.8 \text{ kg}; v_0 = 1.0 \text{ m/s}$.

SOLVE The initial kinetic energy of the box is:

$$K_0 = \tfrac{1}{2}mv_0^2 = \tfrac{1}{2}(1.8 \text{ kg})(1.0 \text{ m/s})^2 = 0.90 \text{ J}$$

When the box reaches 5 cm, it has increased its kinetic energy by the work done by the force, so its new kinetic energy is $K = (0.90 \text{ J}) + (0.438 \text{ J}) = 1.3 \text{ J}$ and the speed of the box is:

$$v = \sqrt{\frac{2K}{m}} = \sqrt{\frac{2(1.3 \text{ J})}{(1.8 \text{ kg})}} = 1.2 \text{ m/s}$$

Similarly, the speed of the box at 10 cm is:

$$v = \sqrt{\frac{2K}{m}} = \sqrt{\frac{2((0.90 \text{ J}) + (1.75 \text{ J}))}{(1.8 \text{ kg})}} = 1.7 \text{ m/s}$$

and the speed of the box at 15 cm is:

$$v = \sqrt{\frac{2K}{m}} = \sqrt{\frac{2((0.90 \text{ J}) + (3.5 \text{ J}))}{(1.8 \text{ kg})}} = 2.2 \text{ m/s}$$

REFLECT Because speed is proportional only to the square root of kinetic energy, or in this case the square root of the work done by the force, the speeds at the end of each interval are increased by less than 1.0 m/s, the initial speed of the box.

73. ORGANIZE AND PLAN The gravitational potential energy equals the work done by gravity but with the opposite sign.
Known: $m = 60 \text{ kg}; \Delta y = 4390 \text{ m}$.

SOLVE We calculate the gravitational potential energy using Equation 5.13:

$$\Delta U = mg\Delta y = (60 \text{ kg})(9.80 \text{ m/s}^2)(4390 \text{ m}) = 2.6 \text{ MJ}$$

REFLECT Expressed in food calories, this energy is only about 600 food calories! See answer to Problem 5.1.

75. ORGANIZE AND PLAN The potential energy stored on a spring equals the work you do on the spring.
Known: $k = 125 \text{ N/m}; d = 0.125 \text{ m}$.

SOLVE The work required to compress the spring is:

$$W_{compress} = \tfrac{1}{2}kd^2 = \tfrac{1}{2}(125 \text{ N/m})(0.125 \text{ m})^2 = 0.977 \text{ J}$$

The work required to extend the spring is:

$$W_{extend} = \tfrac{1}{2}k(-d)^2 = \tfrac{1}{2}(125 \text{ N/m})(-0.125 \text{ m})^2 = 0.977 \text{ J}$$

So the potential energy is the same in both cases, meaning there's no change in potential energy!
REFLECT When you let go of a compressed (or extended) spring, it will oscillate back and forth between compression and extension, reaching the same amplitude in each oscillation. (Although eventually the oscillation will die out due to drag and frictional forces.)

77.   **ORGANIZE AND PLAN**   The unit conversion is straightforward. The gravitational potential energy equals the work done by gravity but with the opposite sign.

Known: $W_a = 120$ kcal; $W_b = 130$ kcal; $m = 62$ kg; $h = 125$ m.

SOLVE   (a) The conversion from food calories to joules is 1 kcal = 4.186 kJ. The energy in the breakfast cereal is:

$$W_a = (120 \text{ kcal}) = (120 \text{ kcal})(4.186 \text{ kJ/kcal}) = 502 \text{ kJ}$$

(b) To climb the hill, the person has to raise his or her potential energy by an amount equal to the work done by gravity but with the opposite sign:

$$W_b = mgh = (62 \text{ kg})(9.80 \text{ m/s}^2)(125 \text{ m}) = 76 \text{ kJ}$$

Expressed in food calories, this change in potential energy is:

$$W_b = (76 \text{ kJ}) = \frac{(76 \text{ kJ})}{(4.186 \text{ kJ/kcal})} = 18 \text{ kcal}$$

With a glass of 1% milk containing 130 kcal of energy, the number of glasses of milk the person has to drink to climb the hill is:

$$\frac{(18 \text{ kcal})}{(130 \text{ kcal/glass})} = 0.14 \text{ glasses of 1\% milk}$$

REFLECT   How many 125 m = 410 ft hills do you climb every day? These numbers illustrate how easy it is to consume more calories than one expends.

79.   **ORGANIZE AND PLAN**   Raising the weight does mechanical work equal to your raising force multiplied with the displacement. This work is also the change in potential energy of the weight. Because of the conversion from food energy to mechanical energy, for each 1 kcal of mechanical energy produced, you burn 5 kcal of food energy, so to burn 100 kcal you need to do 20 kcal of mechanical work. The unit conversion is 1 kcal = 4.186 kJ.

Known: $F = 20.0$ N; $\Delta x = 45$ cm; $W = (20\%)(100 \text{ kcal}) = 20 \text{ kcal} = 83 \text{ kJ}$.

SOLVE   The mechanical work done in raising the weight once is calculated from Equation 5.1:

$$W_1 = F\Delta x = (20.0 \text{ N})(45 \text{ cm}) = 9.0 \text{ J}$$

So the number of raises required to equal 83 kJ of mechanical work (and burn 100 kcal of food energy) is:

$$\frac{W}{W_1} = \frac{(83 \text{ kJ})}{(9 \text{ J})} = 9,300$$

REFLECT   Even if you raise the weight once a second without interruption it would take you over two and a half hours to raise the weight 9,300 times. This is not a reasonable workout session!

81.   **ORGANIZE AND PLAN**   The total mechanical energy is the sum of potential and kinetic energy. The total mechanical energy is constant, so when the golf ball hits the ground, its kinetic energy equals the total mechanical energy (because the potential energy is chosen to be zero at the ground). From the kinetic energy we can calculate the speed.

Known: $m = 45.9$ g; $h_0 = 23.4$ m; $v_0 = 31.2$ m/s.

SOLVE   (a) At every point in its flight, the total mechanical energy of the golf ball is:

$$E = K + U = \tfrac{1}{2}mv^2 + mgh$$

Insert values for when the golf ball is 23.4 m above the ground to calculate the total mechanical energy:

$$E = K_0 + U_0 = \tfrac{1}{2}mv_0^2 + mgh_0 = \tfrac{1}{2}(45.9 \text{ g})(31.2 \text{ m/s})^2 + (45.9 \text{ g})(9.80 \text{ m/s})(23.4 \text{ m}) = 32.9 \text{ J}$$

(b) When the ball hits the ground, all of this energy has been converted to kinetic energy, so we can calculate the ball's speed:

$$v = \sqrt{\frac{2K_{ground}}{m}} = \sqrt{\frac{2E}{m}} = \sqrt{\frac{2(32.9 \text{ J})}{(45.9 \text{ J})}} = 37.8 \text{ m/s}$$

REFLECT   We expect the speed to be greater when the ball hits the ground then when its 23.4 m up, because the work by gravity has accelerated the ball.

83.   **ORGANIZE AND PLAN**   The man does work on the medicine ball equal to the force times the displacement. By conservation of energy, all this work is converted into kinetic energy. Knowing the kinetic energy we can calculate the speed. To stop the ball, the other man must do an equal amount of work, but with the opposite sign.
Known: $m = 5.0$ kg; $F_x = 138$ N; $\Delta x = 0.50$ m.

SOLVE   (a) The man that launches the ball does work:

$$W = F_x \Delta x = (138 \text{ N})(0.50 \text{ m}) = 69 \text{ J}$$

This work is converted to kinetic energy, so the speed of the ball when it leaves the man's hands is:

$$v = \sqrt{\frac{2K}{m}} = \sqrt{\frac{2W}{m}} = \sqrt{\frac{2(69 \text{ J})}{(5.0 \text{ kg})}} = 5.3 \text{ m/s}$$

(b) The other man must do work equal to -69 J to stop the ball.

REFLECT   The work to stop the ball is negative because the force and the displacement are in opposite directions. Even though the work is negative, it will use up muscle energy.

85.   **ORGANIZE AND PLAN**   The roller coaster converts kinetic energy into gravitational potential energy, keeping the total mechanical energy constant, when it goes uphill. If we subtract the potential energy of the elevation increase from the original kinetic energy, we get the kinetic energy of the roller coaster at the given elevation. From this we can easily calculate the speed.
Known: $v_0 = 19.2$ m/s; $\Delta y = 12.2$ m.

SOLVE   If we define original elevation of the roller coaster as having zero potential energy, the total mechanical energy of the roller coaster is simply its kinetic energy given by Equation 5.10:

$$E = K_0 = \tfrac{1}{2}mv_0^2$$

When the roller coaster has increased its elevation by $\Delta y = 12.2$ m, its potential energy increase is given by Equation 5.13:

$$\Delta U = mg\Delta y$$

Since the total mechanical energy stays constant, the kinetic energy must have decreased and is now:

$$K = E - U = \tfrac{1}{2}mv_0^2 - mg\Delta y$$

By using Equation 5.10 a second time, we can calculate the speed v of the roller coaster at this elevation:

$$K = \tfrac{1}{2}mv^2 = \tfrac{1}{2}mv_0^2 - mg\Delta y$$
$$\tfrac{1}{2}v^2 = \tfrac{1}{2}v_0^2 - g\Delta y$$
$$v = \sqrt{v_0^2 - 2g\Delta y} = \sqrt{(19.2 \text{ m/s})^2 - 2(9.80 \text{ m/s}^2)(12.2 \text{ m})} = 11.4 \text{ m/s}$$

REFLECT   Note that it was not necessary to know the mass of the roller coaster. It was also not necessary to calculate the exact values of the potential or kinetic energies (which we couldn't have done without knowing the mass).

87.   **ORGANIZE AND PLAN**   The bowling ball will first fall freely, converting gravitational potential energy to kinetic energy. When it contacts the spring, the spring will do work on the bowling ball to slow it down, and the bowling ball will do work on the spring to compress it. Gravity will continue to do work on the bowling ball as the spring compresses. The spring will be maximally compressed when the speed of the bowling ball is zero, meaning its kinetic energy is zero. This means that the spring is maximally compressed when the change in gravitational potential energy of the bowling ball equals the stored potential energy on the spring (but with the opposite sign; the total energy of the system is constant).
Known: $k = 1340$ N/m; $m = 7.27$ kg; $h = 1.75$ m.

SOLVE   As usual we define a positive *y*-direction to be upward, and we choose $y = 0$ to be the end of the spring when the spring is at its equilibrium length. This is the point where the bowling ball and spring first make contact.

Let's call the maximum compression of the spring Δy. The stored energy in the maximally compressed spring is given by Equation 5.16:

$$\Delta U_{spring} = \tfrac{1}{2}k\Delta y^2$$

The change in gravitational potential energy for the bowling ball when the spring is maximally compressed is:

$$\Delta U_g = -mg(h + \Delta y)$$

The total energy of the system is constant, so the spring is maximally compressed when:

$$\Delta U_{spring} + \Delta U_g = 0$$
$$\tfrac{1}{2}k\Delta y^2 - mg(h + \Delta y) = 0$$

This is a quadratic equation we can solve for the compression Δy (compare to solution of Problem 55):

$$\Delta y = \frac{mg}{k} + \sqrt{\left(\frac{mg}{k}\right)^2 + \frac{2mgh}{k}}$$

Insert our known values to calculate the maximum compression:

$$\Delta y = \frac{(7.27\ \text{kg})(9.80\ \text{m/s}^2)}{(1340\ \text{N/m})} + \sqrt{\left(\frac{(7.27\ \text{kg})(9.80\ \text{m/s}^2)}{(1340\ \text{N/m})}\right)^2 + \frac{2(7.27\ \text{kg})(9.80\ \text{m/s}^2)(1.75\ \text{m})}{(1340\ \text{N/m})}} = 0.488\ \text{m}$$

REFLECT   Note that we don't have to calculate the kinetic energy of the bowling ball in this problem, because the only moment we ask about is one when we know the kinetic energy is zero.

89. **ORGANIZE AND PLAN**   Initially the rubber ball has gravitational potential energy relative to the ground, but no kinetic energy. When the ball is dropped, the potential energy is converted to kinetic energy, keeping the total mechanical energy of the system constant. Just before the ball hits the ground, all of the initial potential energy has been converted to kinetic energy. In the bounce, 25% of the energy is lost (to heat) and the ball rebounds upward with kinetic energy equal to 75% of the initial total mechanical energy. Kinetic energy is converted into gravitational potential energy as the ball travels upward, and when the ball has reached its maximum height, all the kinetic energy the ball had after the bounce has been converted into potential energy.

Known: $h_{before} = 2.4\ \text{m}$; $E_{after}/E_{before} = 0.75$.

SOLVE   (a) The total mechanical energy before the bounce is the initial gravitational potential energy of the ball:

$$E_{before} = \Delta U_{before} = mgh_{before}$$

When the ball hits the ground, the kinetic energy equals the total mechanical energy. From the kinetic energy we can calculate the speed:

$$v = \sqrt{\frac{2K_{before}}{m}} = \sqrt{\frac{2E_{before}}{m}} = \sqrt{\frac{2mgh}{m}} = \sqrt{2gh} = \sqrt{2(9.80\ \text{m/s}^2)(2.4\ \text{m})} = 6.9\ \text{m/s}$$

(b) After the bounce, when the ball reaches its maximum height, the total mechanical energy again equals the gravitational potential energy of the ball:

$$E_{after} = \Delta U_{after} = mgh_{after}$$

Since we know the ratio of the total mechanical energy before and after the bounce, we can calculate the maximum height of the ball after the bounce:

$$h_{after} = \frac{E_{after}}{mg} = \frac{0.75 E_{before}}{mg} = 0.75 h_{before} = 0.75(2.4\ \text{m}) = 1.8\ \text{m}$$

REFLECT   We did not need to know the mass of the ball to calculate either answer.

91. **ORGANIZE AND PLAN**   As shown in the figure below, the three forces act on the car. The force of gravity mg directed downward, the normal force from the ground, and the frictional force from the ground. The acceleration of the car is due to the net force parallel to the ground, which is the component of gravity parallel to the ground reduced by the frictional force. We can calculate the frictional force from the normal force and the coefficient of friction. The normal force is equal to the component of gravity normal to the ground. With all forces known, we

can calculate the work done by friction and gravity by multiplying the forces with the displacement (the rolling distance). We can calculate the final speed of the car from the kinetic energy, which is the work done by gravity reduced by the work done by friction.

Known: $m = 980$ kg; $\theta = 3.6°$; $\mu_r = 0.030$; $\Delta x = 35$ m.

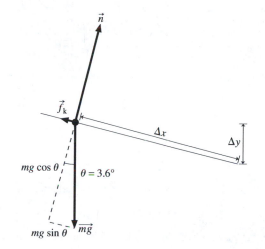

SOLVE   (a) The normal force from the ground is:

$$n = mg\cos\theta = (980\text{ kg})(9.80\text{ m/s}^2)\cos(3.6°) = 9.6\text{ kN}$$

The frictional force is:

$$f_r = -\mu_r n = -(0.030)(9.6\text{ kN}) = -0.29\text{ kN}$$

The work done by friction is:

$$W_f = f_r\Delta x = (-0.29\text{ kN})(35\text{ m}) = -10\text{ kJ}$$

(b) The work done by gravity is:

$$W_g = mg\sin\theta\Delta x = (980\text{ kg})(9.80\text{ m/s}^2)\sin(3.6°)(35\text{ m}) = 21\text{ kJ}$$

(c) The kinetic energy of the car at the bottom of the hill is the work done by gravity reduced by the work done by friction:

$$K = W_g + W_f = (21\text{ kJ}) + (-10\text{ kJ}) = 11\text{ kJ}$$

The final speed of the car is:

$$v = \sqrt{\frac{2K}{m}} = \sqrt{\frac{2(11\text{ kJ})}{(980\text{ kg})}} = 4.7\text{ m/s}$$

REFLECT   The work done by gravity is always equal to the change in the gravitational potential energy (but with the opposite sign):

$$W_g = -\Delta U = -mg\Delta y$$

What's the change in elevation Δy here? As shown in the figure below, $\Delta y = -\Delta x\sin\theta$, so:

$$W_g = -\Delta U = -mg\Delta y = mg\Delta x\sin\theta$$

which is the same result as above.

93.   **ORGANIZE AND PLAN**   It simplifies the problem to separate the motion into two components: horizontal speed v_x, and vertical speed v_y. The stored potential energy in the spring gets converted into kinetic energy associated with the horizontal speed. The downward vertical speed comes from conversion of gravitational potential energy to kinetic energy, but the downward motion is a simple acceleration problem that we don't need use energy arguments to solve; all we want to know from the vertical motion is how long it takes for the pellet to drop 1.20 m. Once we know the time of the fall, we know how far the pellet has traveled horizontally.

Known: $k = 42.0$ N/m; $\Delta y = -1.20$ m; $d = 5.00$ cm; $m = 25.0$ g.

SOLVE The stored potential energy in the spring compressed by $d = 5.00$ cm is:

$$\Delta U = \tfrac{1}{2}kd^2 = \tfrac{1}{2}(42.0 \text{ N/m})(5.00 \text{ cm})^2 = 52.5 \text{ mJ}$$

When the spring is released, this potential energy is converted into kinetic energy associated with the horizontal speed of the pellet. When all of the potential energy has been converted, the final horizontal speed is:

$$v_x = \sqrt{\frac{2K}{m}} = \sqrt{\frac{2\Delta U}{m}} = \sqrt{\frac{2(52.5 \text{ mJ})}{(25.0 \text{ g})}} = 2.05 \text{ m/s}$$

Next we need to find out how long the pellet travels at this horizontal speed, i.e., how long it takes for the pellet to drop 1.20 m. Using Equation 2.12:

$$\Delta y = -\tfrac{1}{2}gt^2$$

$$t = \sqrt{\frac{2(-\Delta y)}{g}} = \sqrt{\frac{2(1.20 \text{ m})}{(9.80 \text{ m/s}^2)}} = 0.495 \text{ s}$$

This means the pellet strikes the floor at a distance from the edge of the table which is:

$$x = v_x t = (2.05 \text{ m/s})(0.495 \text{ s}) = 1.01 \text{ m}.$$

REFLECT Separating the motion into two orthogonal components is a useful strategy when solving many problems. In particular, because kinetic energy is proportional to the square of the speed, and the speed can be calculated as the sum of the squares of the horizontal and vertical components of the velocity, kinetic energy can be thought of as having a horizontal part and a vertical part:

$$K = \tfrac{1}{2}mv^2 = \tfrac{1}{2}m\left(\sqrt{v_x{}^2 + v_y{}^2}\right)^2 = \tfrac{1}{2}mv_x{}^2 + \tfrac{1}{2}mv_y{}^2$$

95. **ORGANIZE AND PLAN** The vertical component of the cat's speed must be at least large enough that the initial kinetic energy associated with the vertical speed equals the change in gravitational potential energy.
Known: $\Delta y = 1.15$ m; $\theta = 75°$.

SOLVE The required change in gravitational potential energy is:

$$\Delta U = mg\Delta y$$

The initial kinetic energy K_y associated with the vertical component of the cat's speed must equal (or be larger than) this change in potential energy. This means the minimum vertical speed is:

$$v_y = \sqrt{\frac{2K_y}{m}} = \sqrt{\frac{2\Delta U}{m}} = \sqrt{\frac{2mg\Delta y}{m}} = \sqrt{2g\Delta y} = \sqrt{2(9.80 \text{ m/s}^2)(1.15 \text{ m})} = 4.7 \text{ m/s}$$

Leaving the floor at an angle θ, the vertical component of the speed is related to the total speed by:

$$v_y = v\sin\theta$$

This means the cat's minimum speed must be:

$$v = \frac{v_y}{\sin\theta} = \frac{(4.7 \text{ m/s})}{\sin 75°} = 4.9 \text{ m/s}$$

REFLECT We have assumed that the distance between the floor and the center of mass of the cat when the cat leaves the floor is the same as the distance between the top of the dresser and the center of mass of the cat when the cat lands. In reality, there is a difference between these two distances which reduces the required speed slightly. Center of mass is discussed in Chapter 6.

97. **ORGANIZE AND PLAN** The maximum acceleration occurs when the spring force is maximum and that happens when the spring is maximally compressed, so there are two unknowns: the spring constant k which we are asked to calculate, and the maximum compression x of the spring. We need two equations to find them. Newton's second law relates the force of the spring to the acceleration, and the force also equals spring constant multiplied by compression. That gives us one equation for k and x. The change in gravitational potential energy of the elevator

car must equal the stored potential energy in the spring when the spring is maximally compressed (because there is no kinetic energy at this moment, and the total mechanical energy is conserved). That gives us the second equation.
Known: $m = 480$ kg; $h = 11.8$ m; $a_{max} = 4g$

SOLVE   Newton's second law gives us the maximum allowed spring force:

$$F_{max} = ma_{max} = 4mg$$

Then the maximum allowed compression x of the spring is:

$$x = \frac{F_{max}}{k} = \frac{4mg}{k}$$

which we can rewrite as an expression for the spring constant:

$$k = \frac{4mg}{x}$$

This is our first equation relating the compression x to the spring constant k. Next we look at energy. The change in gravitational potential energy of the elevator car dropping from its maximum height to its position when the spring is maximally compressed is:

$$\Delta U_g = -mg(h + x)$$

The total mechanical energy $E = K + \Delta U_g + \Delta U_s$ is conserved (constant), and the kinetic energy K is zero both then the elevator cable breaks and when the spring is maximally compressed, so the change in stored potential energy in the spring when the spring is maximally compressed is:

$$\Delta U_s = \tfrac{1}{2}kx^2 = -\Delta U = mg(h + x)$$

This is our second equation relating the compression x to the spring constant k. Use the first equation to substitute for k in the second equation:

$$\tfrac{1}{2}kx^2 = mg(h + x)$$
$$\tfrac{1}{2}\frac{4mg}{x}x^2 = mg(h + x)$$
$$2mgx = mg(h + x)$$
$$x = h$$

The maximum allowed spring compression is equal to the elevator car's maximum height over the spring. We can now use the first equation to calculate the spring constant:

$$k = \frac{4mg}{x} = \frac{4mg}{h} = \frac{4(480\text{ kg})(9.80\text{ m/s}^2)}{(11.8\text{ m})} = 1.59 \text{ kN/m}$$

REFLECT   We had two unknowns and two equations. We could have chosen to use the first equation to substitute for x in the second equation, and then solve directly for the spring constant k. Try it! Both methods will give the correct answer, although solving directly for k means a bit more algebra.

99.   **ORGANIZE AND PLAN**   Power is work per unit time, so if we multiply power by time we get the work.
Known: $P = 8.5$ kW; $t = 30$ s.

SOLVE   Multiply power by time to calculate the work:

$$W = Pt = (8.5\text{ kW})(30\text{ s}) = 0.26 \text{ MJ}$$

REFLECT   What if the power had not been constant, but fluctuated over the 30 s time interval? If the average power was 8.5 kW, the answer would have been the same. In a power versus time graph, the work is the area under the curve.

101.   **ORGANIZE AND PLAN**   From the density we can calculate the mass m of water that rush over the falls in one minute. From the mass and the height of the fall we can calculate the change in potential energy. Once we know the change in energy we can calculate the power, because power is energy delivered per unit time.
Known: $\Delta y = 100$ m; $V = 550 \times 10^6$ m³; $t = 1$ min; $\rho = 1000$ kg/m³.

SOLVE   The mass of water going through the fall in one minute is:

$$m = V\rho = \left(550 \times 10^6 \text{ m}^3\right)\left(1000 \text{ kg/m}^3\right) = 550 \times 10^9 \text{ kg}$$

The change in potential energy of this water is:

$$\Delta U = mg\Delta y = \left(550 \times 10^9 \text{ kg}\right)\left(9.80 \text{ m/s}^2\right)\left(100 \text{ m}\right) = 5.4 \times 10^{14} \text{ J}$$

The total power in the waterfall is:

$$P = \frac{\Delta U}{t} = \frac{\left(5.4 \times 10^{14} \text{ J}\right)}{\left(1 \text{ min}\right)} = 9.0 \text{ TW}$$

REFLECT   The power output of a typical nuclear power station is about 1 GW per reactor. That means the Victoria Falls power is equivalent of about 9000 nuclear reactors!

103.   **ORGANIZE AND PLAN**   The work is the change in gravitational potential energy required to lift the total mass. Once we know the change in energy we can calculate the required power, because power is energy delivered per unit time.

Known: $\theta = 30°$; $m_{\text{person}} = 75$ kg; $m_{\text{chair}} = 22$ kg; $d = 5.6$ m; $t = 12$ s

SOLVE   (a) The total mass to lift is:

$$m = m_{\text{person}} + m_{\text{chair}} = \left(75 \text{ kg}\right) + \left(22 \text{ kg}\right) = 97 \text{ kg}$$

The lifting height, i.e., the vertical distance to lift is:

$$\Delta y = d\sin\theta = \left(5.6 \text{ m}\right)\sin 30° = 2.8 \text{ m}$$

The work done in lifting the person and the chair equals the change in gravitational potential energy:

$$\Delta U = mg\Delta y = \left(97 \text{ kg}\right)\left(9.80 \text{ m/s}^2\right)\left(2.8 \text{ m}\right) = 2.7 \text{ kJ}$$

(b) The power the motor must deliver to make this lift in 12 s is:

$$P = \frac{\Delta U}{t} = \frac{\left(2.7 \text{ kJ}\right)}{\left(12 \text{ s}\right)} = 0.22 \text{ kW}$$

REFLECT   The power we calculate is the required average power. Whether the power output fluctuates or is constant does not affect how much work is done over a period of time. It's the average power over the time period that matters.

105.   **ORGANIZE AND PLAN**   The power of the engine equals energy delivered per unit time, so if we multiply the power with the time, we get the energy delivered. This energy (ignoring drag and frictional forces) will equal the kinetic energy of the car (or more precisely the change in kinetic energy, but the car is initially at rest with zero kinetic energy). From the kinetic energy we can calculate the speed.

Known: $m = 1320$ kg; $P_{\text{engine}} = 280$ hp; $P_{\text{motion}} / P_{\text{engine}} = 40\%$; $t = 4.0$ s.

SOLVE   The power that can be converted into motion is:

$$P_{\text{motion}} = 0.40 P_{\text{engine}} = 0.40\left(280 \text{ hp}\right) = 0.40\left(280 \text{ hp}\right)\left(746 \text{ W/hp}\right) = 84 \text{ kW}$$

The maximum kinetic energy delivered in 4.0 s is:

$$K = P_{\text{motion}} t = \left(84 \text{ kW}\right)\left(4.0 \text{ s}\right) = 0.33 \text{ MJ}$$

The maximum speed of the car after 4.0 s is:

$$v = \sqrt{\frac{2K}{m}} = \sqrt{\frac{2\left(0.33 \text{ MJ}\right)}{\left(1320 \text{ kg}\right)}} = 22 \text{ m/s}$$

REFLECT   Are there any other "efficiencies" to consider? If the tires slip against the road surface, the power that can be converted into motion is reduced further.

107.   **ORGANIZE AND PLAN**   If we multiply the power, i.e., the rate at which the man expends energy while running, with the time spent running per week, we get the extra energy used per week. Then we simply divide by 7 to get the extra daily energy required to maintain a constant weight. The man's normal food consumption does not enter into the problem.

Known: $d = 4 \times 8$ km $= 32$ km; $P = 450$ W; $v = 12$ km/h.

SOLVE   Every week, the man runs 32 km. At a speed of 12 km/h, the time the man spends on running each week is:

$$t = \frac{d}{v} = \frac{(32 \text{ km})}{(12 \text{ km/h})} = 9600 \text{ s}$$

Expending energy at a rate of 450 W = 450 J/s, the extra weekly energy consumption due to running is:

$$W_{\text{weekly}} = Pt = (450 \text{ W})(9600 \text{ s}) = 4.3 \text{ MJ}$$

So to maintain his weight, the man should consume extra food energy each day equal to:

$$W_{\text{daily}} = \frac{W_{\text{weekly}}}{7} = \frac{(4.3 \text{ MJ})}{7} = 0.6 \text{ MJ}$$

REFLECT   While the normal food consumption of the man did not enter into the problem, it allows us to calculate the required increase in food consumption in percent (about 7%).

109.   **ORGANIZE AND PLAN**   The work done by gravity is the gravitational force multiplied by the displacement in the direction of the force. For the graph we are interested in the instantaneous power, which is force multiplied by velocity in the direction of the force. The velocity of the falling apple is easy to calculate.

Known: $m = 0.150$ kg; $h = 2.60$ m.

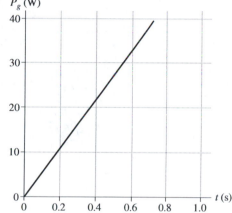

SOLVE   (a) The work done by gravity is:

$$W_g = -mgy$$

where as usual we have defined the positive y-direction as upward, and where $y = 0$ is the starting elevation of the apple. When the apple reaches the ground, i.e., $y = -h$, the work done by gravity is:

$$W_g = -mgy = mgh = (0.150 \text{ kg})(9.80 \text{ m/s}^2)(2.60 \text{ m}) = 3.82 \text{ J}$$

(b) To calculate the instantaneous power supplied by gravity as function of time, we need to multiply the gravitational force $F_y = -mg$ by the velocity $v_y = -gt$ of the apple. The instantaneous power is:

$$P_g = F_y v_y = (-mg)(-gt) = mg^2 t$$

To produce our graph, we also need to know the time at which the apples reaches the ground, i.e., the time when $y = -h$. The vertical position of the apple is given by Equation 2.12:

$$y = -\tfrac{1}{2}gt^2$$

so the time the apple reaches the ground is:

$$t_{\text{ground}} = \sqrt{\frac{-2y}{g}} = \sqrt{\frac{2h}{g}} = \sqrt{\frac{2(2.60 \text{ m})}{(9.80 \text{ m/s}^2)}} = 0.728 \text{ s}$$

Now we can make a graph of the instantaneous power supplied by gravity as a function of time over the entire fall from. The power starts at zero at $t = 0$ and grows linearly with time until it reaches

$$P_{g,\text{ground}} = mg^2 t_{\text{ground}} = (0.150 \text{ kg})(9.80 \text{ m/s}^2)^2 (0.728 \text{ s}) = 10.5 \text{ W}$$

at $t = t_{\text{ground}} = 0.728$ s. The graph is shown in the figure below.

(c) We can use the graph to calculate the average power. The average power of a time interval is the area under the curve, divided by the time interval. If our time interval of interest is the entire fall time, the average power is the area of the triangle under the curve, divided by the fall time:

$$\overline{P}_g = \frac{\frac{1}{2}(10.5 \text{ W})(2.75 \text{ s})}{(2.75 \text{ s})} = \frac{1}{2}(10.5 \text{ W}) = 5.25 \text{ W}$$

Multiplying the average power with the fall time we get:

$$\overline{P}_g t_{\text{ground}} = (5.25 \text{ W})(0.728 \text{ s}) = 3.82 \text{ J}$$

which equals the work done by gravity.

REFLECT   It should be fairly obvious that the average power is half the final instantaneous power.

111.   **ORGANIZE AND PLAN**   The work to lift the blood from feet to brain is the change in the gravitational potential energy. The average power is the work divided by the time it takes to do the work.

Known: $V = 5.0$ L; $\rho = 1.05$ mg/L; $t = 1.0$ min; $\Delta y = 1.85$ m; $P_{\text{actual}} = 6.0$ W.

SOLVE   (a) The work to the lift the blood from feet to brain is:

$$W = \Delta U = mg\Delta y = \rho V g \Delta y = (1.05 \text{ g/mL})(5.0 \text{ L})(9.80 \text{ m/s}^2)(1.85 \text{ m}) = 95 \text{ J}$$

(b) The average power the heart expends in this process is:

$$P = \frac{W}{t} = \frac{(95 \text{ J})}{(1.0 \text{ min})} = 1.6 \text{ W}$$

(c) There are at least two reasons why the actual power consumption of the heart is larger than what we just calculated. First of all, the heart is unlikely to be 100% efficient in converting the energy it uses into pumping the blood. Second, there is more than the change in gravitational potential energy to overcome. Blood is a viscous fluid just like almost all other fluids, so some drag forces from the walls of the blood vessels are expected. Viscosity is discussed in Chapter 10.

REFLECT   What percentage of a person's daily food energy consumption is the heart responsible for?

113.   **ORGANIZE AND PLAN**   The problem text states the amount of energy consumed per minute and how much energy to burn off, so we can calculate how long time the person needs to spend walking. Then we multiply the time with the walking speed to get the distance.

Known: $m = 175$ lbs; $v = 4.5$ mph; $P = 7.0$ kcal/min; $W = 125$ kcal.

SOLVE   Power is energy per unit time, so time is energy divided by power:

$$t = \frac{W}{P} = \frac{(125 \text{ kcal})}{(7.0 \text{ kcal/min})} = 18 \text{ min}$$

Distance is speed multiplied with time:

$$x = vt = (4.5 \text{ mph})(18 \text{ min}) = 1.3 \text{ miles}$$

Since the speed was given in mph, it's reasonable to give the answer in miles. Converting 1.3 miles to km, the answer in SI units is 2.2 km.

REFLECT   We didn't need to know the weight of the person.

115.   ORGANIZE AND PLAN   In a force-versus-position graph, the work is the area under the curve. When the curve is below the x-axis, the force does negative work equal to the area above the curve.
Known: Force-versus-position.

SOLVE   (a) The area under the graph from 0 to 7.0 m consists of two triangles and one rectangle. The total area is:

$$W_a = \tfrac{1}{2}(6.0\ \text{N})(3.0\ \text{m}) + (6.0\ \text{N})(1.0\ \text{m}) + \tfrac{1}{2}(6.0\ \text{N})(3.0\ \text{m}) = 24\ \text{J}$$

(b) The work done by the force as the object moves from 0 to 12.0 m is the work done in part (a) minus the work which is the area above the curve between 7.0 m and 12.0 m. The latter consist of two triangles and one rectangle. The total work from 0 to 12.0 m is:

$$W_b = W_a - \tfrac{1}{2}(4.0\ \text{N})(2.0\ \text{m}) - (4.0\ \text{N})(1.0\ \text{m}) - \tfrac{1}{2}(4.0\ \text{N})(2.0\ \text{m}) = 12\ \text{J}$$

REFLECT   In all portions of the graph, we can write the force in the form $F = kx + F_0$, but with different values on the spring constants k and the offset forces F_0.

117.   ORGANIZE AND PLAN   The froghopper's velocity has a horizontal component and a vertical component. We can calculate the initial horizontal and vertical speeds from the total speed and the angle given to us. The kinetic energy associated with the vertical speed is gradually converted to gravitational potential energy as the froghopper's height above ground increases. When all this kinetic energy has been converted to potential energy, the kinetic energy is zero meaning the vertical speed is zero and the froghopper has reached its maximum height above ground. From the change in potential energy we can calculate the height above ground. To calculate the spring constants of the legs we need to know the compression, which is given to us, and the initial stored energy. The latter is the total kinetic energy of the froghopper (from both horizontal and vertical speeds) as it leaps. The spring constant of each leg is half the total spring constant, because the legs act in parallel.
Known: $x = \tfrac{1}{3} \times 6.1$ mm; $m = 12.3$ mg; $v = 2.8$ m/s; $\theta = 58°$.

SOLVE   (a) The initial vertical speed of the froghopper is:

$$v_y = v\sin\theta = (2.8\ \text{m/s})\sin 58° = 2.4\ \text{m/s}$$

The kinetic energy associated with this initial vertical speed is:

$$K_y = \tfrac{1}{2}mv_y{}^2 = \tfrac{1}{2}(12.3\ \text{mg})(2.4\ \text{m/s})^2 = 35\ \mu\text{J}$$

When the froghopper has reached its maximum height above ground, all of this kinetic energy has been converted to a change in the gravitational potential energy, and we can calculate how high the froghopper leaps from this change:

$$\Delta y = \frac{\Delta U}{mg} = \frac{K_y}{mg} = \frac{(35\ \mu\text{J})}{(12.3\ \text{mg})(9.80\ \text{m/s}^2)} = 29\ \text{cm}$$

(b) The total, initial kinetic energy of the frog hopper is:

$$K = \tfrac{1}{2}mv^2 = \tfrac{1}{2}(12.3\ \text{mg})(2.8\ \text{m/s})^2 = 48\ \mu\text{J}$$

If all of this kinetic energy is stored on a spring compressed by one-third of the froghopper's body length, the spring constant of this spring would be:

$$k_{\text{spring}} = \frac{2W}{x^2} = \frac{2K}{x^2} = \frac{2(48\ \mu\text{J})}{\left(\tfrac{1}{3} \times 6.1\ \text{mm}\right)^2} = 23\ \text{N/m}$$

Since the spring hopper has two legs that act as parallel springs, the spring constant of each leg is half this value:

$$k_{\text{leg}} = \frac{k_{\text{spring}}}{2} = \frac{(23\ \text{N/m})}{2} = 12\ \text{N/m}$$

REFLECT   We have ignored air resistance in this problem? Could that be important?

119.   **ORGANIZE AND PLAN**   If the block is at rest, its kinetic energy is zero, so what we want to find out for part (a) is if the kinetic energy is ever zero, and if so, where. The initial kinetic energy of the block is straightforward to calculate. In a force-versus-position graph, the work is the area under the curve. When the curve is below the x-axis, as it is between $x = 0$ and $x = 1.33$ cm, the force does negative work equal to the area above the curve, reducing the kinetic energy. We need to find the x-value between $x = 0$ and $x = 1.33$ cm for which the sum of the initial kinetic energy plus the work (which is negative) equals zero.

Known: Force-versus-position. $m = 2.0$ kg; $v_0 = 5.0$ m/s.

SOLVE   The initial kinetic energy of the block is:

$$K_0 = \tfrac{1}{2}mv_0^2 = \tfrac{1}{2}(2.0\ \text{kg})(5.0\ \text{m/s})^2 = 25\ \text{J}$$

Consequently, if the kinetic energy is ever going to be zero, the work done by the force between between $x = 0$ and $x = 1.33$ cm must be less than or equal to -25 J (because this is the only portion where the force does negative work, slowing the block down, elsewhere the work is positive). The area under the curve between $x = 0$ and $x = 1.33$ cm is:

$$W = -\frac{(40\ \text{N})(1.33\ \text{m})}{2} = -27\ \text{J}$$

From which we conclude that the block's kinetic energy is zero at some point in the interval between $x = 0$ and $x = 1.33$ cm . The force in this portion of the graph is linear and can be written similar to Hooke's law, except with an added constant offset:

$$F = kx + F_0$$

where $F_0 = -40$ N is the offset force for zero displacement and $k = 30$ N/m so that the force reaches 20 N by $x = 2$ m. The negative work done by the force is the area above the curve. This area consists of a rectangle with base x and height $-F$ and a triangle with the same base x and height $F - F_0$. Consequently, the work done by the force F in this portion of the graph is:

$$W = -\left(-Fx + \tfrac{1}{2}(F - F_0)x\right) = -\left(-(kx + F_0)x + \tfrac{1}{2}kx^2\right) = \tfrac{1}{2}kx^2 + F_0 x$$

We want to know where the sum of the initial kinetic energy and this work equals zero, i.e., for which value of x that $K_0 + W = 0$. This gives us a quadratic equation:

$$K_0 + \tfrac{1}{2}kx^2 + F_0 x = 0$$

which we can solve for x:

$$x = \frac{-F_0 \pm \sqrt{F_0^2 - 2kK_0}}{k}$$

Insert known values to calculate x:

$$x = \frac{-F_0 \pm \sqrt{F_0^2 - 2kK_0}}{k} = \frac{-(-40\ \text{N}) \pm \sqrt{(-40\ \text{N})^2 - 2(30\ \text{N/m})(25\ \text{J})}}{(30\ \text{N/m})} =$$

$$= (1.33\ \text{m}) \pm (0.33\ \text{m}) = 1.0\ \text{m}$$

The answer to part (a) is that the block is at rest at $x = 1.0$ m. The answer to part (b) is that there is no such position, because the force on the block is always negative (to the left). Once the block has reached $x = 1.0$ m, it turns around and goes to the left.

REFLECT   The result for the work done by the force F should be fairly obvious if we regard the force F as two separate forces. One force is a regular spring force kx and the other a regular constant force F_0. The total work done by both forces is simply the sum of the work done by each individual force.

121.   **ORGANIZE AND PLAN**   If the passengers should not experience an upward force that exceeds $4g$, then the centripetal force should not exceed $3mg$, in order that the sum of the centripetal force plus the gravitational force does not exceed the limit. This puts a limit to the speed at point B. From the speed we can calculate the minimum height of point B from conservation of total mechanical energy. Since we are given the height of point C we can calculate the speed at point C, also from conversation of total mechanical energy.

Known: $m = 1500$ kg; $h_A = 40$ m; $v_A = 3$ m/s; $F_{r,\text{max}} = 3mg$; $R = 15$ m; $h_C = 20$ m.

SOLVE   (a) The centripetal force is given by Equation 4.9 and should not exceed *3mg*:

$$F_r = \frac{mv^2}{R} \leq 3mg$$

This puts a limit on the speed of the roller coaster at point B:

$$v \leq \sqrt{3gR} = \sqrt{3(9.80 \text{ m/s}^2)(15 \text{ m})} = 21 \text{ m/s}$$

Conservation of total mechanical energy gives us the minimum value of *h* that keeps the speed below 21 m/s and satisfies the requirement of the problem text:

$$\Delta K + \Delta U = 0$$
$$(K_B - K_A) + (U_B - U_A) = 0$$
$$\left(\tfrac{1}{2}mv_B^2 - \tfrac{1}{2}mv_A^2\right) + (mgh - mgh_A) = 0$$
$$h = h_A - \frac{1}{2g}\left(v_B^2 - v_A^2\right) \geq (40 \text{ m}) - \frac{1}{2(9.80 \text{ m/s}^2)}\left((21 \text{ m/s})^2 - (3 \text{ m/s})^2\right) = 18 \text{ m}$$

(b) Conservation of total mechanical energy gives us the speed of the roller coaster at point C:

$$\Delta K + \Delta U = 0$$
$$(K_C - K_A) + (U_C - U_A) = 0$$
$$\left(\tfrac{1}{2}mv_C^2 - \tfrac{1}{2}mv_A^2\right) + (mgh_C - mgh_A) = 0$$
$$v_C = \sqrt{v_A^2 - 2g(h_C - h_A)} = \sqrt{(3 \text{ m/s})^2 - 2(9.80 \text{ m/s})((20 \text{ m}) - (40 \text{ m}))} = 20 \text{ m/s}$$

REFLECT   With the known value of the roller coaster's mass we could have calculated intermediate values of several quantities, including centripetal force, kinetic energies, and potential energies.

When we talk about experiencing a certain number of *g*'s, what we refer to is acceleration, not force.

123.   **ORGANIZE AND PLAN**   Because the golf ball starts and ends its journey at the same elevation, i.e, with the same potential energy, the entire change in kinetic energy must be due to the air resistance. The work by the air resistance equals the change in kinetic energy.

Known: $m = 45.9$ g; $v_0 = 42.6$ m/s; $v_1 = 31.9$ m/s.

SOLVE   The change in kinetic energy and the work done by air resistance was:

$$W_f = \Delta K = \tfrac{1}{2}mv_1^2 - \tfrac{1}{2}mv_0^2 = \tfrac{1}{2}(45.9 \text{ g})(31.9 \text{ m/s})^2 - \tfrac{1}{2}(45.9 \text{ g})(42.6 \text{ m/s})^2 = -18.3 \text{ J}$$

REFLECT   Drag forces always do negative work.

125.   **ORGANIZE AND PLAN**   As usual we separate the motion into two components: horizontal speed v_x, and vertical speed v_y. The stored potential energy in the spring gets converted into kinetic energy associated with the horizontal speed. The downward vertical speed comes from conversion of gravitational potential energy to kinetic energy, keeping the total mechanical energy of the system constant. From kinematics we can calculate how long it takes for the pellet to drop 1.20 m. Once we know the time of the fall, we know how far the pellet has traveled horizontally.

Known: $k = 72.0$ N/m; $\Delta y = -1.20$ m; $d = 3.20$ cm; $m = 15.0$ g.

SOLVE   (a) The stored potential energy in the spring compressed by $d = 3.20$ cm is:

$$\Delta U_s = \tfrac{1}{2}kd^2 = \tfrac{1}{2}(72.0 \text{ N/m})(3.20 \text{ cm})^2 = 36.9 \text{ mJ}$$

When the spring is released, this potential energy is converted into kinetic energy associated with the horizontal speed of the pellet. When all of the potential energy has been converted, the horizontal speed is:

$$v_x = \sqrt{\frac{2K_0}{m}} = \sqrt{\frac{2\Delta U_s}{m}} = \sqrt{\frac{2(36.9 \text{ mJ})}{(15.0 \text{ g})}} = 2.22 \text{ m/s}$$

(b) The total mechanical energy is conserved, so the kinetic energy when the pellet hits the ground is the sum of the initial gravitational potential energy relative to ground and the initial spring potential energy:

$$K = E = \Delta U_g + \Delta U_s = -mg\Delta y + \Delta U_s = -(15.0 \text{ g})(9.80 \text{ m/s}^2)(-1.20 \text{ m}) + (36.9 \text{ mJ}) = 213 \text{ mJ}$$

From this we can calculate the speed when the pellet hits the ground:

$$v = \sqrt{\frac{2K}{m}} = \sqrt{\frac{2(213 \text{ mJ})}{(15.0 \text{ mg})}} = 5.33 \text{ m/s}$$

(c) Next we need to find long it takes for the pellet to drop 1.20 m. Using Equation 2.12:

$$\Delta y = -\tfrac{1}{2}gt^2$$

$$t = \sqrt{\frac{2(-\Delta y)}{g}} = \sqrt{\frac{2(1.20 \text{ m})}{(9.80 \text{ m/s}^2)}} = 0.495 \text{ s}$$

This means the pellet travels horizontally a distance:

$$x = v_x t = (2.22 \text{ m/s})(0.495 \text{ s}) = 1.10 \text{ m}$$

REFLECT   Once the spring has done its work, the total mechanical energy of the system is always the sum of the pellets kinetic energy and its potential energy above ground.

6 MOMENTUM AND COLLISIONS

CONCEPTUAL QUESTIONS

1. **ORGANIZE AND PLAN** We're asked to compare the average force of the wall on a ball, depending on whether it sticks to the wall or rebounds off the wall. In our coordinate system, we'll take the positive direction to be toward the wall. We'll also assume that the time Δt it takes the ball to come to a stop and stick to the wall is the same as the time the ball is in contact with the wall as it rebounds.

SOLVE Force is proportional to the change in momentum according to

$$\vec{F}\Delta t = \Delta \vec{p}$$

If the ball sticks to the wall, its final momentum is zero, and the change in momentum is

$$\Delta \vec{p} = 0 - \vec{p}_i = -\vec{p}_i$$

However, if the ball rebounds off the wall, the final momentum is in the negative direction and the change in momentum is larger"

$$\Delta \vec{p} = \vec{p}_f - \vec{p}_i > 0 - \vec{p}_i \text{ where } \vec{p}_i > 0$$

Given identical values for Δt, a larger change in momentum requires a larger force. Therefore, average force is greater if the ball rebounds off the wall.

REFLECT For an elastic collision, conservation of momentum requires that the magnitude of the change in momentum is twice as large as the initial momentum. This is like the difference between catching a basketball and being hit with a basketball that bounces off you, if both are thrown at the same speed.

3. **ORGANIZE AND PLAN** Let's consider the explosive bolts that hold the first stage of a rocket to the second stage. A bolt has mass m. We'll use the frame of reference of the moving rocket.

SOLVE With respect to the moving rocket, the explosive bolts have zero velocity. Therefore,

$$\vec{p} = m\vec{v} = 0 \frac{\text{kg} \cdot \text{m}}{\text{s}}$$

Now the bolts explode to release the second stage of the rocket. Kinetic energy is not a vector quantity and does not depend on the directions of the fragments. Suppose a bolt simply splits into two fragments of equal mass $\frac{1}{2}m$.

The center of mass of the two fragments still has $\vec{v} = 0$. To conserve momentum, the two fragments must travel in opposite directions at the same speed $v \neq 0$ with respect to the rocket. Each fragment has kinetic energy

$$K_{\text{fragment}} = \frac{1}{2}\left(\frac{1}{2}mv^2\right)$$

Since there are two fragments, the total kinetic energy is

$$K = \frac{1}{2}mv^2$$

Because velocity is not zero, kinetic energy must be positive.

Now, if we say a system of particles has zero kinetic energy, this implies that each particle must have a speed of zero with respect to its reference frame. Therefore, the center of mass of the system of particles also has a speed and velocity of zero. Thus in this case the momentum of the system of particles must be zero.

REFLECT The example of two fragments can be expanded to any number of fragments. A kinetic energy of zero for a system of particles necessarily means zero momentum. However, the inverse is not true: A momentum of zero does not necessarily mean kinetic energy of zero.

5. **SOLVE**   We consider the system of the rocket and the atoms of the rocket's chemical propellant. It is quite sufficient to show that the propellant and the rocket have initial velocity of zero. Let v_{2f} be the velocity of the rocket. The mass of propellant m_1 is forced out of the rocket by way of a chemical reaction. In order to conserve momentum, the rocket must move the opposite direction, according to

$$m_1 v_{1f} + m_2 v_{2f} = (m_1 + m_2) v_i$$

For $v_i = 0$ before the rocket engine fires,

$$v_{2f} = \frac{m_1 v_{1f}}{m_2}$$

None of the variables represent a quantity related to air, and momentum is conserved. Air is not included in any expression for conservation of momentum for the rocket and its propellant.

REFLECT   In order to argue that the rocket needs air to "push off," we attempt to use Newton's third law. When the exhaust gases exert a force on the air, the air would have to exert a force on the exhaust gases, and the gases would transmit this force to the rocket. This would be like a car accelerating because the pavement exerts force on the tires, and hence on the car. Unlike the tires, the exhaust gases are compressible and this argument fails.

7. **SOLVE**   The formula for kinetic energy as a function of speed is

$$K = \tfrac{1}{2} m v^2$$

The formula for the magnitude of momentum as a function of speed is

$$p = mv$$

Kinetic energy increases as the square of the speed. Momentum increases only as the first power of speed. Kinetic energy increases faster with increasing speed.

REFLECT   Automobile collisions are more dangerous in terms of personal injury and property damage at higher speeds than at lower speeds. Momentum increases as a linear function of speed. But at higher speeds, there is much more kinetic energy available to do work on the car. It is the work done on the car that causes damage.

9. When you "follow through" with a tennis racquet or golf club, you maintain contact with the ball for as long a time interval as possible. Nothing that you do with the club after the ball leaves the head of the club has any effect on the ball.

    SOLVE   The loss of momentum of the golf club $\Delta p = m_1 v_{1i} - m_1 v_{1f}$ is the value of the gain in momentum of the ball $\Delta p = m_2 v_{2f} - m_2 v_{2i}$. The only loss in momentum of the club occurs during the short time interval in which it is in contact with the ball. Let's ignore the fact that the club is being swung in a large circle and assume that the club moves in a straight line during this time. When we consider that the initial velocity of the ball is zero,

$$m_1 \vec{v}_{1i} - m_1 \vec{v}_{1f} = m_2 \vec{v}_{2f} - 0 = \Delta \vec{p} = \vec{J} = \vec{F} \Delta t$$

A strong follow-through maximizes Δt and hence the final velocity of the ball.

REFLECT   It isn't possible for a player to exert enough force for a good shot on a tennis ball or golf ball solely by "pushing" it with the racquet or the club. Instead, the player relies on a racquet or club with substantial mass. In addition, the racquet or club moves at a high speed, giving it a large initial momentum.

11. **ORGANIZE AND PLAN**   It takes a certain impulse $\vec{J} = \vec{F} \Delta t$ to stop the egg. The key to this question is in deciding how to reduce the speed and hence the momentum of your egg in as long a time interval as possible.

    SOLVE   Physical principles: There is a minimum force required to crack the egg. Less force is required to crack the egg on a sharp edge. You can achieve the impulse required to stop the egg by increasing either the applied force or the time interval. Design characteristics: Use elastic material or bands to control the rate of acceleration. Use a flexible surface to distribute the force evenly over the surface of the egg.

    REFLECT   There are variations on the "egg drop" in which you are allowed to encapsulate your egg in a container designed to "cushion" the egg on landing.

13. **SOLVE**   The requirement in the high jump is for the athlete's body to clear the bar without knocking it off the supports. If we model the athlete's body as a point, then the center of mass clearly must clear the bar. If instead we model the athlete's body as, say, a sphere with finite radius, then the center of mass must clear the height of the bar

plus one radius. However, the athlete's body is actually a flexible extended object and can assume a U-shaped position. In this case, no part of the athlete's body is present at the center of mass. The athlete can thus roll over the bar in this position, and the center of mass can pass under the bar without the athlete touching the bar.

REFLECT   Dick Fosbury perfected this high-jump style and used it to set a new world Olympic record at the Summer Games in 1968. Using the now-standard "Fosbury Flop," the athlete goes over the bar arching his or her back, with the center of mass remaining below the bar during the entire jump.

15.   **ORGANIZE AND PLAN**   For any dimension N, the center of mass of a system of objects is displaced from the origin of that axis by the expression $N_{cm} = \dfrac{1}{M} \sum_{i=1}^{n} m_i n_i$.

SOLVE   For a system of two objects, there is no mass that does not lie on the line between the two objects. The weighted average mass must therefore be on the line that connects the two. Here's a further illustration: For a system of three objects, we'll declare the plane defined by the three objects to be the x-y plane in a three-axis Cartesian coordinate system. Suppose we claim that the center of mass of this system does not lie on the x-y plane. Then the value Z_{cm} must not be equal to zero. If our three objects have masses m_1, m_2 and m_3, we see that

$$Z_{cm} = \frac{1}{M} \sum_{i=1}^{3} m_i z_i = \frac{m_1 z_1 + m_2 z_2 + m_3 z_3}{m_1 + m_2 + m_3}$$

Since the position of any point on the x-y plane can be stated as an ordered triple $(x, y, 0)$ and all three of our points are on the x-y plane, $z_1 = z_2 = z_3 = 0$. Therefore,

$$Z_{cm} = \frac{m_1 z_1 + m_2 z_2 + m_3 z_3}{m_1 + m_2 + m_3} = \frac{m_1(0) + m_2(0) + m_3(0)}{m_1 + m_2 + m_3} = 0.$$

The center of mass does not have $Z_{cm} \neq 0$ and therefore lies on the x-y plane.

Each object has zero displacement from the line connecting the two objects. And the center of mass of three objects has to lie in the plane determined by the three.

REFLECT   If our systems of two or three objects had a center of mass not on their centerline or plane, respectively, then we could place the objects over the edge of a table top and expect them to remain balanced there, with no object over the table and nothing between the objects and the floor except air!

17.   **ORGANIZE AND PLAN**   Newton's cradle, shown in Figure CQ6.17 in the text, is a close approximation of elastic collisions between balls. In elastic collisions, kinetic energy is conserved. Kinetic energy is never gained during a collision. We'll use the subscript 1 to apply to the single ball and the subscript 2 to apply to the two balls.

SOLVE   We assume that all the balls in Newton's cradle are identical with mass m. The balls that are pulled back and released have velocity v at the instant of collision. Suppose the collisions between balls are elastic. We propose a hypothesis that if we pull back and release two balls, only one ball is ejected from the opposite end, but with twice the velocity, that is, $2v$. In this situation, momentum is conserved:

$$p_2 = (2m)v = p_1 = m(2v)$$

Let's see what happens to kinetic energy:

$$K = \frac{p^2}{2m}$$

$$p_2 = \sqrt{2(2m)\tfrac{1}{2}(2m)v_2^2}$$

$$p_1 = \sqrt{2m\left(\tfrac{1}{2}mv_1^2\right)}$$

But since momentum is conserved,

$$p_2 = p_1 = \sqrt{2(2m)\tfrac{1}{2}(2m)v_2^2} = \sqrt{2m\left(\tfrac{1}{2}mv_1^2\right)}$$

$$4m^2 v_2^2 = m^2 v_1^2 \quad \text{and}$$

$$v_1 = 2v_2$$

So according to our hypothesis,

$$K_2 = \tfrac{1}{2}\left(2\ m\right)v_2^2 = mv_2^2 \text{ and}$$

$$K_1 = \tfrac{1}{2}\ m\left(2v_2\right)^2 = 2\ mv_2^2$$

If this were to be the case, then the kinetic energy of the system would increase and the final kinetic energy of the single ball would be twice the initial kinetic energy of the two balls. Since there is no input of energy to the system, that result is impossible.

REFLECT   In fact, any increase in the velocity of the single ball will cause the final kinetic energy of the system to be higher than the initial kinetic energy. Kinetic energy can only be gained during an explosion, in which there is a source of stored potential energy within the system. This is not an explosion. In a practical demonstration of Newton's cradle, a small amount of kinetic energy is lost due to friction. Therefore, the two balls to which momentum and kinetic energy are transferred don't go quite as high as the original two balls.

Multiple-Choice Problems

19.   **ORGANIZE AND PLAN**   This is an impulse-momentum problem. The car must lose all its momentum $p = mv$ in order to come to a stop. The applied force accomplishes this in a period of time according to the impulse-momentum theory $J = \Delta\vec{p} = \vec{F}\Delta t$.

Known: $m = 960$ kg; $v_0 = 25$ m/s; $v = 0$; $\Delta t = 15$ s.

SOLVE   Since the car slows to a stop,

$$\Delta p = p_f - p_i = F\Delta t$$

$$F = \frac{p_f - p_i}{\Delta t} = \frac{m\left(v_f - v_i\right)}{\Delta t} = \frac{\left(960\ \text{kg}\right)\left(25\ \text{m/s}\right)}{15\ \text{s}} = 1600\ \text{N}$$

The correct answer is (a).

REFLECT   The smaller the force, the longer it takes the object to stop. The units of impulse are the same as those of momentum, $\dfrac{\text{kg} \cdot \text{m}}{\text{s}}$.

21.   **ORGANIZE AND PLAN**   Here we are supposed to find momentum after a free fall. We can use conservation of mechanical energy and the relationship we know between momentum and kinetic energy, $K = \dfrac{p^2}{2m}$.

Known: $m = 5.00$ kg ; $\Delta y = 12.0$ m.

SOLVE   The boulder's gravitational potential energy is

$$U_g = mg\Delta y$$

When the boulder strikes the ground, all its potential energy has been converted to kinetic energy,

$$K = \tfrac{1}{2}mv^2 = mg\Delta y$$

Since $K = \dfrac{p^2}{2m}$,

$$p = \sqrt{2mK} = \sqrt{2m\left(mg\Delta y\right)} = \sqrt{2m^2 g\Delta y}$$

$$p = \sqrt{2\left(5.00\ \text{kg}\right)^2\left(9.80\ \text{m/s}^2\right)\left(12.0\ \text{m}\right)} = 76.7\ \text{kg} \cdot \text{m/s}$$

The correct answer is (b).

REFLECT   We can check our result against what we know about free fall: $v_y^2 = v_{y0}^2 + 2g\Delta y$, so $v_y = \sqrt{2g\Delta y}$, which gives $p = mv_y = \sqrt{2m^2 g\Delta y}$ which is the expression for p given above.

23. **ORGANIZE AND PLAN**   In this problem, the velocity and momentum change direction and therefore change sign. We'll use $\Delta p = p - p_0$ and calculate the momenta using the ball's initial and final velocity. We'll set up our coordinate system so that the ball strikes the wall in the negative direction and rebounds in the positive direction.

Known: $m = 75$ g; $v_0 = -12$ m/s; $v = 10.$m/s

SOLVE   First, we change the mass of the ball to kilograms:

$$m = 75 \text{ g}\left(\frac{1 \text{ kg}}{1000 \text{ g}}\right) = 0.075 \text{ kg}$$

The change in momentum is

$$\Delta p = p - p_0 = mv - mv_0 = m(v - v_0)$$

$$\Delta p = 0.075 \text{ kg}\left(10.\text{m/s} - (-12 \text{ m/s})\right) = 1.65 \text{ kg} \cdot \text{m/s}$$

The correct answer is (d).

REFLECT   Notice the double negative when we evaluate. We are subtracting the negative initial velocity. If we had set our coordinate system in the opposite direction with the ball striking the wall in the positive directions, Δp would be negative.

25. **ORGANIZE AND PLAN**   In this perfectly inelastic collision, the gliders stick together and move as one unit. As in Problem 24, we'll use subscripts 1 and 2 for the heavier and lighter gliders, respectively. Since the lighter glider is initially at rest, we can use $m_1 v_{1xi} = (m_1 + m_2)v_{xf}$ to find v_{xf}.

Known: $m_1 = 177$ g; $m_2 = 133$ g.

SOLVE   Rearranging to get v_{xf},

$$m_1 v_{1xi} = (m_1 + m_2)v_{xf}$$

$$v_{xf} = \frac{m_1 v_{1xi}}{m_1 + m_2} = 0.200 \text{ m/s}$$

The correct answer is (a).

REFLECT   We carefully note that it is not necessary to convert mass to kilograms in this problem. The masses cancel, resulting in a dimensionless ratio to be multiplied by v_{1xi}. If we needed to calculate K, we'd have to make this conversion.

27. **ORGANIZE AND PLAN**   Two moving wads of putty combine to become one. That makes this a perfectly inelastic collision. We are to find the final velocity of the combined mass. We can use $\vec{v}_f = \dfrac{m_1\vec{v}_{1i} + m_2\vec{v}_{2i}}{m_1 + m_2}$. We aren't told the masses of the wads, only that they are equal.

Known: $m_1 = m_2 = m$; $\vec{v}_{1i} = 2.50$ m/s in the x-direction; $\vec{v}_{2i} = 2.50$ m/s in the y-direction.

SOLVE   We'll set our coordinate system so that the first wad is moving in the positive x-direction and the second wad is moving perpendicular to the first in the positive y-direction. Since we have a two-dimensional problem, we'll use vector space notation:

$$\vec{v}_{1i} = 2.50 \text{ m/s}\,\hat{i} \text{ and } \vec{v}_{2i} = 2.50 \text{ m/s}\,\hat{j}$$

Substituting known values into the chosen equation,

$$\vec{v}_f = \frac{m_1\vec{v}_{1i} + m_2\vec{v}_{2i}}{m_1 + m_2} = \frac{m(2.50 \text{ m/s}\,\hat{i}) + m(2.50 \text{ m/s}\,\hat{j})}{2m} = 1.25 \text{ m/s}\,\hat{i} + 1.25 \text{ m/s}\,\hat{j}$$

Since we want speed, not velocity, we only need the magnitude of the velocity:

$$v = \sqrt{2(1.25 \text{ m/s})^2} = 1.77 \text{ m/s}$$

The correct answer is (d).

REFLECT   For objects of equal mass colliding at right angles, the final speed will be $\sqrt{2}/2$ times the initial speed.

29. **ORGANIZE AND PLAN** We'll establish our coordinate system with the origin at the zero mark of the 2.00-m stick.

We' use the formula for center of mass in one dimension, $X_{cm} = \frac{1}{M}\sum_{i=1}^{n} m_i x_i$ to find the center of mass.

Known: $m_1 = 0.45$ kg; $x_1 = 0.80$ m; $m_2 = 0.60$ kg; $x_2 = 1.10$ m; $m_3 = 1.15$ kg; $x_3 = 1.60$ m.

SOLVE
Using the formula for center of mass,

$$X_{cm} = \frac{1}{M}\sum_{i=1}^{n} m_i x_i = \frac{m_1 x_1 + m_2 x_2 + m_3 x_3}{m_1 + m_2 + m_3}$$

$$X_{cm} = \frac{(0.45 \text{ kg})(0.80 \text{ m}) + (0.60 \text{ kg})(1.10 \text{ m}) + (1.15 \text{ kg})(1.60 \text{ m})}{0.45 \text{ kg} + 0.60 \text{ kg} + 1.15 \text{ kg}} = 1.30 \text{ m}$$

The center of mass is 1.30m from the origin, the end of the 2.00-m stick.

The correct answer is (c).

REFLECT We can use the same principle, obtain the same answer, and simplify the math by declaring the origin to be the location of one of the masses. Try this using the location of the 0.60kg mass as the origin, but be careful of the signs.

31. **ORGANIZE AND PLAN** In this problem, both balls are initially moving, and we'll use subscripts 1 and 2 to refer to them. Since we don't know that the collision is elastic, we can rely on conservation of momentum, but not conservation of kinetic energy. We'll use $m_1 v_{1i} + m_2 v_{2i} = m_1 v_{1f} + m_2 v_{2f}$ to find the velocity of the second ball v_{2f}. In our coordinate system, the 0.20kg ball will be moving in the positive direction, that is, to the right.

Known: $m_1 = 0.20$ kg; $m_2 = 0.40$ kg; $v_{1i} = 3.0$ m/s; $v_{2i} = -2.0$ m/s; $v_{1f} = -3.0$ m/s.

SOLVE Rearranging to find v_{2f},

$$m_1 v_{1i} + m_2 v_{2i} = m_1 v_{1f} + m_2 v_{2f}$$

$$\frac{m_1 v_{1i} + m_2 v_{2i} - m_1 v_{1f}}{m_2} = v_{2f}$$

$$v_{2f} = \frac{(0.20 \text{ kg})(3.0 \text{ m/s}) + (0.40 \text{ kg})(-2.0 \text{ m}) - (0.20 \text{ kg})(-3.0 \text{ m})}{0.40 \text{ kg}} = 1.00 \text{ m/s}$$

Since the value is positive, the ball is traveling to the right.

The correct answer is (b).

REFLECT Since the first ball "bounces off" the second ball and changes velocity to the same speed in the opposite direction, it transfers more momentum to the second ball than if it remained at rest after the collision.

Problems

33. **ORGANIZE AND PLAN** This is an impulse-momentum problem where the bicycle and rider start from rest. We'll use $J = F\Delta t$ to find the force required.

Known: $v_i = 0$ m/s; $v_f = 15$ m/s; $\Delta t = 8.0$ s; $m = 105$ kg.

SOLVE Using the definition of acceleration,

$$a_x = \frac{v_f - v_i}{\Delta t}$$

Using the definition of impulse,

$$J = F\Delta t = \Delta p$$

Since

$$\Delta p = p_f - p_i$$

and

$$p_i = 0 \text{ kg} \cdot \text{m/s},$$

$$J = p_f - 0 = mv_f = F\Delta t$$

$$F = \frac{mv_f}{\Delta t} = \frac{(105 \text{ kg})(15 \text{ m/s})}{8.0 \text{ s}} = 2.0 \times 10^2 \text{ N}$$

$$p = mv = (105 \text{ kg})(15 \text{ m/s}) = 1600 \text{ kg m/s}$$

REFLECT   As an alternative, we could have used a kinematic equation to find $a = \Delta v / \Delta t$, then Newton's second law to find force. The units of $mv / \Delta t$ are kg•m/s² which is equivalent to newtons.

35.   **ORGANIZE AND PLAN**   We are asked to find the magnitude of the momentum from mass and speed. We'll use the definition of momentum $p = mv$.

Known: $m = 64$ kg; $v = 7.3$ m/s.

SOLVE   Using the definition of momentum,

$$p = mv = (64 \text{ kg})(7.3 \text{ m/s}) = 470 \text{ kg•m/s}$$

REFLECT   A person with a mass of 64kg "weighs" about 140lb on a spring scale. This is a reasonable weight. A person in good physical condition can run 100.m in 10-11 seconds (9.1 – 10 m/s). So a speed of 7.3 m/s is also reasonable.

37.   **ORGANIZE AND PLAN**   Using the impulse-momentum theorem, we can find the impulse imparted to the baseball by the moving bat. We'll use $J = \Delta p$ and $p = mv$. In our coordinate system, velocity will be negative as the ball approaches the bat.

Known: $v_i = -34.5$ m/s.; $v_f = 39.2$ m/s.; $m = 145$ g.

SOLVE   First, convert mass to kilograms:

$$m = 145 \text{ g}\left(\frac{1 \text{ kg}}{1000 \text{ g}}\right) = 0.145 \text{ kg}$$

Then, impulse is

$$J = \Delta p = m(v_f - v_i) = (0.145 \text{ kg})(39.2 \text{ m/s} - (-34.5 \text{ m/s})) = 10.7 \text{ kg•m/s}$$

REFLECT   If we said the ball was pitched in the positive direction and batted in the negative direction, the magnitude of the impulse would be the same but the sign would be negative instead.

39.   **ORGANIZE AND PLAN**   We'll use dimensional analysis to show that $1 \text{N} \cdot \text{s} = 1 \text{kg} \cdot \text{m/s}$. Then we'll use $J = \Delta p$ and $p = mv$ to find the speed of the rocket could have when the engine stops firing.

Known: $J = 7.5 \text{ N} \cdot \text{s}$; $m = 140$ g; $v_i = 0$ m/s.

SOLVE   (a) According to Newton's second law,

$$F = ma$$

In units of measure,

$$\text{N} = (\text{kg})(\text{m/s}^2)$$

When we multiply newtons by seconds,

$$J = \text{N} \cdot \text{s} = (\text{kg} \cdot \text{m/s}^2)(\text{s}) = \text{kg} \cdot \text{m/s}$$

This is the unit of measure for momentum that we have been using.

(b) First we convert mass to kilograms.

$$m = 140 \text{ g}\left(\frac{1 \text{ kg}}{1000 \text{ g}}\right) = 0.140 \text{ kg}$$

Then we see that

$$J = m(v_f - v_i)$$

and since $v_i = 0$,

$$v_f = \frac{J}{m} = \frac{7.5 \text{ N} \cdot \text{s}}{0.140 \text{ kg}} = 53.6 \text{ m/s}$$

REFLECT   In practice, any rocket loses mass as propellant is consumed, and its acceleration increases with firing time. The speed we calculated here is the greatest speed this rocket could have under the conditions specified. Suppose that the rocket starts with a mass greater than 140 g; the final velocity will be smaller than we calculated. There are other ways of calculating the speed of a rocket that are beyond the scope of the present chapter.

41.   **ORGANIZE AND PLAN**   We can use Newton's second law $F = ma$ and the impulse momentum theorem $\Delta p = F\Delta t$ to find the average force and impulse. Since the flea starts at rest, the impulse is equal to the final momentum.

Known: $\Delta t = 1.2 \text{ ms}$; $a_y = 100 \text{ g} = 980 \text{ m/s}^2$; $m = 220 \text{ mg}$ MNS check

SOLVE   First, we convert to SI base units:

$$\Delta t = 1.2 \text{ ms} \left(\frac{1 \text{ s}}{1000 \text{ ms}} \right) = 1.2 \times 10^{-3} \text{ s}$$

$$m = 220 \text{ mg} \left(\frac{1 \text{ kg}}{1 \times 10^6 \text{ mg}} \right) = 2.20 \times 10^{-4} \text{ kg}$$

(a) Using Newton's second law,

$$F = ma = \left(2.20 \times 10^{-4} \text{ kg} \right) \left(980. \text{m/s}^2 \right) = 0.216 \text{ N}$$

(b) The impulse of the ground on the flea is

$$J = F\Delta t = \left(0.216 \text{ N} \right) \left(1.2 \times 10^{-3} \text{ s} \right) = 2.59 \times 10^{-4} \text{ kg} \cdot \text{m/s}$$

(c) The change in momentum of the flea is

$$J = \Delta p = p_f - p_i$$

and since $p_i = 0 \text{kg} \cdot \text{m/s}$,

$$\Delta p = p_f = mv_f = ma_y \Delta t = \left(2.20 \times 10^{-4} \text{ kg} \right) \left(980. \text{m/s}^2 \right) \left(1.2 \times 10^{-3} \text{ s} \right) = 2.59 \text{ kg} \cdot \text{m/s}$$

REFLECT   The units for impulse and momentum work out correctly, so we have confidence in our answer. The answer to (c) is a restatement of (a) and (b) combined. If a flea had the body length of a human, it would be able to clear tall buildings in a single bound.

43.   **ORGANIZE AND PLAN**   We'll use the impulse momentum theorem, $\vec{F}\Delta t = \vec{p}$ to calculate the final momentum of the puck. We'll treat the momenta in the x- and y-directions separately and express the result in vector-space notation. In (b), when the puck has an initial velocity, we have to take into account the value of $\vec{p}_i$ in $\Delta\vec{p} = m\vec{v}$.

Known: (a) $v_i = 0$; $\vec{F}_{net} = 0.340 \text{ N}\hat{i} + 0.240 \text{ N}\hat{j}$; $m = 170 \text{ g}$; $\Delta t = 4.50 \text{ s}$.

SOLVE   First, we convert mass to kilograms:

$$m = 170 \text{ g} \left(\frac{1 \text{ kg}}{1000 \text{ g}} \right) = 0.170 \text{ kg}$$

(a) In the x-direction,

$$\vec{J}_x = \Delta\vec{p}_x = \vec{p}_{xf} - \vec{p}_{xi} = \vec{p}_{xf} = \vec{F}_x \Delta t = \left(0.340 \text{ N}\hat{i} \right) \left(4.50 \text{ s} \right)$$

$$\Delta\vec{p}_x = 1.53 \text{kg} \cdot \text{m/s}\hat{i}$$

In the y-direction,

$$\vec{J}_y = \Delta\vec{p}_y = \vec{p}_{yf} - \vec{p}_{yi} = \vec{p}_{yf} = \vec{F}_y \Delta t = \left(0.240 \text{ N}\hat{j} \right) \left(4.50 \text{ s} \right)$$

$$\Delta\vec{p}_y = 1.08 \text{ kg} \cdot \text{m/s}\hat{j}$$

$$\vec{p}_f = \vec{p}_{xf} + \vec{p}_{yf} = 1.53 \text{ kg} \cdot \text{m/s}\hat{i} + 1.08 \text{ kg} \cdot \text{m/s}\hat{j}$$

(b) Now we do the same thing we did in (a), but now we take into account the initial velocity:

In the x-direction,

$$\vec{p}_{xf} = \vec{p}_{xi} + \vec{J}_x = m\vec{v}_{xi} + \vec{J}_x = (0.175 \text{ kg})(2.90 \text{ m/s})\hat{i} + (0.340 \text{ N})(4.50 \text{ s})\hat{i}$$

$$\vec{p}_{xf} = 2.04 \text{ kg} \cdot \text{m/s}\hat{i}$$

In the y-direction,

$$\vec{p}_{yf} = \vec{p}_{yi} + \vec{J}_y = m\vec{v}_{yi} + \vec{J}_y = (0.175 \text{ kg})(1.35 \text{ m/s})\hat{ij} + (0.240 \text{ N})(4.50 \text{ s})\hat{j}$$

$$\vec{p}_{xf} = 1.32 \text{ kg} \cdot \text{m/s}\hat{j}$$

$$\vec{p}_f = \vec{p}_{fx} + \vec{p}_{fy} = 2.04 \text{ kg} \cdot \text{m/s}\hat{i} + 1.32 \text{ kg} \cdot \text{m/s}\hat{j}$$

REFLECT   Impulse has the same units of measure as momentum, so we can correctly add values of $\text{kg} \cdot \text{m/s}$ to values of $\text{N} \cdot \text{s}$.

45.   **ORGANIZE AND PLAN**   We'll use the kinematic equation $v_{yf}^2 = v_{yi}^2 + 2a_y\Delta y$ to find the speed of each ball as it strikes the ground (Galileo showed us the speeds would be the same!). Then we can use $\vec{p} = m\vec{v}$ to find the momentum for each ball. After that, we'll use $K = p^2/2m$ to find kinetic energy.

Known: $\Delta y = -19.0 \text{ m}$; $a_y = -g = -9.80 \text{ m/s}^2$; $m_1 = 1.5 \text{ kg}$; $m_2 = 4.5 \text{ kg}$; $v_{iy} = 0 \text{ m/s}$.

SOLVE   (a) Let's find the final velocity , which is independent of mass:

$$v_{yf}^2 = v_{yi}^2 + 2a_y\Delta y$$

Since $v_{yi} = 0$,

$$v_{yf} = \sqrt{2a_y\Delta y} = \sqrt{2(-9.80 \text{ m/s}^2)(-19 \text{ m})} = 19.3 \text{ m/s}$$

Now we can find the momentum of each ball. For the 1.50kg ball,

$$\vec{p} = m\vec{v} = (1.50 \text{ kg})(-19.3 \text{ m/s}) = -28.9 \text{ kg} \cdot \text{m/s (downward)}$$

For the 4.50kg ball,

$$\vec{p} = m\vec{v} = (4.50 \text{ kg})(19.3 \text{ m/s}) = 86.8 \text{ kg} \cdot \text{m/s downward}$$

(b) Now to find the kinetic energy for each ball:

For the lighter ball,

$$K = \frac{p^2}{2m} = \frac{(28.9 \text{ kg} \bullet \text{m/s})^2}{2(1.5 \text{ kg})} = 279 \text{ J}$$

For the heavier ball,

$$K = \frac{p^2}{2\ m} = \frac{(86.8 \text{ kg} \bullet \text{m/s})^2}{2(4.5 \text{ kg})} = 837 \text{ J}$$

REFLECT   We could have equated gravitational potential energy atop the 19.0m building with the kinetic energy after free fall, calculated momentum, and then found the speed of each ball. The formulas representing the laws of classical mechanics are intertwined and are always consistent.

47.   **ORGANIZE AND PLAN**   In this problem the runner's direction continuously changes. She runs a circular track, so we have to use vector addition to find the change in momentum at any point. Since we're only interested in the magnitude of the change in momentum, we'll just take the absolute value.

Known: $r = 63.7 \text{ m}$; $m = 64 \text{ kg}$.

SOLVE   (a) She starts with

$$\vec{p}_i = m\vec{v}_i = (64 \text{ kg})(5.3 \text{ m/s})\hat{j} = (339 \text{ kg} \cdot \text{m/s})\hat{j}$$

When she has gone halfway around the circle, she is traveling at the same speed but in the opposite direction, so

$$\vec{p}_f = m\vec{v}_f = (64 \text{ kg})(-5.3 \text{ m/s})\hat{j} = (-339 \text{ kg} \cdot \text{m/s})\hat{j}$$

$$\Delta p = |p_f - p_i| = |-339 \text{ kg} \cdot \text{m/s} - 339 \text{ kg} \cdot \text{m/s}| = 678 \text{ kg} \cdot \text{m/s}$$

(b) Again, her initial momentum is $(339 \text{ kg} \cdot \text{m/s})\hat{j}$. When she had traveled one-quarter of the way around the circle, her velocity vector has changed direction by 90°, and

$$\vec{p}_f = (-339 \text{ kg} \cdot \text{m/s})\hat{i}$$

So now,

$$\vec{p}_f = (-339 \text{ kg} \cdot \text{m/s})\hat{i}$$

and

$$\Delta\vec{p} = (-339 \text{ kg} \cdot \text{m/s})\hat{i} - (339 \text{ kg} \cdot \text{m/s})\hat{j}$$

$$\Delta p = \left|\sqrt{(-339 \text{ kg} \cdot \text{m/s})^2 + (-339 \text{ kg} \cdot \text{m/s})^2}\right| = 479 \text{ kg} \cdot \text{m/s}$$

(c) When the runner has come one full circle, her momentum has the same direction and magnitude as initially, and the change is zero:

$$\Delta p = |339 \text{ kg} \cdot \text{m/s} - 339 \text{ kg} \cdot \text{m/s}| = 0 \text{ kg} \cdot \text{m/s}$$

REFLECT We have to remember to use vector addition of components when doing calculations with momentum.

49. **ORGANIZE AND PLAN** We'll set up our coordinate system so that the side cushion is the x-axis. We'll resolve the billiard ball's velocity into x- and y- components based on the 30.°angle of incidence. Then we'll multiply each component by mass m to get the x- and y-components of momentum. We'll find change in momentum by subtracting initial momentum components from final. Then we'll repeat this process for the reduced final speed. Knowing the time during impact, we can equate momentum to impulse to find the components of the force of the cushion on the ball.

Known: $\theta = 30.°$; $m = 160.\text{g}$; $\Delta t = 25 \text{ ms}$; (a) $v_i = 1.67 \text{ m/s}$; $v_f = 1.67 \text{ m/s}$; (b) $v_i = 1.67 \text{ m/s}$; $v_f = 1.42 \text{ m/s}$.

SOLVE First convert to SI units:

$$m = 160.\text{g}\left(\frac{1 \text{ kg}}{1000 \text{ g}}\right) = 0.160 \text{ kg}$$

$$\Delta t = 25 \text{ ms}\left(\frac{1 \text{ s}}{1000 \text{ ms}}\right) = 0.025 \text{ s}$$

(a)

$$\Delta\vec{v} = \vec{v}_f - \vec{v}_i = (1.67\cos 30° \text{ m/s})\hat{i} + (1.67\sin 30° \text{ m/s})\hat{j} - \left[(1.67\cos 30° \text{ m/s})\hat{i} - (1.67\sin 30° \text{ m/s})\hat{j}\right]$$

$$\Delta\vec{v} = (1.67 \text{ m/s})\hat{j}$$

$$\vec{p} = m\Delta\vec{v} = (0.160 \text{ kg})(1.67 \text{ m/s})\hat{j} = (0.267 \text{ kg} \bullet \text{m/s})\hat{j}$$

(b)

$$\Delta\vec{v} = \vec{v}_f - \vec{v}_i = (1.42\cos 30° \text{ m/s})\hat{i} + (1.42\sin 30° \text{ m/s})\hat{j} - \left[(1.67\cos 30° \text{ m/s})\hat{i} - (1.67\sin 30° \text{ m/s})\hat{j}\right]$$

$$\Delta\vec{v} = (0.217 \text{ m/s})\hat{i} + (1.545 \text{ m/s})\hat{j}$$

$$\Delta\vec{p} = m\Delta\vec{v} = (0.160 \text{ kg})\left[(0.217 \text{ m/s})\hat{i} + (1.545 \text{ m/s})\hat{j}\right] = (-0.0347 \text{ kg} \cdot \text{m/s})\hat{i} + (0.247 \text{ kg} \cdot \text{m/s})\hat{j}$$

(c) $\Delta\vec{p} = \vec{F}\Delta t$

$$\vec{F} = \frac{\Delta\vec{p}}{\Delta t} = \frac{(0.267 \text{ kg} \bullet \text{m/s})\hat{j}}{0.025 \text{ s}} = (10.7 \text{ N})\hat{j}$$

There is no x-component of the force of the cushion on the ball when the initial and final speeds are the same, but when the speed is reduced by the collision,

$$\vec{F} = \frac{\Delta \vec{p}}{\Delta t} = \frac{(-0.0347 \text{ kg} \cdot \text{m/s})\hat{i}}{0.025 \text{ s}} + \frac{(0.2472 \text{ kg} \cdot \text{m/s})\hat{j}}{0.025 \text{ s}} = (-1.39 \text{ N})\hat{i} + (9.89 \text{ N})\hat{j}$$

REFLECT   The answers are reasonable: In (a), the x-component of velocity and hence momentum remains the same. There should be no force in the x-direction. In (b), the change in velocity is smaller, and the force was also smaller, as we expect. Also, there is a small x-component of force in the negative direction to slow the ball down.

51.   **ORGANIZE AND PLAN**   We are given a graph of force versus time in figure P6.51 in the text. Using data from curve 1, we calculate average force between $t - 0$ s and $t = 0.5$ s. We can use the arithmetic mean because there is a linear relationship between F and t. Then we calculate impulse from $\vec{J} = \vec{F}\Delta t$. We can use average impulse, even if the force changes, to find change in momentum. Since the chunk initially had zero momentum and we know its mass, we can find velocity from $p = mv$. We'll use subscripts 1 and 2 for the time intervals.

Known: $m = 120$g; 1st time interval: $\Delta t = 0.50$ s; $F_i = 0$ N; $F_f = 6$ N (from graph); 2nd time interval: $\Delta t = 0.50$ s; $F = 6$ N.

SOLVE   First we convert mass to kilograms:

$$m = 120.\text{g}\left(\frac{1 \text{ kg}}{1000 \text{ g}}\right) = 0.120 \text{ kg}$$

(a) Then, using values from the graph, we calculate the average force

$$\overline{F} = \frac{F_i + F_f}{2} = \frac{0 \text{ N} + 6 \text{ N}}{2} = 3 \text{ N}$$

Now we can multiply average force for this interval by time to get impulse.

$$\overline{J}_1 = \overline{F}_1 \Delta t = (3 \text{ N})(0.50 \text{ s}) = 1.50 \text{ N} \cdot \text{s}$$

For the second time interval, the force is constant and

$$\overline{J}_2 = \overline{F}_2 \Delta t = (6 \text{ N})(0.50 \text{ s}) = 3.00 \text{ N} \cdot \text{s}$$

The total impulse is

$$J_{\text{total}} = \overline{J}_1 + \overline{J}_2 = 1.5 \text{ N} \cdot \text{s} + 3.0 \text{ N} \cdot \text{s} = 4.5 \text{ N} \cdot \text{s}$$

(b) Using the impulse from (a), we can find final velocity. Since $p_i = 0$

$$\Delta p = p_f - p_i = p_f$$

$$J_{\text{total}} = \Delta p = p_f = mv_f = \frac{4.5 \text{ N} \cdot \text{s}}{0.120 \text{ kg}} = 38 \text{ m/s}$$

REFLECT   Since $\vec{J} = \vec{F}\Delta t$, we can see that impulse is represented by the area under curve 1 between the upper and lower limits. In (a) this is the area of a triangle. In (b), it is the area of a rectangle. Notice that we could still solve for v_f as long as we were given v_i.

53.   **ORGANIZE AND PLAN**   In this problem we add the momenta of many rubber bullets to equal the magnitude of the momentum of the charging rhinoceros. Since we are given the mass and velocity of each bullet, we can use $J = \Delta p = m\Delta v$ to find the impulse of each bullet. We are given the time during which the bullets are fired and their rate of firing, which allows us to calculate the number of bullets and therefore their total momentum, $(n_{\text{bullet}})(J_{\text{bullet}}) = \Delta p_{\text{total}}$. Using conservation of momentum and knowing the speed of the rhino, we can use $\Delta p_{\text{bullets}} + m_{\text{rhino}}v_{\text{rhino}} = 0$ to find the mass of the rhino. The velocity of the rhino is negative, opposite the velocity of the bullets.

Known: $m_{\text{bullet}} = 20.$g; $v_{\text{rhino}} = -0.81$ m/s; rate $= 15$ bullets/s; $v_{\text{bullet}} = 73$ m/s; $\Delta t = 34$s.

SOLVE   First, convert mass to kilograms:

$$m_{\text{bullet}} = 20.\text{g}\left(\frac{1 \text{ kg}}{1000 \text{ g}}\right) = 0.020 \text{ kg}$$

(a)Now we find the impulse of each bullet, assuming the bullets start at rest.

$$J_{bullet} = \Delta p = mv_{bullet} = (0.020 \text{ kg})(73 \text{ m/s}) = 1.46 \text{ N·s}$$

Using the rate and total time,

$$(\text{rate})(\Delta t)(J_{bullet}) + m_{rhino}v_{rhino} = 0$$

$$m_{rhino} = \frac{-(\text{rate})(\Delta t)(J_{bullet})}{v_{rhino}} = \frac{-(15 \text{ bullets/s})(34 \text{ s})(1.46 \text{ N·s/bullet})}{-0.81 \text{ m/s}} = 919 \text{ kg}$$

REFLECT   That's about a one-ton rhinoceros! If the velocity of the bullets is positive, the velocity of the rhino must be negative. We note that we have to fire a very large number of bullets (510) with a total mass of about 10 kg. The individual rubber bullets will not harm the rhino, but one 10 kg "cannonball" traveling at 73 m/s might do so.

55.   **ORGANIZE AND PLAN**   The two pieces of the meteoroid will move in opposite directions. We must calculate the mass of the second piece. We'll use conservation of momentum where the initial momentum of the system is zero, $m_1v_1 = -m_2v_2$. Subscripts 1 and 2 refer to the two pieces of the meteoroid.
Known: $m_{total} = 130.\text{g}$; $m_1 = 55$ g; $v_1 = 0.65$ m/s.

SOLVE   First, convert mass to kilograms:

$$m_{total} = 130.\text{g}\left(\frac{1 \text{ kg}}{1000 \text{ g}}\right) = 0.130 \text{ kg}$$

Likewise,

$$m_1 = 55 \text{ g} = 0.055 \text{ kg}$$

Then we find the mass of the second piece,

$$m_2 = m_{total} - m_1 = 0.130.\text{kg} - 0.055 \text{ kg} = 0.075 \text{ kg}$$

Now,

$$m_1v_1 = -m_2v_2$$

$$v_2 = \frac{-m_1v_1}{m_2} = \frac{-(0.055 \text{ kg})(0.65 \text{ m/s})}{0.075} = -0.48 \text{ m/s}$$

REFLECT   Since the meteoroid is initially at rest, the two pieces must move in opposite directions to conserve momentum. We notice that in this particular problem we would not have had to convert mass to kilograms since we are only using the ratio of masses. It's good practice to do so, to minimize errors in other types of calculations.

57.   **ORGANIZE AND PLAN**   Here momentum is conserved and both objects are moving after the collision. We'll assume that the club and the ball are moving in the same straight line just before and after the collision. We're to find the speed of the head of the golf club. We'll use $m_1v_{1i} + m_2v_{2i} = m_1v_{1f} + m_2v_{2f}$ where subscript 1 refers to the club and subscript 2 refers to the ball. The club initially moves in the positive x-direction.
Known: $m_1 = 250.\text{g}$; $m_2 = 45.7$ g; $v_{1i} = 24.2$ m/s; $v_{2i} = 0$ m/s; $v_{2f} = 37.6$ m/s.

SOLVE   First we convert mass to kilograms:

$$m_1 = 250.\text{g}\left(\frac{1 \text{ kg}}{1000 \text{ g}}\right) = 0.250 \text{ kg}$$

Likewise,

$$m_2 = 0.0457 \text{ kg}$$

For conservation of momentum,

$$m_1v_{1i} + m_2v_{2i} = m_1v_{1f} + m_2v_{2f}$$

$$v_{1f} = \frac{m_1v_{1i} + m_2v_{2i} - m_2v_{2f}}{m_1} = \frac{(0.250 \text{ kg})(24.2 \text{ m/s}) + 0 \text{ kg·m/s} - (0.0457 \text{ kg})(37.6 \text{ m/s})}{0.250 \text{ kg}}$$

$$v_f = 17.3 \text{ m/s}$$

The club head is moving at $17.3\,\text{m/s}$ in its original direction.

REFLECT   When a golfer swings the club, the club head "follows through" after colliding with the ball and ends up over the golfer's shoulder, so the positive final direction of the club head is reasonable. Only part of the head's momentum is imparted to the ball.

59.   **ORGANIZE AND PLAN**   This is a perfectly inelastic collision. Since we have only one final speed, v_f, we can use $m_1 v_{1i} + m_2 v_{2i} = (m_1 + m_2) v_f$ to find that final speed.

Known: $m_1 = 1030$ kg; $m_2 = 1140$ kg; $v_{1i} = 3.4$ m/s; $v_{2i} = 0$ m/s.

SOLVE   For a perfectly inelastic collision,

$$m_1 v_{1i} + m_2 v_{2i} = (m_1 + m_2) v_f$$

Isolating v_{1f},

$$v_f = \frac{m_1 v_1 + m_2 v_2}{m_1 + m_2} = \frac{(1030 \text{ kg})(3.4 \text{ m/s}) + 0 \text{ kg} \cdot \text{m/s}}{1030 \text{ kg} + 1140 \text{ kg}} = 1.6 \text{ m/s}$$

REFLECT   The combined mass travels in the same (positive) direction as the incoming object, which is reasonable. Notice the similarity between the formula for v_f and the formula we learned for center of mass. These are both weighted averages (see pages 137–138 in the text).

61.   **ORGANIZE AND PLAN**   In this elastic collision we are given the initial velocity of only one skater. This means we'll be able to find the final velocities as expressions in terms of the initial velocity of the incoming object which is unknown [Did we intend to include a value for the velocity of the incoming skater? -LLS] We'll use the formulas derived from both conservation of momentum and of mechanical energy:

$$v_{1xf} = \left(\frac{m_1 - m_2}{m_1 + m_2} \right) v_{1xi} \text{ and } v_{2xf} = \left(\frac{2m_1}{m_1 + m_2} \right) v_{1xi}$$

Known: $m_1 = 60.0$ kg; $m_2 = 87.5$ kg; $v_{2xi} = 0$ m/s.

SOLVE   To find an expression of the final velocity of the first skater,

$$v_{1xf} = \left(\frac{m_1 - m_2}{m_1 + m_2} \right) v_{1xi} = \frac{(60.0 \text{ kg}) - (87.5 \text{ kg})}{(60.0 \text{ kg}) + (87.5 \text{ kg})} v_{1xi}$$

$$v_{1xf} = -0.186 v_{1xi}$$

For the second skater,

$$v_{2xf} = \left(\frac{2m_1}{m_1 + m_2} \right) v_{1xi} = \frac{2(60.0 \text{ kg})}{60.0 \text{ kg} + 87.5 \text{ kg}} v_{1xi}$$

$$v_{2xf} = 0.81 \, v_{1xi}$$

REFLECT   It makes sense that the less massive ball would "bounce off" the more massive stationary ball. If the stationary ball had the lesser mass, then both balls would travel in the positive direction after the collision.

63.   **ORGANIZE AND PLAN**   Here we're asked for algebraic rather than numeric solutions. We'll start with $K_i = K_f$ for conservation of kinetic energy of the two particles, and $\vec{p}_i = \vec{p}_f$ for conservation of momentum. We'll identify the two particles with subscripts 1 and 2.

Known: m_1; m_2; $v_{1xi}\hat{i}$; $v_{2xi}\hat{i}$; $v_{1xf}\hat{i}$; $v_{2xf}\hat{i}$.

SOLVE   (a) First, for conservation of momentum,

$$\vec{p}_i = \vec{p}_f$$

$$m_1 v_{1xi}\hat{i} + m_2 v_{2xi}\hat{i} = m_1 v_{1xf}\hat{i} + m_2 v_{2xf}\hat{i}$$

For conservation of mechanical energy,

$$v_{1xi}^2 \quad \tfrac{1}{2}m_1 v_{1xi}^2 + \tfrac{1}{2}m_2 v_{2xi}^2 = \tfrac{1}{2}m_1 v_{1xf}^2 + \tfrac{1}{2}m_2 v_{2xf}^2$$

(b) For the algebraic solution, we are only concerned about the magnitude and sign of the momentum. From conservation of energy,

$$\tfrac{1}{2}m_1 v_{1xi}^2 + \tfrac{1}{2}m_2 v_{2xi}^2 = \tfrac{1}{2}m_1 v_{1xf}^2 + \tfrac{1}{2}m_2 v_{2xf}^2$$

Rearranging and factoring,

$$m_1\left(v_{1xi}^2 - v_{1xf}^2\right) = m_2\left(v_{2xf}^2 - v_{2xi}^2\right)$$

From conservation of momentum,

$$m_1\left(v_{1xi} - v_{1xf}\right) = m_2\left(v_{2xf} - v_{2xi}\right)$$

Dividing the former by the latter,

$$\frac{m_1\left(v_{1xi}^2 - v_{1xf}^2\right)}{m_1\left(v_{1xi} - v_{1xf}\right)} = \frac{m_2\left(v_{2xf}^2 - v_{2xi}^2\right)}{m_2\left(v_{2xf} - v_{2xi}\right)}$$

We expand the difference of squares and cancel masses and like terms:

$$v_{1xi} + v_{1xf} = v_{2xf} + v_{2xi}$$

Rearranging,

$$v_{1xi} - v_{2xi} = v_{2xf} - v_{1xf} = -\left(v_{1xf} - v_{2xf}\right)$$

(c) The interpretation of (b) is that in an elastic collision the difference between the initial speeds is equal to the difference between the final speeds.

REFLECT These two equations provide a tool for working with elastic collisions.

65. **ORGANIZE AND PLAN** We'll calculate an expression for final velocity in terms of initial velocity of the incoming vehicle, which we know to be correct from conservation of momentum, $p_i = p_f$. Then we will do the same from the formula for kinetic energy, $K = \tfrac{1}{2}mv^2$, supposing that it is also conserved. Then we'll show that the final velocities cannot be equal. This will demonstrate that kinetic energy is not conserved for this perfectly inelastic collision.

Known: $m_1 = m_2 = m$; $v_{2xi} = 0\,\text{m/s}$.

SOLVE For conservation of momentum,

$$m_1 v_{1xi} + m_2 v_{2xi} = \left(m_1 + m_2\right)v_{fx}$$

Since masses are equal and $v_{2xi} = 0$,

$$v_{1xi} = 2v_{xf}$$

If kinetic energy is also conserved, then

$$\tfrac{1}{2}m_1 v_{1xi}^2 + 0 = \tfrac{1}{2}\left(m_1 + m_2\right)v_{xi}^2$$

Again, since masses are equal,

$$\tfrac{1}{2}mv_{1xi}^2 = \tfrac{1}{2}(2\ m)v_{xf}^2 \text{ and}$$

$$v_{1xi}^2 = 2v_{xf}^2$$

Squaring the expression we obtained from conservation of momentum, which we know to be true,

$$v_{1xi}^2 = 4v_{xf}^2$$

Since

$$2v_{xf}^2 \neq 4v_{xf}^2$$

kinetic energy is explicitly not conserved.

REFLECT This is a general proof that kinetic energy is never conserved for a perfectly inelastic collision.

67.   **ORGANIZE AND PLAN**   We are given the ballistic pendulum, shown in Figure P6.67 in the text. A bullet embeds itself in the wooden block, transferring momentum to the block. We'll consider conservation of momentum and conservation of mechanical energy in the ballistic pendulum.

Known: m_{bullet} ; m_{block}; h.

SOLVE   The first part of the process involves the bullet embedding itself in the wood block. During this pert, momentum is conserved, but kinetic energy is not conserved because it is a perfectly inelastic collision. This first step happens so fast that the block hasn't enough time to move very far.

Now that the bullet has transferred its momentum, the block, with bullet in it has velocity and hence kinetic energy. During this second part of the process, kinetic energy is conserved but momentum is not, because of the external forces of gravity and the cord supporting the wood block. The kinetic energy of the bullet/block system is determined by the velocity it has as a result of the first part. MNS check

REFLECT   The height to which the pendulum rises is then dependent on the speed of the block after the bullet strikes it. The key to the use of the ballistic pendulum is the conversion of the initial momentum of the bullet to the velocity of the block/bullet system.

69.   **ORGANIZE AND PLAN**   This is an algebraic problem. We are not asked for a numeric solution. We'll use conservation of momentum to calculate the speed of the bullet/block system just after the bullet stops. In this perfectly inelastic collision, $mv = (m+M)V$, since the block's original speed was zero. Finally we'll calculate kinetic energy $K = \frac{1}{2}mv^2$, convert to gravitational potential energy $U_g = mgh$, and express height as a function of v.

Known: mass of bullet $= m$; mass of block $= M$; velocity of bullet $= v$; velocity of bullet/block system $= V$; height to which pendulum swings $= h$.

SOLVE   (a) From conservation of momentum,

$$mv = (m+M)V$$

$$V = \frac{mv}{(m+M)}$$

From conservation of mechanical energy,

$$\tfrac{1}{2}mv^2 = mgh$$

$$V^2 = 2gh = \frac{m^2v^2}{(m+M)^2} \text{ so}$$

$$v = \frac{\sqrt{2(m+M)^2 gh}}{m}$$

REFLECT   This is the equation a forensic scientist uses to calculate the velocity of the bullet from the laboratory observation of the pendulum.

71.   **ORGANIZE AND PLAN**   We'll treat this as a perfect inelastic collision between eagle and jackrabbit. We'll use $m_1 v_{1i} + m_2 v_{2i} = (m_1 + m_2)v_f$ to find final velocity. Once we know final velocity, we can use $K = \frac{1}{2}mv^2$ to find initial and final kinetic energy and to determine kinetic energy loss. Finally, we recalculate the final velocity for the eagle/jackrabbit system as before, but using an initial velocity for the rabbit. We'll use subscripts 1 for the eagle and 2 for the rabbit.

Known: $m_1 = 6.0$ kg; $m_2 = 5.0$ kg; $v_{1xi} = 322$ km/h; for (a) $v_{2xi} = 0$ m/s; for (b) $v_{2xi} = -2.0$ m/s.

SOLVE   First we convert the eagle's speed to m/s :

$$v_{1xi} = 322 \frac{\text{km}}{\text{h}}\left(\frac{1\text{ h}}{3600\text{ s}}\right)\left(\frac{1000\text{ m}}{\text{km}}\right) = 89.4 \text{ m/s}$$

(a) for this inelastic collision,

$$m_1 v_{1i} + m_2 v_{2i} = (m_1 + m_2)v_f$$

Since $v_{2xi} = 0$ m/s,

$$v_{xf} = \frac{m_1 v_{1xi}}{m_1 + m_2} = \frac{(6.0\text{ kg})(89.4\text{ m/s})}{6.0\text{ kg} + 5.0\text{ kg}} = 48.8\text{ m/s}$$

(b) The initial kinetic energy for the eagle/rabbit system was

$$K_i = \tfrac{1}{2}mv_{1xi}^2 = \tfrac{1}{2}(6.0\text{ kg})(89.4\text{ m/s})^2 = 2.40\times10^4\text{ J}$$

The final kinetic energy is

$$K_f = \tfrac{1}{2}(6.0\text{ kg} + 5.0\text{ kg})(48.8\text{ m/s})^2 = 1.31\times10^4\text{ J}$$

The fraction of kinetic energy lost is

$$\frac{K_f - K_i}{K_i} = \frac{1.31\times10^4\text{ J} - 2.40\times10^4\text{ J}}{2.40\times10^4\text{ J}} = -0.454\text{ decimal fraction}$$

(c) If the rabbit is running toward the eagle [silly rabbit!] at -2.00 m/s, the final speed of the eagle-rabbit system is

$$v_{xf} = \frac{m_1 v_{1xi} + m_2 v_{2xi}}{m_1 + m_2} = \frac{(6.0\text{ kg})(89.4\text{ m/s}) + (5.0\text{ kg})(-2.0\text{ m/s})}{6.0\text{ kg} + 5.0\text{ kg}} = 47.9\text{ m/s}$$

REFLECT   Unlike in animated cartoons, the force the eagle exerts on its prey in the short interval of the collision is enough to cause physiological damage to the prey. We see that the initial velocity of the rabbit in (c) is so small compared to the eagle's that is has little effect on final speed.

73.   **ORGANIZE AND PLAN**   This is a perfectly elastic collision, so both momentum and kinetic energy are conserved. We're given the masses of the bat and ball, but velocity information for the ball only. We'll set our coordinate system so that the ball is pitched in the negative direction and rebounds from the bat in the positive direction. We'll write equations for the conservation of momentum and the conservation of energy. We can express the change in speed of the bat, $v_{2f} - v_{2i}$ in terms of known variables. We can also find the squares of the speeds of the bat, $v_{2f}^2 - v_{2i}^2$. We can factor the difference of the squares and divide by the difference of the variables. This will give us two equations in v_{2f} and v_{2i} which will allow us to solve for both.

Known: $m_1 = 145\text{ g} = 0.145\text{ kg}$; $m_2 = 916\text{ g} = 0.916\text{ kg}$; $v_{1i} = 32\text{ m/s}$; $v_{1f} = -40.\text{ m/s}$.

SOLVE   From conservation of momentum:

$$m_1 v_{1i} + m_2 v_{2i} = m_1 v_{1f} + m_2 v_{2f}$$

$$m_1\left(v_{1i} - v_{1f}\right) = m_2\left(v_{2f} - v_{2i}\right)$$

$$\frac{m_1\left(v_{1i} - v_{1f}\right)}{m_2} = \left(v_{2f} - v_{2i}\right) = \frac{0.145\text{ kg}\left((-32\text{ m/s}) + (40.\text{m/s})\right)}{0.916\text{ kg}} = -11.40\text{ m/s}$$

From conservation of energy,

$$\tfrac{1}{2}m_1 v_{1i}^2 + \tfrac{1}{2}m_2 v_{2i}^2 = \tfrac{1}{2}m_1 v_{1f}^2 + \tfrac{1}{2}m_2 v_{2f}^2$$

$$\frac{m_1\left(v_{1i}^2 - v_{1f}^2\right)}{m_2} = v_{2f}^2 - v_{2i}^2$$

$$\left(v_{2f} - v_{2i}\right)\left(v_{2f} + v_{2i}\right) = \frac{0.145\text{ kg}\left((-32\text{ m/s})^2 + (40.\text{m/s})^2\right)}{0.916\text{ kg}} = -91.18\text{ m}^2/\text{s}^2$$

These give us

$$v_{2f} + v_{2i} = 8.00\text{ m/s}$$

$$v_{2f} - v_{2i} = -11.40\text{ m/s}$$

Solving,

$$v_{2f} = -1.70\text{ m/s}$$

$$v_{21} = 9.70\text{ m/s}$$

REFLECT   The units work out correctly. This batted ball is a bit unusual because it is not much faster than a well-pitched ball. Notice that the bat rebounds backward! If the initial speed of the bat had been greater, more momentum could have been transferred to the ball, and the batter could have "followed through" with the bat.

75.   **ORGANIZE AND PLAN**   The gravitational potential energy of the ball U_g is converted to kinetic energy as it falls. Ignoring air resistance, the ball only loses energy during its inelastic collision with the floor. It rebounds from the floor with an amount of kinetic energy specified by the coefficient of restitution. This kinetic energy is once again converted to U_g. We're to find its rebound height after the third bounce.

Known: COR = 0.82.

SOLVE   Since

$$\text{COR} = \frac{K_f}{K_i} = \frac{\frac{1}{2}mv_f^2}{\frac{1}{2}mv_i^2} = \frac{mgh_f}{mgh_i} = \frac{h_f}{h_i}$$

We see that it is also the ratio of final to initial height.

The final height of one bounce n is the initial height for the next bounce $n+1$. For three bounces, the final height becomes

$$h_i(\text{COR})^3 = h_f = (0.82)^3 h_i = 0.55 h_i$$

REFLECT   The COR of an unknown material may be determined by dropping it from a known height on a hard surface and observing the height to which it rebounds. The famous Wham-O Superball® reportedly had a COR of 0.92. [Time magazine 10-22-1965]

77.   **ORGANIZE AND PLAN**   In this problem, we'll write equations for the conservation of momentum and the loss of kinetic energy. For momentum, we'll express one velocity in terms of the other. Then we'll substitute this expression into the equation for kinetic energy. This gives us a quadratic equation that we can solve using the quadratic formula.

Known $m_1 = m_2 = m$; $v_{1i} = 0.95$ m/s; $v_{2i} = 0$ m/s; COR = 0.90.

SOLVE   Since the masses are equal, mass cancels in both equations and we can work with velocities alone. From conservation of momentum,

$$v_{1i} = v_{1f} + v_{2f}$$

From conservation of kinetic energy,

$$0.90v_{1i}^2 = v_{1f}^2 + v_{2f}^2$$

$$v_{1f} + v_{2f} = 0.95 \text{ m/s}$$

$$v_{1f}^2 + v_{2f}^2 = 0.81225$$

$$v_{1f} = 0.95 - v_{2f}$$

$$v_{2f}^2 = 0.81225 - v_{1f}^2$$

Substituting,

$$v_{2f}^2 = 0.81225 - \left(0.95 - v_{2f}\right)^2$$

Expanding the difference of squares and rearranging to the general form of a quadratic equation, we get

$$v_{2f}^2 - 0.95 \, v_{2f} + 0.040125 = 0$$

Evaluating with the quadratic formula, we get

$$v_{2f} = 0.906 \text{ m/s or}$$

$$v_{2f} = 0.0443 \text{ m/s}$$

Choosing $v_{2f} = 0.906 \text{ m/s} \cong 0.91 \text{ m/s}$, we find that

$$v_{1f} = 0.0443 \text{ m/s} \cong 0.044 \text{ m/s}$$

REFLECT   We notice that the quadratic formula gives us both roots, that is, values for both v_{2f} and v_{1f}. This is because the masses are the same. We chose $v_{1f} = 0.044$ m/s because the leading ball must travel faster than the trailing ball. Otherwise we would have one solid ball traveling through a second slower ball.

79. **ORGANIZE AND PLAN** We'll set up our coordinate system so one wad of putty is traveling in the positive x-direction and the other in the positive y-direction. We'll use vector space notation to keep track of the vector quantities $\vec{v}$ and $\vec{p}$. We can find the final velocity by using the formula for momentum of a perfectly inelastic collision.

Known: $m_1 = m_2 = m$; $\vec{v}_{1i} = (1.45 \text{ m/s})\hat{i}$; $\vec{v}_{2i} = (1.45 \text{ m/s})\hat{j}$.

SOLVE For a perfectly inelastic collision,

$$m_1 \vec{v}_{1i} + m_2 \vec{v}_{2i} = (m_1 + m_2)\vec{v}_f$$

$$\vec{v}_f = \frac{m_1 \vec{v}_{1i} + m_2 \vec{v}_{2i}}{m_1 + m_2}$$

Since the masses are equal,

$$\vec{v}_f = \frac{\vec{v}_{1i} + \vec{v}_{2i}}{2} = \frac{(1.45 \text{ m/s})\hat{i} + (1.45 \text{ m/s})\hat{j}}{2} = (0.725 \text{ m/s})\hat{i} + (0.725 \text{ m/s})\hat{j}, \text{ or}$$

$$v_f = 2\frac{\sqrt{(1.45 \text{ m/s})^2 + (1.45 \text{ m/s})^2}}{} = 1.03 \text{ m/s}$$

$$\theta = \tan^{-1}\left(\frac{1.45}{1.45}\right) = 45° \text{ above the positive } x\text{-axis}$$

REFLECT Unlike a motion problem in which we add velocity vectors for one mass, here we have to take into account that we have two masses that become joined.

81. **ORGANIZE AND PLAN** In this problem we need to find the velocity of the second car. The problem is made a bit easier because the masses are the same. We set up a coordinate system with the first car traveling in the positive x-direction. We'll use conservation of momentum for a perfectly inelastic collision and cancel the masses. We'll use subscript 1 for the eastbound car and 2 for the northbound car.

Known: $m_1 = m_2 = m$; $v_{1i} = 25 \text{mi/h} = 11.1 \text{m/s}$ in the x-direction; $\theta = 54°$ N of E, or above the positive x-axis.

SOLVE For a perfectly inelastic collision,

$$m_1 \vec{v}_{1i} + m_2 \vec{v}_{2i} = (m_1 + m_2)\vec{v}_f$$

Since the masses are equal,

$$\vec{v}_{1i} + \vec{v}_{2i} = 2\vec{v}_f$$

$$\frac{\vec{v}_{1i}}{2} + \frac{\vec{v}_{2i}}{2} = \vec{v}_f$$

Since $\vec{v}_f$ is the resultant velocity vector, $v_{1i}/2$ is the x-component and $v_{2i}/2$ is the y-component.

$$\tan(54°) = \frac{v_{2i}/2}{v_{1i}/2}$$

The second car was speeding.

$$v_{2i} = v_{1i}\tan(54°) = 25 \text{ mi/h}(1.38) = 34 \text{ mi/h}$$

REFLECT Since the cars were of equal mass, we can tell intuitively that the northbound car was going faster than the eastbound car.

83. **ORGANIZE AND PLAN** For this elastic collision with one ball at rest, we'll write equations for conservation of momentum in both the x-direction and y-direction. The equal masses will cancel. We know that second ball is at a right angle to the first, so we'll solve two equations in two unknowns, v_{1f} and v_{2f}. It doesn't matter which ball goes off at a 45° angle, so we'll declare that to be ball 1.

Known: $m_1 = m_2 = m$; $\vec{v}_{1i} = 1.65$ m/s in the positive x-direction ; $\vec{v}_{2i} = 0$ m/s; $\theta = 45°$.

SOLVE In the x-direction,

$$m_1 \vec{v}_{1i} + m_2 \vec{v}_{2i} = m_1 \vec{v}_{1f} + m_2 \vec{v}_{2f}$$

Since $v_{2i} = 0$ and the masses are equal,

$$\vec{v}_{1i} = \vec{v}_{1f} + \vec{v}_{2f}$$

Let ball 1 go off at 45° after the collision. In the x-direction,

$$v_{1f}\cos(45°) + v_{2f}(\cos\theta) = 1.65\,\text{m/s}$$

In the y-direction,

$$0 = v_{1f}\sin(45°) + v_{2f}(\sin\theta)$$

Since we have an elastic collision with one object at rest,

$$\theta = 90° - 45° = 45° \text{ below the } x\text{-axis, or } -45°$$

Then,

$$v_{1f}\cos(45°) + v_{2f}\cos(-45°) = 1.65 \text{ m/s}$$

$$v_{1f}\sin(45°) + v_{2f}\sin(-45°) = 0 \text{ m/s}$$

Solving these two simultaneous equations,

$$v_{1f} = 1.16 \text{ m/s at 45° above the positive } x\text{-axis and}$$

$$v_{2f} = 1.16 \text{ m/s at 45° below the positive } x\text{-axis}$$

REFLECT   If this were an inelastic collision, then we would need to know one other piece of information, such as an angle or the final speed of a ball, in order to solve.

85.   **ORGANIZE AND PLAN**   We'll use $\vec{p}_{h\nu,i} + \vec{p}_{e,i} = \vec{p}_{h\nu,f} + \vec{p}_{e,f}$ for this elastic collision where subscript e is for the electron and subscript $h\nu$ is for the photon. We'll use a conventional coordinate system and work in vector space notation.

Known: $\vec{p}_{h\nu,i} = (1.0\times10^{-21} \text{ kg}\cdot\text{m/s})\hat{i}$; $\vec{p}_{e,i} = 0 \text{ kg}\cdot\text{m/s}$; $\vec{p}_{h\nu,f} = (1.8\times10^{-22} \text{ kg}\cdot\text{m/s})\hat{i} - (3.1\times10^{-28} \text{ kg}\cdot\text{m/s})\hat{j}$.

SOLVE

$$\vec{p}_i = \vec{p}_f$$

$$\vec{p}_{h\nu,i} + \vec{p}_{e,i} = \vec{p}_{h\nu,f} + \vec{p}_{e,f}$$

Since $\vec{p}_{e,i} = 0$

$$\vec{p}_{e,f} = \vec{p}_{h\nu,i} - \vec{p}_{h\nu,f} = (1.0\times10^{-21} \text{ kg}\cdot\text{m/s})\hat{i} - \left[(1.8\times10^{-22} \text{ kg}\cdot\text{m/s})\hat{i} - (3.1\times10^{-28} \text{ kg}\cdot\text{m/s})\hat{j}\right]$$

$$\vec{p}_{e,f} = (8.2\times10^{-22} \text{ kg}\cdot\text{m/s})\hat{i} + (3.1\times10^{-28} \text{ kg}\cdot\text{m/s})\hat{j}$$

REFLECT   Even though a photon has no mass, relativistic mechanics tells us that its momentum is its energy divided by the speed of light.

87.   **ORGANIZE AND PLAN**   Since the initial velocity of the mortar shell is expressed using level ground (the Earth) as the frame of reference, we'll continue to use this frame of reference. We'll calculate the x- and y-components of the final velocity of the third piece with respect to the Earth.

Known: $v_{i,\text{shell}} = 35.0$ m/s at 60° above horizontal; $m_1 = 10.\text{kg}$; $v_{1f} = 38.0$ m/s in the positive x-direction;

$m_2 = 10.\text{kg}$; $v_{2f} = 11.5$ m/s in the positive y-direction.

SOLVE   At the top of the trajectory, the unexploded shell has a speed of

$$v_x = v_i\cos 60° = 17.5 \text{ m/s}.$$

This means that the initial momentum to be conserved in the x-direction is

$$\vec{p}_i = mv_x = (25\text{kg})(17.5 \text{ m/s})\hat{i}$$

so

$$\vec{v}_{3fx} = \frac{(25 \text{ kg})(17.5 \text{ m/s}) - (10 \text{ kg})(38 \text{ m/s})}{5 \text{ kg}} = (11.5 \text{ m/s})\hat{i}$$

In the y-direction, there is no initial momentum, and the only final momenta are those of the 10 kg piece that travels vertically upward, and that of the third piece:

$$\vec{v}_{3yx} = \frac{0-(10 \text{ kg})(11.5 \text{ m/s})}{5 \text{ kg}} = (-23.0 \text{ m/s})\hat{j}$$

The final velocity of the third piece is then

$$\vec{v}_{3f} = (11.5 \text{ m/s})\hat{i} - (23.0 \text{ m/s})\hat{j}$$

with respect to the Earth.

REFLECT   To an observer on Earth standing at a right angle to the path of the shell, this third piece would appear to be traveling down and forward.

89.   **ORGANIZE AND PLAN**   We'll declare the meterstick to be the x-axis. We're trying to find the center of mass in the x-direction. We'll use $X_{cm} = \frac{1}{M}\sum_{i=1}^{n} m_i x_i$. Distance is measured from the origin, the zero end of the meterstick.

Known: $m_1 = 0.250$ kg; $m_2 = 0.500$ kg; $x_1 = 0.200$ m; $x_2 = 0.500$ m.

SOLVE   Using the formula for center of mass,

$$X_{cm} = \frac{1}{M}\sum_{i=1}^{n} m_i x_i = \frac{m_{1x}x_1 + m_2 x_2}{m_1 + m_2} = \frac{(0.250 \text{ kg})(0.200 \text{ m}) + (0.500 \text{ kg})(0.500 \text{ m})}{0.250 \text{ kg} + 0.500 \text{ kg}}$$

$$X_{cm} = 0.400 \text{ m}$$

REFLECT   It doesn't matter where we choose to start measuring. If we declare this point to be between the masses, however, one displacement will be negative and we must take into account the sign.

91.   **ORGANIZE AND PLAN**   We'll use the center of mass formula $X_{cm} = \frac{1}{M}\sum_{i=1}^{n} m_i x_i$ for the two-body Sun-Jupiter system as we did in Problem 90. Then we'll compare the location of X_m to the radius of Sun.

Known: $m_{\text{Sun}} = 1.989 \times 10^{30}$ kg; $m_{\text{Jupiter}} = 1.899 \times 10^{27}$ kg; $x_{\text{Jupiter}} = 7.786 \times 10^8$ km $= 7.786 \times 10^{11}$ m; $x_{\text{Sun}} = 0$; $r_{\text{Sun}} = 6.96 \times 10^8$ m.

SOLVE   Calculating the center of mass,

$$X_{cm} = \frac{1}{M}\sum_{i=1}^{n} m_i x_i = \frac{0 + (1.899 \times 10^{27} \text{ kg})(7.786 \times 10^{11} \text{ m})}{1.989 \times 10^{30} \text{ kg} + 1.899 \times 10^{27} \text{ kg}} = 7.43 \times 10^8 \text{ m}$$

Comparing this to Sun's radius of 6.96×10^8 m we find the center of mass to be just above its surface, by about 7% the length of its radius.

REFLECT   Jupiter is about 300 times more massive than Earth and about 5 times more distant from Sun. This has a significant effect on the location of the center of mass compared to that in Problem 90. It is still very close to Sun.

93.   **ORGANIZE AND PLAN**   In this two-body system, we already know the center of mass and must find the location of one of the objects. It makes sense to measure from the pivot, but we'll have to watch the signs. We'll use $X_{cm} = \frac{1}{M}\sum_{i=1}^{n} m_i x_i$ and solve for one of the distance values. We'll let x_2 be the positive distance from the pivot point to the child.

Known: $m_1 = 28$ kg; $x_1 = -2.8$ m; $m_2 = 38$ kg; $X_{cm} = 0$ (that is, at the pivot).

SOLVE   Rearranging,

$$X_{cm} = \frac{1}{M}\sum_{i=1}^{n} m_i x_i = \frac{m_1 x_1 + m_2 x_2}{m_1 + m_2}$$

$$X_{cm}(m_1 + m_2) = m_1 x_1 + m_2 x_2$$

Since

$$X_{cm} = 0,$$

$$-m_1 x_1 = m_2 x_2$$

$$x_2 = \frac{-m_1 x_1}{m_2} = \frac{-(28 \text{ kg})(-2.8 \text{ m})}{38 \text{ kg}} = 2.2 \text{ m}$$

REFLECT   It's reasonable to expect the heavier child to sit closer to the pivot point in order to balance. We'll see this again when we study torque.

95.   ORGANIZE AND PLAN   In part (a), we'll use $V_{cm,x} = \frac{1}{M}\sum_{i=1}^{n} m_i v_{ix}$ to find the velocity of the center of mass. In part

(b) we need to find both final velocities when the objects have equal mass and are initially in motion. From conservation of momentum, we'll use $v_{1xi} + v_{2xi} = v_{1xf} + v_{2xf}$. From conservation of mechanical energy, we'll

use $v_{1xi} - v_{2x1} = v_{2xf} - v_{1xf}$. This gives us two simultaneous equations and our solution for both velocities.

Known: $v_{1xi} = 0.350 \text{ m/s}$; $v_{2xi} = 0.250 \text{ m/s}$.

SOLVE   (a) For the center of mass velocity,

$$V_{cm,x} = \frac{1}{M}\sum_{i=1}^{n} m_i v_{ix}$$

For two particles of equal mass, this reduces to

$$V_{cm,x} = \frac{mv_{1xi} + mv_{2xi}}{2m} = \frac{v_{1xi} + v_{2xi}}{2} = \frac{0.350 \text{ m/s} + 0.250 \text{ m/s}}{2} = 0.300 \text{ m/s}$$

(b) From conservation of momentum we have

$$v_{1xi} + v_{2xi} = v_{1xf} + v_{2xf} = 0.600 \text{ m/s}$$

From conservation of mechanical energy we have

$$v_{1xi} - v_{2x1} = v_{2xf} - v_{1xf} = 0.100 \text{ m/s}$$

This gives us the two simultaneous equations

$$v_{1xf} + v_{2xf} = 0.600 \text{ m/s}$$

$$v_{2xf} - v_{1xf} = 0.100 \text{ m/s}$$

Solving, we get

$$v_{2xf} = 0.350 \text{ m/s}$$

and

$$v_{1xf} = 0.250 \text{ m/s}$$

We see that due to the collision, the objects "trade" velocities.

(c) Now we check the center-of-mass velocity. Since the masses are equal they cancel:

$$V_{cm,x} = \frac{m_1 v_{1xf} + m_2 v_{2fx}}{2m} = \frac{0.250 \text{ m/s} + 0.350 \text{ m/s}}{2} = 0.300 \text{ m/s}$$

REFLECT   The center-of-mass velocity is weighted more toward the massive object. In this case, the masses are equal. There is no weighting and the result is simply the arithmetic mean.

97.   ORGANIZE AND PLAN   We're given the x and y coordinates of two soccer players with respect to the origin, the corner of the field. These coordinates give us the x and y distances for use in the

formulas $X_{cm} = \frac{m_1 x_1 + m_2 x_2}{m_1 + m_2}$ and $Y_{cm} = \frac{m_1 y_1 + m_2 y_2}{m_1 + m_2}$ to get the coordinates for the center of mass of the two-player

system.

Known: $m_1 = 59.0 \text{ kg}$; $m_2 = 71.5 \text{ kg}$; from the given coordinates we have $x_1 = 24.3 \text{ m}$; $y_1 = 35.9 \text{ m}$; $x_2 = 78.8 \text{ m}$;

$y_2 = 21.5 \text{ m}$.

SOLVE   For the x-position of the center of mass,

$$X_{cm} = \frac{m_1 x_1 + m_2 x_2}{m_1 + m_2} = \frac{(59.0 \text{ kg})(24.3 \text{ m}) + (71.5 \text{ kg})(78.8 \text{ m})}{59.0 \text{ kg} + 71.5 \text{ kg}} = 54.2 \text{ m}$$

Likewise for the y-position of the center of mass,

$$Y_{cm} = \frac{m_1 y_1 + m_2 y_2}{m_1 + m_2} = \frac{(59.0 \text{ kg})(35.9 \text{ m}) + (71.5 \text{ kg})(21.5 \text{ m})}{59.0 \text{ kg} + 71.5 \text{ kg}} = 28.0 \text{ m}$$

The center-of-mass coordinates are $(54.2 \text{ m}, 28.0 \text{ m})$.

REFLECT   The center of mass is on the line connecting the two players, closer to the more massive player.

99.   **ORGANIZE AND PLAN**   We're given the information shown in Figure P6.99 in the text. For part (a), we'll use trigonometry to find the distance the center of mass of the legs rises. For part (b) we have to use the weighted masses of the legs and the remainder of the body as two different masses. We're not concerned about how far the center of mass moves in the x-direction. We'll use $Y_{cm} = \frac{1}{M} \sum_{i=1}^{n} m_i y_i$ for the center of mass in the y-direction. We let M be the total body mass and h be the height the center of mass rises.

Known: $\theta = 50.0°$; $L_{\text{legs}} = 95 \text{ cm}$; $m_{\text{legs}} = 0.345 \, M$; $m_{\text{body}} = 0.655 \, M$.

SOLVE   First we convert the length of the legs to meters:

$$L_{\text{legs}} = 95 \text{ cm} \left(\frac{1 \text{ m}}{100 \text{ cm}} \right) = 0.95 \text{ m}$$

(a) Then we use trigonometry to find h :

$$h = \frac{(0.95 \text{ m})}{2} \sin 50.0° = 0.36 \text{ m}$$

(b) Now we can use the answer in (a) to find the center of mass of the entire body:

$$Y_{cm} = \frac{(m_{\text{legs}})(h) + (m_{\text{body}})(0)}{M} = \frac{(0.345 \, M)(0.36 \text{ m})}{M} = 0.12 \text{ m}$$

(c) The only external force acting on the body is the normal force of the floor on the body, opposing gravity.

REFLECT   The body's center of mass also moves slightly toward the head. The problem does not ask this, but we have all the information we need to calculate it. Stored chemical energy in glucose in the blood is oxidized to produce muscle contraction. This results in force on the floor. The reactive force of the floor on the body produces the leg rises.

101.   **ORGANIZE AND PLAN**   We're to prove that the center of mass accelerated with an applied force $\vec{F}_{\text{net}}$ according to Newton's second law, $\vec{F}_{\text{net}} = M\vec{A}_{cm}$. We'll start out with the formula for the velocity of the center of mass. We'll divide by time to get acceleration, then rearrange to finish the proof. This is an algebraic solution, with no numeric answer required.

Known: applied force $= \vec{F}_{\text{net}}$; $V_{cm} \frac{1}{M} \sum_{i=1}^{n} m_i v_i$.

SOLVE   Acceleration is change in velocity with time, so starting with

$$V_{cm} \frac{1}{M} \sum_{i=1}^{n} m_i v_i$$

We divide by Δt to get

$$\frac{V_{cm}}{\Delta t} = \frac{1}{M} \sum_{i=1}^{n} \frac{m_i v_i}{\Delta t}$$

Now, since

$$\vec{F}\Delta t = mv$$

$$\vec{F} = \frac{mv}{\Delta t}, \text{ and}$$

$$\sum_{i=1}^{n} F_i = \sum_{i=1}^{n} \frac{m_1 v_1}{\Delta t}$$

$$\vec{F}_{net} = \sum_{i=1}^{n} F_i = \sum_{i=1}^{n} \frac{m_1 v_1}{\Delta t}, \text{ and}$$

$$M = \sum_{i=1}^{n} m_i, \text{ then}$$

$$\vec{F}_{net} = M\vec{A}_{cm}$$

REFLECT   We remember that net force, $\vec{F}_{net}$ is the weighted vector sum of all the forces acting on all the particles.

103.   **ORGANIZE AND PLAN**   To find the force, we'll use $\vec{F}_{net} = m\vec{a}$. First we'll find the changes in velocity in both directions, then divide each by time to get acceleration. Then knowing the mass of the electron, we calculate force.
Known: $v_i = (6.2\times10^6 \text{ m/s})\hat{i} - (5.8\times10^5 \text{ m/s})\hat{j}$; $v_f = (-3.7\times10^6 \text{ m/s})\hat{i} - (15.8\times10^6 \text{ m/s})\hat{j}$; $\Delta t = 3.0 \text{ s}$;
$m_e = 9.109\times10^{-31} \text{ kg}$.

SOLVE   In the *x*-direction,

$$\vec{a}_x = \frac{\vec{v}_{xf} - \vec{v}_{xi}}{\Delta t} = \frac{(-3.7\times10^6 \text{ m/s})\hat{i} - (6.2\times10^6 \text{ m/s})\hat{i}}{3.0 \text{ s}} = -3.30\times10^6 \text{ m/s}^2$$

In the *y*-direction,

$$\vec{a}_y = \frac{\vec{v}_{yf} - \vec{v}_{yi}}{\Delta t} = \frac{(-15.8\times10^6 \text{ m/s})\hat{i} - (-5.8\times10^5 \text{ m/s})\hat{i}}{3.0 \text{s}} = -5.07\times10^6 \text{m/s}^2$$

$$\vec{a} = (-3.30\times10^6 \text{ m/s}^2)\hat{i} - (5.07\times10^6 \text{ m/s}^2)\hat{j}$$

$$\vec{F} = m\vec{a} = 9.109\times10^{-31} \text{ kg}(-3.30\times10^6 \text{ m/s}^2)\hat{i} - (5.07\times10^6 \text{ m/s}^2)\hat{j}$$

$$\vec{F} = (-3.01\times10^{-24} \text{ N})\hat{i} - (4.62\times10^{-24} \text{ N})\hat{j}$$

REFLECT   This is just for one particle. If there was a group of particles, we'd just find the center of mass velocity and treat it like the velocity of a particle as we did here.

105.   **ORGANIZE AND PLAN**   We'll declare the incoming ball to have velocity in the positive *x*-direction. According to Newton's third law, when the ball hits the stationary racket, the racket exerts a force on the ball, but the ball also exerts a force on the racket (and the player holding it). We're to find the force of the racket on the ball and the recoil speed of the racket/player system. We'll use the momentum-impulse theorem, $F\Delta t = \Delta p$.

Known: $m_1 = 68.0 \text{ kg}$; $m_2 = 57.0 \text{ g} = 0.057 \text{ kg}$; $v_{2i} = 50.0 \text{ m/s}$; $v_{2f} = -12.5 \text{ m/s}$;
$v_{1i} = 0 \text{ m/s}$ $\Delta t = 35 \text{ ms} = 0.035 \text{ s}$.

SOLVE   (a) From the impulse-momentum theorem,

$$\Delta p = F\Delta t$$

$$F = \frac{\Delta p}{\Delta t} = \frac{m(v_f - v_i)}{\Delta t} = \frac{0.057 \text{ kg}(-12.5 \text{ m/s} - 50.0 \text{ m/s})}{0.035 \text{ s}} = -102 \text{ N}$$

(b) The same magnitude of force acts in the opposite direction on the racket/player system. The impulse is:

$$J = F\Delta t = (102 \text{ N})(0.035 \text{ s}) = 3.57 \text{ N} \cdot \text{s}$$

$$v_f = \frac{J}{m} = \frac{3.57 \text{N} \cdot \text{s}}{68.0 \text{ kg}} = 0.0525 \text{ m/s}$$

REFLECT   The recoil velocity of the player is small because her mass is large compared to that of the ball. However, she's in the air, and this velocity backward will affect her balance. MNS check

107.   **ORGANIZE AND PLAN**   This is an explosion problem with zero initial momentum. The two particles will move apart from one another. We'll use $(m_1 + m_2)v_i = m_1 v_{f1} + m_2 v_{f2}$ to find the final velocity of the second piece, v_{2f}.
Known: $v_i = 0$; $m_1 = 92 \text{ mg} = 9.2\times10^{-5} \text{ kg}$; $m_2 = 71 \text{ mg} = 7.1\times10^{-5} \text{ kg}$; $v_{1f} = 48 \text{ cm/s} = 0.48 \text{m/s}$.

SOLVE   Using conservation of momentum,

$$(m_1 + m_2)v_i = m_1 v_{f1} + m_2 v_{f2}$$

$$v_i = 0$$

and

$$m_2 v_2 = -m_1 v_{1f}$$

$$v_{2f} = \frac{-m_1 v_{1f}}{m_2} = \frac{-(9.2 \times 10^{-5}\ \text{kg})(0.48\ \text{m/s})}{7.1 \times 10^{-5}\ \text{kg}} = -0.62\ \text{m/s}$$

The other piece moves in the opposite direction at $0.62\ \text{m/s}$.

REFLECT   The units come out to be m/s, so we are confident in our answer. A popcorn kernel does act like a bomb! Water vaporizes, pressurizing the inside of the kernel until the hull bursts.

109.   **ORGANIZE AND PLAN**   First, we'll use kinematic equations to calculate Δt, assuming constant acceleration. Next we'll calculate the loss in momentum in the bullet and equate that to impulse $\Delta p = J$. Knowing Δt from (a) allows us to calculate force from impulse. Next we'll consider the momentum transmitted to the block using $J = \Delta p = \Delta(mv)$. We'll assume the block does not have time to move very far. Lastly, we'll calculate acceleration of the block to find the force of friction and use $F_k = F_n \mu_k$ to calculate the coefficient of friction.
Known: $m_{\text{block}} = 4.0\ \text{kg}$; $m_{\text{bullet}} = 20.\text{g} = 0.020\ \text{kg}$; $\Delta x_{\text{bullet}} = 20.\text{cm} = 0.20\ \text{m}$; $v_{\text{bullet},i} = 800.\ \text{m/s}$; $v_{\text{bullet},f} = 425\ \text{m/s}$; $\Delta x_{\text{block}} = 81\text{cm} = 0.81\text{m}$; $v_{\text{block},f} = 0$.

SOLVE   (a) $v_{\text{bullet},f}^2 = v_{\text{bullet},i}^2 + 2a\Delta x_{\text{bullet}}$

$$a = \frac{v_{\text{bullet},f}^2 - v_{\text{bullet},i}^2}{2\Delta x_{\text{bullet}}} = \frac{(425\ \text{m/s})^2 - (800\ \text{m/s})^2}{2(0.20\ \text{m})} = -1.148 \times 10^6\ \text{m/s}^2$$

$$\Delta t = \frac{425\ \text{m/s} - 800\ \text{m/s}}{-1.148 \times 10^6\ \text{m/s}^2} = 3.27 \times 10^{-4}\ \text{s}$$

(b) Now we calculate loss in momentum and force:

$$J = \Delta p = m_{\text{bullet}}\left(v_{\text{bullet},f} - v_{\text{bullet},i}\right) = (0.020\ \text{kg})(425\ \text{m/s} - 800\ \text{m/s}) = -7.5\ \text{N} \cdot \text{s}$$

$$F = \frac{J}{\Delta t} = \frac{7.5\ \text{kg} \cdot \text{m/s}}{3.27 \times 10^{-4}\ \text{s}} = 2.30 \times 10^4\ \text{N}$$

(c) The loss in momentum of the bullet we calculated in (b) is transmitted to the block.

$$J = \Delta p = m\Delta v = m_{\text{block}}\left(v_{\text{block},f} - v_{\text{block},i}\right)$$

$$v_{\text{block},f} = \frac{J}{m_{\text{block}}} = \frac{7.5\ \text{kg} \cdot \text{m/s}}{4.0\ \text{kg}} = 1.875\ \text{m/s}$$

(d) Now, using the final velocity of the block from (c) as its initial velocity in a kinematic equation,

$$v_{\text{block},f}^2 = v_{\text{block},i}^2 + 2a\Delta x_{\text{block}}$$

Since the block comes to rest,

$$a = \frac{-v_{\text{block},f}^2}{2\Delta x_{\text{block}}} = \frac{-(1.875\ \text{m/s})^2}{2(0.81\ \text{m})} = -2.17\ \text{m/s}^2$$

From Newton's second law,

$$F_k = m_{\text{block}}a = -(4.0\ \text{kg})(2.17\ \text{m/s}^2) = -8.68\ \text{N}$$

$$\mu_k = \left|\frac{F_k}{F_n}\right| = \left|\frac{F_k}{mg}\right| = \left|\frac{-8.68\ \text{N}}{(4.0\ \text{kg})(9.80\ \text{m/s}^2)}\right| = 0.22$$

REFLECT   This might be a competitor for a ballistic pendulum. However, we can accurately and reproducibly measure height. It's difficult to produce two surfaces that have precisely the same coefficient of friction.

111.   **ORGANIZE AND PLAN**   Compare this to Problem 104. Here we add the initial momentum of the snow, which is zero, to the momentum of the toboggan. Then we divide by the total mass to find final velocity using $m_1 v_{1i} + m_2 v_{2i} = (m_1 + m_2) v_f$.

Known: $m_1 = 8.6$ kg; $m_2 = 12$ kg; $v_{1i} = 23$ km/h; $v_{2i} = 0$.

SOLVE   First we convert 23 km/h to 6.39 m/s. We'll round to the correct number of significant digits at the end of the problem. Then we use the formula for a perfectly inelastic collision:

$$m_1 v_{1i} + m_2 v_{2i} = (m_1 + m_2) v_f$$

$$v_f = \frac{m_1 v_{1i} + 0}{m_1 + m_2} = \frac{(8.6 \text{ kg})(6.39 \text{ m/s})}{8.6 \text{ kg} + 12 \text{ kg}} = 2.7 \text{ m/s}$$

REFLECT   Momentum is a vector quantity. Even though the falling snow has momentum in the *y*-direction, that momentum is transferred to the snowy ground. It does not contribute to momentum in the *x*-direction.

113.   **ORGANIZE AND PLAN**   This is a combination of a kinematic problem and an inelastic collision problem. First we use kinematic equations to find and hence momentum, then after the collision with the stationary target car, we convert momentum back to velocity. In part (b) we only use the kinematic equations to find final velocity. We'll tilt the axes as we do with an inclined plane problem.

Known: $m_1 = 950.$ kg ; $m_2 = 1240$ kg; $\Delta x = 36$ m; $\theta = 2.5°$; for (h) $v_i = 0$ m/s.

SOLVE   (a) $a_x = g \sin \theta$

$$v_f^2 = v_i^2 + 2 a_x \Delta x$$

Since $v_i = 0$ m/s,

$$v_f = \sqrt{2 \, g \sin \theta \Delta x} = \sqrt{2(9.80)(\sin 2.5°)(18 \text{ m})} = 3.92 \text{ m/s}$$

Halfway down the slope, the car collides with the second car. Their combined momentum is then

$$m_1 v_{1i} + m_2 v_{2i} = (m_1 + m_2) v_f \text{ and since the second car was stationary,}$$

$$v_f = \frac{m_1 v_1}{m_1 + m_2} = \frac{(3.92)(950.\text{kg})}{950 \text{ kg} + 1240 \text{ kg}} = 1.70 \text{ m/s}$$

Now, still ignoring friction, and using this initial velocity of 1.70 m/s for the rest of this part,

$$v_f = \sqrt{v_i^2 + 2 a_x \Delta x} = \sqrt{(1.70 \text{ m/s})^2 + 2(9.80 \text{ m/s}^2)(\sin 2.5°)(18 \text{ m})} = 4.28 \text{ m/s}$$

(b) Now suppose there is no stationary second car. We find the final speed of the first car as before, but using the entire slope of 36 m.

$$v_f = \sqrt{2 g \sin \theta \Delta x} = \sqrt{2(9.80)(\sin 2.5°)(36 \text{ m})} = 5.55 \text{ m/s}$$

REFLECT   The collision provides some means of "braking." This is like the ball in a mechanical pinball machine striking barriers as it rolls on its way down to the flippers or the "drain."

115.   **ORGANIZE AND PLAN**   This is a two-body momentum problem, the reverse of a perfectly inelastic collision. We'll need to calculate the astronaut's final speed from conservation of momentum using $(m_1 + m_2) v_i = m_1 v_{f1} + m_2 v_{f2}$.

Once we know the speed, we can calculate the time as a constant speed kinematic problem. In part (b) we'll recalculate using an initial speed of the astronaut. We'll use the spacecraft is the frame of reference with positive direction being in the direction from the astronaut toward the spacecraft.

Known: $m_1 = 128$ kg; $m_2 = 1.10$ kg; $\Delta x = 15.0$ m; (a) $v_{1i} = 0$; $v_{2i} = 0$; $v_{2f} = -5.40$ m/s; (b) $v_{1i} = -2.85$ cm/s $= -2.85 \times 10^{-2}$ m/s ; $v_{2f} = -5.4285$ m/s.

SOLVE   When the astronaut is stationary with respect to the spacecraft,

$$(m_1 + m_2) v_i = m_1 v_{f1} + m_2 v_{f2}$$

$$m_1 v_{f1} = -m_2 v_{f2}$$

$$v_{1f} = \frac{-m_2 v_{2f}}{m_1} = \frac{-(1.10 \text{ kg})(-5.40 \text{ m/s})}{128 \text{ kg}} = 0.0464 \text{ m/s}$$

At that rate, the time it will take to return to the spacecraft is

$$\Delta t = \frac{\Delta x}{v_v} = \frac{15.0 \text{ m}}{0.0464 \text{ m/s}} = 323 \text{ s}$$

The astronaut will make it back to the craft in about 5½ minutes.

(b) Now, when the astronaut (with wrench) is traveling away from the craft at -2.85×10^{-2} m/s conservation of momentum becomes

$$v_{1f} = \frac{(m_1 + m_2)v_i - m_2 v_{2f}}{m_1}$$

$$v_{1f} = \frac{(128 \text{ kg} + 1.1 \text{ kg})(-0.0285 \text{ m/s}) - (1.10 \text{ kg})(-5.40 \text{ m/s})}{128 \text{ kg}} = 0.0179 \text{ m/s}$$

Repeating the math, at this rate, time back to the craft will be 838 s or about 14 minutes

REFLECT   Astronauts in this situation have to be careful to throw the wrench away from their center of mass in a direct line with the spacecraft. Otherwise they will end up spinning in space. We'll visit this again in Chapter 8. If the astronaut were going just a little faster away from the craft, it would be impossible to reverse direction!

117.   **ORGANIZE AND PLAN**   There are four segments to the model: head, torso, legs and arms. We know each of their masses as a fraction of total body mass M. We'll use the floor as a datum and calculate where the center of mass is for each segment with respect to the floor. We'll let the y-positions be the centers of mass for the various segments.

Known: $m_1 = 0.069\,M$; $m_2 = 0.460\,M$; $m_3 = 0.346\,M$; $m_4 = 0.125\,M$; $y_1 = 167.5$ cm; $y_2 = 125$ cm; $y_3 = 47.5$ cm $y_4 = 122.5$ cm.

SOLVE   (a) From the floor,

$$Y_{cm} = \frac{1}{M}\sum_{i=1}^{n} m_i y_i$$

For the four segments,

$$Y_{cm} = \frac{m_1 y_1 + m_2 y_2 + m_3 y_3 + m_4 y_4}{M}$$

$$Y_{cm} = \frac{(0.069\,M)(167.5 \text{ cm}) + (0.460\,M)(125 \text{ cm}) + (0.346\,M)(47.5 \text{ cm}) + (0.125\,M)(122.5 \text{ cm})}{M}$$

$$Y_{cm} = 100.8 \text{ cm}$$

(b) Now, if the model raises its arms, the only change in the calculation is the new position of the center of mass of the arms, which becomes 187.5cm. Recalculating, the new overall center of mass is

$$Y_{cm} = 108.9 \text{ cm}$$

REFLECT   Most of the mass of the body is in the legs and torso. Raising the arms only increases the body's center of mass by about 8 cm. We did not convert centimeters to meters because the values were not being used to calculate other quantities.

119.   **ORGANIZE AND PLAN**   Using the data in problem 117, we see that the starting position in the jumping jack is like (b) in 117, with the exception of the separation of the feet. This is a big exception, because the trigonometry involved also lowers the centers of mass of the other three segments as well!

Known: Angle between leg and vertical $= 60°/2 = 30°$; data from problem 17.

SOLVE   We calculate the new center of mass of the legs as

$$y_{legs} = (47.5 \text{ cm})\cos 30° = 41.1 \text{ cm}$$

This means that the legs only reach to a vertical height of 82.3 cm. This is 12.7cm lower than before, so we reduce the centers of mass of the other three segments by this same amount, with the following result:

	Y,cm
Head	154.8
Torso	112.3
Legs	41.1
Arms	174.8

Using this data for the first position of the jumping jack, we find

(a) $Y_{cm} = \dfrac{(0.069\,M)(154.8\text{ cm}) + (0.460\,M)(112.3\text{ cm}) + (0.346\,M)(41.1\text{ cm}) + (0.125\,M)(174.8\text{ cm})}{M}$

$$Y_{cm} = 98.4 \text{ cm}$$

(b) Then we see that the second position in the jumping jack is exactly like (a) in problem 117, with $Y_{cm} = 100.8$ cm. This second position has a center of mass 2.4 cm higher than the first.

REFLECT   Raising one's arms overhead raises the body's center of mass less than placing one's feet apart 60° lowers it. So we are more stable, with a lower center of mass, with our arms up and feet apart.

121.   **ORGANIZE AND PLAN**   We're asked to derive equations 6.12 and 6.13:

$$v_{1xf} = \left(\frac{m_1 - m_2}{m_1 + m_2}\right) v_{1xi} \qquad (6.12)$$

$$v_{2f} = \left(\frac{2m_1}{m_1 + m_2}\right) v_{1i} \qquad (6.13)$$

We start with equations for conservation of momentum $m_1 v_{1i} + m_2 v_{2i} = m_1 v_{1f} + m_2 v_{2f}$ and conservation of energy, $\frac{1}{2} m_1 v_{1i}^2 + \frac{1}{2} m_2 v_{2i}^2 = \frac{1}{2} m_1 v_{1f}^2 + \frac{1}{2} m_2 v_{2f}^2$. We rearrange, factor out m_1 and m_2 and subtract the equations giving us the result in problem 63. Then we substitute the result, $v_{2f} = v_{1i} + v_{1f} - v_{2i}$ into $m_1\left(v_{1i} - v_{1f}\right) = m_2\left(v_{2f} - v_{2i}\right)$ and solve first for v_{1f}, then for v_{2f}.

SOLVE   From conservation of momentum,

$$m_1 v_{1i} + m_2 v_{2i} = m_1 v_{1f} + m_2 v_{2f}$$

From conservation of energy,

$$\tfrac{1}{2} m_1 v_{1i}^2 + \tfrac{1}{2} m_2 v_{2i}^2 = \tfrac{1}{2} m_1 v_{1f}^2 + \tfrac{1}{2} m_2 v_{2f}^2$$

Rearranging and factoring out m_1 and m_2,

$$m_1\left(v_{1i} - v_{1f}\right) = m_2\left(v_{2f} - v_{2i}\right)$$

$$m_1\left(v_{1i}^2 - v_{1f}^2\right) = m_2\left(v_{2f}^2 - v_{2i}^2\right)$$

Dividing and rearranging, we get

$$v_{1i} - v_{21} = v_{2f} - v_{1f}$$

Solving for v_{2f}, substituting into

$$m_1\left(v_{1i} - v_{1f}\right) = m_2\left(v_{2f} - v_{2i}\right) \text{ and rearranging, we get}$$

$$v_{1f} = \left(\frac{m_1 - m_2}{m_1 + m_2}\right) v_{1i} + \left(\frac{2m_2}{m_1 + m_2}\right) v_{2i}$$

When the target is stationary, this becomes

$$v_{1f} = \left(\frac{m_1 - m_2}{m_1 + m_2}\right) v_{1i}$$

Likewise, by substituting

$$v_{1f} = v_{2f} + v_{2i} - v_{1i} \text{ into}$$

$$m_1\left(v_{1i}-v_{1f}\right)=m_2\left(v_{2f}-v_{2i}\right)$$

We get

$$v_{2f}=\left(\frac{2m_1}{m_1+m_2}\right)$$

when the target is stationary.

REFLECT   The equations we're asked to derive are special cases where the target is stationary. We can use the full version of the equations when the target is moving.

123.   **ORGANIZE AND PLAN**   We're told this is an elastic collision with a stationary target. Since the deuterium target is more massive than the neutron, we know the neutron will recoil in the opposite direction. But here, we don't care about the direction of the recoil, just the speed. When this happens in practice, there are many neutrons moving in many directions, surrounded by deuterium nuclei. We'll use the subscript n for the neutron and D for the deuterium nucleus.

Known: $m_n=1.67\times10^{-27}$ kg ; $m_D=3.34\times10^{-27}$ kg.

SOLVE   (a) From conservation of momentum and energy, we have

$$v_f=\left(\frac{m_n-m_D}{m_n+m_D}\right)v_i=\left(\frac{1.67\times10^{-27}\text{ kg}-3.34\times10^{-27}\text{ kg}}{1.67\times10^{-27}\text{ kg}+3.34\times10^{-27}\text{ kg}}\right)v_i$$

$$v_f=-0.333v_i\text{ or for just the speed,}$$

$$v_f=0.333v_i$$

(b) In order for the neutron to finally have less than 1% of its initial speed,

$$0.01=\left(0.333\right)^n$$

$$n\log\left(0.333\right)=\log\left(0.01\right)$$

$$n=\frac{-2}{\log\left(0.333\right)}=4.19$$

The neutron would have to undergo the next higher integer, or 5 collisions, to reduce its speed to less than 1% of initial.

REFLECT   Slower neutrons are more effective at initialized chain reactions. Enrico Fermi's research group was the first to experiment with this process, using hydrogen-1 nuclei in the water from a fishpond.

7

OSCILLATIONS

CONCEPTUAL QUESTIONS

1. **SOLVE** The motion of a child on a Ferris wheel is periodic because the child repeatedly follows the same trajectory. It is not oscillatory because the child does not go back and forth on the same trajectory.

    REFLECT All oscillatory motion is periodic but not all periodic motion is oscillatory. Oscillatory motion requires the object to move back and forth on the same trajectory. What is interesting is that if you consider the shadow of the child on the ground you would observe oscillatory motion of the shadow if the sun is directly above the Ferris wheel.

3. **SOLVE** Acceleration is zero when the velocity is maximized. In the case of a mass oscillating on a spring, this is when the spring is in its un-stretched position. The velocity is zero when the spring is in it's fully compressed or fully extended position

    REFLECT For one full period, T, if we assume the period begins at $t = 0$ with the mass at rest and a fully extended spring, the velocity is zero at $t = 0$, $t = \dfrac{T}{2}$ and $t = T$. The acceleration is zero when there is no force being exerted by the spring. This occurs at $t = \dfrac{T}{4}$ and at $t = \dfrac{3T}{4}$.

5. **SOLVE** The period of a SHO is given by $T = 2\pi\sqrt{\dfrac{m}{k}}$. If the spring constant k is doubled, the period is decreased by a factor of $\dfrac{1}{\sqrt{2}}$

    REFLECT Increase of the spring constant k implies a stiffer spring (a Slinky has a much lower spring constant than a screen door spring). From experience, stiffer springs will produce faster oscillations than looser springs when the mass is the same for both.

7. **SOLVE** Take two bungee cords, and affix one to one wall and the other to an opposing wall. Affix the other ends of the cords to your utility belt. They should be stretched out. Have a colleague apply a known force from a calibrated device. Measure the displacement for the known force to determine the spring constant, k, of the bungee cord set up. Ask your colleague to push you along the direction of the bungee cord line then let you go. You will oscillate back and forth. Determine the period of oscillation, T. Recall the relationship $T = 2\pi\sqrt{\dfrac{k}{m}}$. Solve for the unknown mass in terms of k and T.

    REFLECT Of course, the bungee cord setup is a little low-tech. The inset on page 152 depicts a more appropriately engineered device called the Body Mass Measurement Device. Even thought it is not called the Bungee Cord Astronaut Mass Assessor the principle is the same.

9. **SOLVE** The net work done by the spring over one complete cycle is zero. When the spring is extended it is applying a force in the direction of the displacement of the mass, hence doing positive work on the mass. When the mass passes the equilibrium extension of the spring the spring applies a force opposite the direction of the displacement of the mass, hence doing negative work on the mass. In other words, the mass is doing work on the spring. After the spring is fully compressed the process starts over.

REFLECT Experience of real masses on springs tells us that the oscillations eventually die out. In that case, the spring will actually do work over each oscillation on the surrounding environment (heating the air and the surface over which the mass is sliding). It is important to recall the assumptions made when describing physical systems. We assume in this question that there are no frictional forces and that all the energy is contained in the mass-spring system.

11.   **SOLVE** If you double the amplitude at small angles the period will not be affected because, in the small amplitude approximation, the period has no dependence upon the amplitude (we note that 2*small is still small). However, as the angle increases, the period begins to exhibit a pronounced dependence upon the amplitude for the following reason:

The small angle approximation comes from the mathematical approximation $\sin\theta \simeq \theta$ for small θ. As the angle increases the true relationship of $\sin\theta < \theta$ becomes more apparent (see figure 7.14). This inequality means that the small angle approximation over-estimates the tangential force on the pendulum at large angles. An over-estimate of the tangential force results in an under-estimate of the period of oscillation. Therefore as the angle increases the period increases (as illustrated by Table 7.2).

For part a, doubling the amplitude from 2° to 4° produces only a very small change in the period (roughly 0.004% change in the period).

For part b, doubling the amplitude from 20° to 40° produces an increase of the period by approximately 2% (obtained via interpolation of values from Table 7.2).

REFLECT The detailed mathematics of the solution for the real pendulum is beyond the scope of this course. However, the small angle approximation illustrated in the pendulum is exemplary for all systems that oscillate close to an equilibrium value (in this case the equilibrium value of the pendulum is when $\theta = 0$). The properties of many systems are described well by this approximation which explains why simple harmonic motion is a ubiquitous inclusion in physics textbooks. By studying the SHO-exemplified by amass on a spring-we learn about many things. This problem illustrates the care that must be taken with such useful approximations.

13.   **SOLVE** The period is proportional to $\sqrt{L}$. If you want to cut the period in half you have to cut the length in one fourth.

REFLECT Proportional reasoning can save you a lot of time. Make sure the answers conform with your intuition (recall Question 10).

15.   **SOLVE** On the way down: During the initial part of the leap, the bungee cord plays no role as the gravitational potential energy of the jumper gets converted into kinetic energy. As the jumper falls and begins to stretch the bungee cord, the kinetic energy as well as the additional gravitational potential energy lost in falling gets converted into the elastic energy of the bungee cord as it extends. When all the gravitational potential energy possessed by the jumper at the beginning of the leap is now stored in a fully extended bungee cord (assuming the jumper just barely touches the water when the cord is extended) the velocity of the jumper stops.

On the way up: The extended bungee cord then accelerates the jumper upward. As the jumper accelerates up, the elastic energy gets converted back to kinetic energy and gravitational potential energy. Assuming no energy losses, once the cord reaches its un-stretched position again, the jumper will have the same speed (opposite velocity) that it had when it began to stretch the cord on the way down. This upward velocity should then result in the jumper reaching the jump point as the kinetic energy is converted to gravitational potential energy.

Repeat above steps

REFLECT Energy losses such as wind resistance and heating of the bungee cord constantly deplete the total energy of the system. This results in the gradual decrease of the amplitude of the oscillation.

MULTIPLE-CHOICE PROBLEMS

17.   **ORGANIZE AND PLAN**   The period T is related to the frequency f by the relationship $T = \dfrac{1}{f}$.

Given information: $f = 200$ Hz

SOLVE   The period is $T = \dfrac{1}{200\text{Hz}} = 0.005\text{Hz}^{-1}$

Choice (c): 0.005 s

REFLECT   The units of Hz are oscillations per second (frequency units) while Hz^{-1} are seconds per oscillations. Recalling the fundamental units for Hz makes it trivial to recall the relationship between T and f.

19.   **ORGANIZE AND PLAN**   We begin by piecing out the times between landmark points in a full oscillation: the definition of a period in terms of the maximum positive displacement, the maximum negative displacement, and zero displacement from $x = 0$ equilibrium point.

The time it takes and oscillator to go from the maximum displacement $x = A$ back to the same position at $x = A$ is one period T. The time it takes to go from

$$x = A \rightarrow -A \text{ is } t_{1/2} = \frac{T}{2}.$$

SOLVE   Because the motion about the $x = 0$ position is symmetrical in time, the time point of the zero displacement lies halfway between the maximum positive displacement and the maximum negative displacement which is

Choice (b): $t = \dfrac{T}{4}$

REFLECT   One can also solve this problem by dealing directly with the position as a function of time. At $t = 0$ the argument in the cosine function is zero corresponding to the maximum displacement. The first time after this when the cosine function is zero corresponds to an argument of $\pi/2$. If we set the argument $\dfrac{2\pi}{T}t = \dfrac{\pi}{2}$ and solve for t we

find the selected answer. In one period this will also occur at $t = \dfrac{3}{4}T$ but that was not an available choice.

21.   **ORGANIZE AND PLAN**   We are asked to find the maximum speed for a SHO given the mass, the spring constant and the amplitude. We can use conservation of energy and reasoning about where the mass is when it attains its maximum velocity to solve this problem.

When fully extended, the velocity of the mass is zero yielding zero kinetic energy. At this point all the energy is in the potential energy of the spring:

$$E = \frac{1}{2}kA^2.$$

When the spring attached to the mass is in its equilibrium position there is no potential energy in the spring and therefore all the energy in the system is in the kinetic energy:

$$K_{x=0} = \frac{1}{2}mv_{max}^2$$

Using the principle of energy conservation and assuming there are only two forms of energy in this system (K and U) the energy when there is no velocity will be the same as the energy when there is no potential energy. This yields:

$$\frac{1}{2}kA^2 = \frac{1}{2}mv_{max}^2$$

Solving for v yields:

$$v_{max} = A\sqrt{k/m}$$

Since we are given A, k and m we can simply plug in the values from here to obtain the maximum velocity.

SOLVE   Substituting in given values:

$$v_{max} = 2\ m\sqrt{\frac{20.0\ N/m}{1.50\ kg}} =$$

Choice (b): 7.30 m/s

REFLECT   The equation derived above is derived in the text (section 7.3). It is always a good idea to go through the relationship and test that the proportionalities jibe with experience. Larger displacements result in larger maximum velocities (check). Stiffer springs result in larger maximum velocities (check). And smaller masses, i.e., masses easier to accelerate, result in larger maximum velocities (check).

23.   **ORGANIZE AND PLAN**   For a SHO with a given period $T = 5.00$ s we are asked to determine the new period if the mass is doubled. In question 4 we determined that the period would increase by a factor of $\sqrt{2}$. So we find the elongated period as follows:

$$T_{2m} = T_m \sqrt{2}$$

SOLVE   Calculating we find:

$$T_{2m} = 5.00\ s\sqrt{2} =$$

Choice (c): 7.07 s

REFLECT   Choices a and b can be dismissed since we know the oscillator will slow down (hence increase the period). Choice d would follow if the period was proportional to the mass which is not the case... It is proportional to the square-root of the mass leaving only choice c by process of elimination.

25.   **ORGANIZE AND PLAN**   The period of oscillation of a mass-spring system is given by the relationship:

$$T = 2\pi\sqrt{\frac{m}{k}}$$

In order increase the period, the ratio m/k must be increased. This can be done by fixing k and increasing m, fixing m and decreasing k, or by increasing or decreasing both in such a way that m/k increases.

SOLVE   Which of the selections are consistent with the fully exhausted list above: a.) increasing the spring constant? assuming mass is fixed... NO b.) increasing the oscillation amplitude? NO amplitude does not affect the period of a SHO. c.) increasing the ratio k/m? NO. m/k must increase. d.) increasing the mass? assuming k is fixed... YES!

Choice (d): Increasing the mass will produce an increased period

REFLECT   Large masses are more resistant to accelerations than small masses under the same force. Since the spring force does not scale with the mass (like it does in the pendulum), the inertial property of mass affects the acceleration.

27.   **ORGANIZE AND PLAN**   The period of a simple pendulum in the small angle limit is given by $T = 2\pi\sqrt{\dfrac{L}{g}}$. We will assume $g = 9.8$ m/s^2 unless otherwise explicitly stated.

SOLVE   Plugging in the given $L = 13.4$ m yields:

$$T = 2\pi\sqrt{\frac{13.4\ m}{9.8\ m/s^2}} =$$

Choice (a): 7.35 s

REFLECT   This is a very tall pendulum, examples of which exists in virtually all science museums. Look for the Foucault pendulum (they represent one of the first demonstrations that the earth is spinning on its axis). Next time you find yourself near one, whip out a pad, paper and your watch and determine the length of the pendulum by measuring the period.

29.   **ORGANIZE AND PLAN**   The relationship for the period of a simple period is given as $T = 2\pi\sqrt{L/g}$ but this expression could also be viewed as an expression for the acceleration due to gravity in terms of the period and length of a pendulum. Manipulating the above equation to isolate g yields:

$$g = \frac{4\pi^2 L}{T^2}$$

We measure the period to be $T = 2.40$ s and the length of the pendulum to be $L = 1.20$ m

SOLVE   Substituting in values yields:

$$g = \frac{4\pi^2 1.20 \text{ m}}{(2.40 \text{ s})^2} = $$

Choice (d): 8.22 m/s^2

REFLECT   The gravitational acceleration on this new planet is less than g_{Earth}. On Earth, this pendulum would have a shorter period ($T_E = 2.2$ s).

PROBLEMS

31.   **ORGANIZE AND PLAN**   To distinguish weather or not a particular motion is periodic you just have to determine if its motion repeats over time.

Oscillatory motion is a subset of periodic motion. Oscillatory motion is periodic motion that traces the same path back and forth over the period of motion.

We obtain the frequency in Hz we need to know the orbital period in seconds. The orbital period of the Earth is 365.25 days.

Converting days to seconds:

$$365.25 \frac{\text{days}}{\text{osc}} * \frac{3600 \text{ s}}{1 \text{ hr}} * \frac{24 \text{ hr}}{\text{day}} = 3.16 \times 10^7 \frac{\text{s}}{\text{osc}}$$

With the period T in hand we obtain the frequency f in Hz: $f = T^{-1}$

SOLVE   Simple orbital motion of the Earth is periodic since the satellite travels around a path that repeats in time. However, it is not oscillatory because the path does not go back and forth along the periodic path (motion is always in one direction around the sun).

The frequency of the orbital motion of the earth around the sun is:

$$f_E = T^{-1} = 3.17 \times 10^{-8} \text{ Hz}$$

REFLECT   Hertz are probably not the most illuminating units for the frequency of the orbital motion (oscillations per year is more natural but you should be able to convert between units). Its also interesting to note that the orbital motion of the earth has many non-oscillatory yet periodic aspects. For example, the rotational axis precesses (like a spinning top) with a period of approximately 26,000 years.

33.   **ORGANIZE AND PLAN**   The frequency is given in units of beats per minute (160 bpm). To convert this frequency to a period between beats in units of seconds we convert bpm to beats per second (aka oscillations per seconds or Hz). The converting bpm yields:

$$1 \frac{\text{beat}}{\text{min}} \rightarrow \frac{\text{beats}}{\text{min}} * \frac{\text{min}}{60 \text{ sec}} = 1.67 \times 10^{-2} \text{ Hz}$$

Consequently, $f = 160$ bpm $= 160 * 1.67 \times 10^{-2} Hz = 2.67$ Hz

The period of one beat is then $T = f^{-1}$.

SOLVE   Substituting the frequency value into the expression for the period:

$$T = 1/2 .67 \text{ Hz} = 0.374 \text{ s}$$

REFLECT   The time between beats is then 0.374 seconds. For comparison, the time between beats for a typical march (like the star spangled banner) is 0.5 seconds. Presto is quite a fast tempo.

35.   ORGANIZE AND PLAN   Given the frequency we determine the period by inverting the frequency. The angular frequency ω (in units of radians per second) is obtained via a unit transformation of the frequency in Hz.

$$\omega = 2\pi \left[rad/osc \right] * f \left[osc/s \right] = 2\pi f \left[rad/sec \right]$$

SOLVE   The period is:

$$T = \frac{1}{16,000 \text{ Hz}} = 6.25 \times 10^{-5} \text{ s}$$

The angular frequency is:

$$\omega = 2\pi * 16 \text{ kHz} = 10^5 \text{ rad/s}$$

REFLECT   The turbines exhibit periodic yet not oscillatory motion. Basic skills absolutely required for dealing with periodic motion is a skill that is getting a lot of practice in these problems; namely the ability to swim freely between frequency, period and angular frequency.

37.   ORGANIZE AND PLAN   If we approach a problem (attempting to obtain a particular functional relationship between the variables) for example how the period relates to other variables in the system, the one thing that must always be true is that the units have to be consistent.

SOLVE   We begin by listing all the reasonably possible variables in the problem. In the case of the mass on a spring the following variables could possibly be important (neglecting air resistance and friction): mass (m)[kg], spring constant (k)[kg/s^2], amplitude of oscillation (A)[m] (units are shown in brackets[]).

However those variables are combined in an expression for the period, they have to yield units of time. The only variable that possesses a unit of time is the spring constant k. However, the seconds are squared and they are in the denominator. Consequently, the period must be proportional to $\sqrt{k^{-1}}$. This produces units of $\left[\dfrac{s}{\sqrt{kg}} \right]$. We need to get rid of the $\sqrt{1/kg}$ units to produce a pure time unit. The only variable in the problem that possesses kg units is the mass. We can multiply $\sqrt{k^{-1}}$ by $\sqrt{m}$ to obtain units of seconds.

Therefore, the only way to combine the variables to get units of seconds is to make

$$T \propto \sqrt{\frac{m}{k}}$$

REFLECT   What is truly remarkable about dimensional analysis (DA) is that dimensions are the only thing considered. In this chapter, the authors have used principle based deduction of relationships between variables, carefully reasoned approximations and derived models that have been born out by experimental results. Dimensional analysis is much more straightforward and you get a lot out of a little work but what you don't get out of DA is the dimensionless, sometimes very important, aspects of the problem. One example is the 2π that turns the proportionality in this problem into an equal sign. Another more interesting example is the small angle approximation in the simple pendulum. The angle measured in radians is effectively dimensionless. That is why we miss the dependence upon the angular amplitude in the DA of the pendulum case.

39.   ORGANIZE AND PLAN   We are given the mass (in grams) and the frequency in seconds and asked to find the spring constant in a mass-spring system.

We will first convert the frequency to period: $T = f^{-1}$

The relationship between the mass, period and the spring constant is:

$$T = 2\pi \sqrt{\frac{m}{k}}$$

Isolating this equation for *k* yields:

$$k = 4\pi^2 \frac{m}{T^2}$$

We will convert the mass of the spider from grams to kg:

$$m = 1.4 \text{ } g = 1.4 \times 10^{-3} \text{ kg}$$

The period of oscillation is:

$$T = \frac{1}{1.1 \text{ Hz}} = 0.909 \text{ s}$$

SOLVE   Inserting values given in the problem yields a spring constant of the web:

$$k = 4\pi^2 \frac{1.4 \times 10^{-3} \text{ kg}}{(0.909 \text{ s})^2} = 0.067 \text{ N/m}$$

REFLECT   We can check the answer by recalculating the frequency using the derived spring constant and mass of the spider. This spring constant for a slinky is approximately 1 N/m. So, the spiderweb has a spring constant 7% of a slinky.

41.   **ORGANIZE AND PLAN**   Given the bungee jumper follows the equation of motion:

$$x(t) = 5 \text{ m} \cos(\omega t)$$

we can find the position at any time as long as we know ω.

In the previous problem we are given the frequency $f = 0.125$ Hz. The relationship between the angular frequency has been derived above and in the text:

$$\omega = 2\pi f$$

So, in this problem $\omega = 2\pi 0.125$ rad/sec $= 0.785$ rad/sec

SOLVE   Plugging in various values of time:

$$x(0.25 \text{ s}) = 5 \text{ m} \cos\ 0.785 \text{ rad/s}\ 0.25 \text{ s} = 4.9 \text{ m}$$
$$x(0.50 \text{ s}) = 5 \text{ m} \cos\ 0.785 \text{ rad/s}\ 0.50 \text{ s} = 4.6 \text{ m}$$
$$x(1.0 \text{ s}) = 5 \text{ m} \cos\ 0.785 \text{ rad/s}\ 1.0 \text{ s} = 3.5 \text{ m}$$

REFLECT   With a frequency of oscillation of 0.125 Hz and corresponding period of 8 s there is not much action happening in the first second of motion. Referring to the figure in problem 40 one second and less barely takes us out of the first hump in the oscillation. Our numerical answers jibe with the graphical results.

43.   **ORGANIZE AND PLAN**   Given the period and the mass of a mass-spring system we can obtain the spring constant from manipulating the relationship:

$$T = 2\pi\sqrt{\frac{m}{k}}$$

Solving for the spring constant:

$$k = 4\pi^2 \frac{m}{T^2}$$

SOLVE   Plugging in values:

$$k = 4\pi^2 \frac{m}{T^2} = 4\pi^2 \frac{0.975 \text{ kg}}{0.25 \text{ s}^2} = 154 \text{ N/m}$$

Reflect   In cases where you have no intuition for the final number one way to check the answer is to assume you are given the result you obtained and recalculate one of the originally given values. Calculating the period using the values of m and k verifies the consistency of the result.

45.   **ORGANIZE AND PLAN**   The period of an SHO is proportional to the square-root of the mass:

$$T \propto \sqrt{m}$$

Let T_0 and m_0 be the period and mass of the mass spring system and let T_1 and m_1 be the period and mass of the system with Δm added.

With the same proportionality constants in each case, we obtain

$$\frac{T_0}{T_1} = \sqrt{\frac{m_0}{m_0}}$$

Isolating f m_1 yields:

$$m_1 = m_0 \frac{T_1^2}{T_0^2}$$

Finally, $\Delta m = m_1 - m_0 = m_0 \left(\frac{T_1^2}{T_0^2} - 1 \right)$

We note that $T_1 = 1.2 * T_0$

SOLVE Plugging in known values:

$$\Delta m = 0.200 \text{ kg} \left(\frac{1.2^2}{1} - 1 \right) = 0.088 \text{ kg}$$

REFLECT The fractional increase of mass required to increase the period by $x\%$ is $\left(1 + \frac{x}{100} \right)^2 - 1$. In this case,

44% increase in mass produces a 20% change in the period. A larger fractional increase in the mass is required because of the square root of the mass is proportional to the first power of period. If the period were proportional to the mass the fractional increase would be identical.

47. **ORGANIZE AND PLAN** We need to plot $x(t) = A \cos \left(\frac{2\pi t}{T} \right)$. For this problem $A = 0.50$ m and $T = 0.720$ s

SOLVE See figure below.

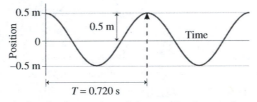

REFLECT The period sets the scale for the time axis and the amplitude sets the scale for the y axis.

49. **ORGANIZE AND PLAN** We plot the given function (shown in figure below)
Simple harmonic motion is characterized by sinusoidal position as a function of time. To determine if the given function represents a simple harmonic oscillator we simply need to determine if it is sinusoidal.
SOLVE

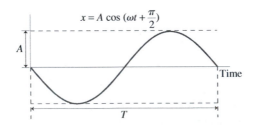

The position of a simple harmonic oscillator as a function of time is sinusoidal. Where this begins on the time axis is only a matter of when you started looking at it. In this problem we start looking when the displacement is zero (since $\cos \frac{\pi}{2} = 0$). In the next instant the position is negative so the oscillator must have had a velocity in the negative direction.

REFLECT Choice of initial condition (or phase angle) of a SHO is irrelevant to the general features of the oscillator.

51. **ORGANIZE AND PLAN** This problem is very similar to problem 48. In this case, however, we are given the mass and period of oscillation and intended to find the displacement d, where d is the gravitationally induced stretch from zero gravity equilibrium position.

When the mass hangs down from the spring, the spring exerts a upward force on the mass equal in magnitude to $\vec{F}_s = kx\hat{j}$. When the mass-spring system is in equilibrium the total force on the mass is zero. For that to occur the force exerted by the spring has to be equal in magnitude but opposite in direction to the gravitational force $\vec{F}_g = -mg\hat{j}$. From this we obtain:

$$k = \frac{mg}{d}$$

Since the spring force is linear in the displacement, the spring constant of the oscillation about the equilibrium point in the gravitational field is the same as the spring constant in the absence of the gravitational field. With $k(d)$ and the mass in hand we obtain the relationship for the period: $T = 2\pi\sqrt{m/k}$

Isolating for k:

$$k = 4\pi^2 \frac{m}{T^2} = \frac{mg}{d}$$

Finally, solving for d yields:

$$d = \frac{gT^2}{4\pi^2}$$

SOLVE   Plugging in values:

$$d = \frac{9.8 \text{ m/s}^2 * (1.04 \text{ s})^2}{4\pi^2} = 0.268 \text{ m}$$

REFLECT   As always we rely on units to ensure that our calculations are not obviously off track. This diligence pays off more as the calculations become more complicated and therefore more error prone.

53.   **ORGANIZE AND PLAN**   If the blocks move together as one mass, the mass spring system simply consists of the sum of the two masses and the spring connected to the bottom mass.

The period is simply:

$$T = 2\pi\sqrt{\frac{m_1 + m_2}{k}}$$

If the coefficient of static friction between the masses is μ, then the maximum frictional force applied to the top block is $\mu * F_N$ (where $F_N = m_2 g$). This force is the maximum force that may be used to accelerate the top block without slipping. The maximum acceleration of the mass-spring system is given by the amplitude of the sinusoidal expression for the acceleration given in equation 7.11:

$$a_{max} = \frac{kA}{m_{total}}$$

This maximum acceleration must be equal to the acceleration of the top block. Using Newtons second law the maximum acceleration that can be applied by the frictional force without slipping is $a_{f\,max} = \mu g$

Equating the maximum acceleration allowed by the frictional force to the acceleration of the mass-spring system yields:

$$\frac{kA_{max}}{m_{total}} = \mu g$$

Solving for A_{max}:

$$A_{max} = \frac{\mu g m_{total}}{k}$$

SOLVE   The value for the oscillation period:

$$T = 2\pi\sqrt{\frac{0.34 \text{ kg}}{25 \text{ N/m}}} = 0.732 \text{ s}$$

Inserting values to obtain A_{max}:

$$A_{max} = \frac{0.14 * 9.8 \text{ m/s}^2 * 0.34 \text{ kg}}{25 \text{ N/m}} = 0.018 \text{ m}$$

REFLECT   For smaller amplitudes, the masses will oscillate as one connected unit. For amplitudes larger than this, the larger block will oscillate out of phase with the bottom block and the properties of the mass-spring system will change. In the zero friction limit the mass in the mass-spring system is just the bottom one. Consequently, an "effective mass" of the mass spring system must decrease when the top block is slipping and therefore the oscillating frequency will increase. This problem becomes very interesting when the top block starts slipping.

55.   **ORGANIZE AND PLAN**   If we assume the oscillator is released from rest at $t = 0$ the oscillator begins with no kinetic energy and all the energy of the system is in potential energy. As the potential energy gets converted to kinetic energy U decreases and K increases. We must find the point in time at which the two values meet in the middle.

SOLVE   On the graph below we plot the potential and kinetic energy as a function of time (in addition to the position plot shown to illustrate the potential confusion in what a period is).

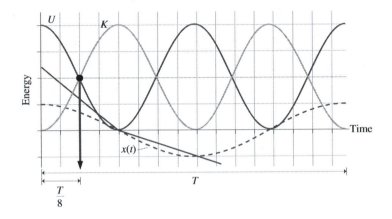

The time between release of the oscillator and the balance between U and K is $\frac{T}{8}$.

REFLECT   It is crucial in this problem to realize that the period of the oscillation is defined as a full cycling of the position. A common mistake is to look at the energy functions and assign the period based on a full cycling of the potential. This would result is a shorter period determination by a factor of 1/2.

57.   **ORGANIZE AND PLAN**   The maximum speed of an oscillator is $v_m = A\omega$ where A is the amplitude and ω is the angular frequency.
The maximum acceleration of an oscillator is $a_m = A\omega^2$.
Both v_m and a_m are given in the problem. We can solve both expressions for ω then equate to obtain:

$$\frac{v_m}{A} = \sqrt{\frac{a_m}{A}}$$

Further manipulation yields:

$$A = \frac{v_m^2}{a_m}$$

SOLVE   Plugging in values:

$$A = \frac{\left(0.95 \text{ m/s}\right)^2}{1.56 \text{ m/s}^2} = 0.579 \text{ m}$$

REFLECT   It is impossible to stress to the college student how important units are especially when the answers cannot be cross-referenced with experience. In addition one can always do consistency checks to rule out calculation errors.

59.   **ORGANIZE AND PLAN**   We are given the following values for a mass-spring system: k, m and A.

The total mechanical energy of the oscillator is: $E = \dfrac{1}{2}kA^2$

When the displacement x_0 is given we can use the conservation of energy and the given values of k, m and A to obtain the velocity.

$$E = \frac{1}{2}mv^2 + \frac{1}{2}kx_0^2$$

Solving for v:

$$v = \pm\sqrt{\frac{2E}{m} - \omega^2 x_0^2}$$

The fraction of total energy that is kinetic energy at a given displacement is 1 minus the fraction of the total energy that exists as potential energy.

$$F_k = 1 - \frac{\dfrac{1}{2}kx_0^2}{\dfrac{1}{2}kA^2} = 1 - \frac{x_0^2}{A^2}$$

SOLVE   Plugging in values into derived equations:

The total mechanical energy: $E = \dfrac{1}{2}\,94.0 \text{ N/m} * (0.560 \text{ m})^2 = 14.7 \text{ J}$

The speed S when $x = -0.210$ m

$$S = |v| = \sqrt{\frac{2E}{m} - \omega^2 x_0^2} = \sqrt{\frac{2 * 14.7 \text{ J}}{1.06 \text{ kg}} - \left(\frac{94.0 \text{ N/m}}{1.06 \text{ kg}}\right) * (0.210 \text{ m})^2} = 4.88 \text{ m/s}$$

The fraction of the total energy partitioned into the kinetic energy is:

$$F_k = 1 - \frac{(0.500 \text{ m})^2}{(0.560 \text{ m})^2} = 0.203$$

REFLECT   Contrast the different methods of solution used to answer the similar problems 58 and 59. In problem 58 we solved the problem without using conservation of energy. We could have solved each problem in either way. The the principle of conservation of energy is often more straight forward.

61.   **ORGANIZE AND PLAN**   The shortest duration of time between maximum velocity and maximum acceleration is $\dfrac{T}{4}$. This time is given as 2.50 s.

With the period in hand and the given value of the spring constant k we obtain the mass using the relation:

$$T = 2\pi\sqrt{\frac{m}{k}}$$

Isolating for m yields:

$$m = \frac{kT^2}{4\pi^2}$$

The maximum velocity is given by: $v_m = A\sqrt{\dfrac{k}{m}}$

The maximum acceleration is given by: $a_m = A\dfrac{k}{m}$

SOLVE

The period $T = 4 * 2.5 \text{ s} = 10 \text{ s}$

The mass is $m = \dfrac{4.0 \text{ N/m} * (10 \text{ s})^2}{4\pi^2} = 10.1 \text{ kg}$

The maximum velocity $v_m = 0.75 \text{ m} * \sqrt{\dfrac{4.0 \text{ N/m}}{10.1 \text{ kg}}} = 0.47 \text{ m/s}$

The maximum acceleration is given by: $a_m = 0.75 \text{ m} * \dfrac{4.0 \text{ N/m}}{10.1 \text{ kg}} = 0.30 \text{ m/s}^2$

REFLECT   This is a very slow oscillation. The given spring constant is comparable to a slinky (which is 1N/m) and a mass which is comparable to about 2.5 gallons of water this slow oscillation makes sense.

63.   **ORGANIZE AND PLAN**   We are given the mass, the spring constant and the speed at a given displacement. The given information allows us to determine the total energy of the mass-spring system directly:

$$E = \frac{1}{2}kx^2 + \frac{1}{2}mv^2$$

The total energy and the spring constant allows us to obtain the amplitude: $E = \dfrac{1}{2}kA^2$. Isolating for A yields:

$$A = \sqrt{2E/k}$$

With the total energy and the position x_0 at an instant we can find the speed at that instant using conservation of energy:

$$v = \sqrt{\frac{2E}{m} - \frac{k}{m}x_0^2}$$

SOLVE   Plugging in values:

The total energy is

$$E = \frac{1}{2}14 \text{ N/m} * (0.22 \text{ m})^2 + \frac{1}{2}0.60 \text{ kg} * (0.95 \text{ m/s})^2 = 1.63 \text{ J}$$

The amplitude of oscillation is $A = \sqrt{2*1.63 \text{ J}/14 \text{ N/m}} = 0.483 \text{ m}$

The oscillators speed at $x = 0.22$ m is:

$$v = \sqrt{\frac{2*1.63 \text{ J}}{0.60 \text{ kg}} - \frac{14 \text{ N/m}}{0.60 \text{ kg}}(0.11 \text{ m})^2} = 2.27 \text{ m/s}$$

REFLECT   The maximum velocity is 2.33 m/s. At x = 0.11 m the mass is less than 1/4 a full amplitude away from the equilibrium position and therefore near to its maximum velocity which is consistent with the final result.

65.   **ORGANIZE AND PLAN**   One full oscillation of the drive produces one rotation of a wheel with diameter D resulting in a πD displacement of the train. Said another way the distance traveled by the train is πDm/osc. A wheel spinning at a frequency of f with a diameter D will therefore produce a velocity of $v = \pi D f$. In this case we are given the required velocity and asked to find the frequency. Isolating for f yields:

$$f = \frac{v}{\pi D}$$

SOLVE   The required frequency of the drive shaft is: $f = \dfrac{26 \text{ m/s}}{\pi 1.42 \text{ m}} = 5.82 \text{ Hz}.$

REFLECT   If you think about the sound a steam locomotive makes the clicking chug a chug of the wheels is consistent with the result. Unit analysis techniques are a tremendous help in constructing solutions to questions not just checking answers. In the above problem we went from m/osc, to osc/s, to m/s to get to the desired expression.

67. **ORGANIZE AND PLAN**   Rotations per minute can be converted into Hz by a simple unit conversion:

$$N \frac{\text{rotatons}}{\text{min}} \frac{1\,\text{min}}{60\text{s}} = \frac{N}{60}\,\text{Hz}$$

Angular frequency in terms of oscillation frequency is $\omega = 2\pi f$.

SOLVE   Converting 600 rpm to Hz yields: $f = 10$ Hz.

Angular frequency is $\omega = 2\pi 10\,\text{Hz} = 62.8\,s^{-1}$

REFLECT   Another example of the power of unit conversion. Convinced that unit analysis is important yet?

69. **ORGANIZE AND PLAN**   The period of a simple pendulum in the small angle limit is $T = 2\pi\sqrt{\dfrac{L}{g}}$. Isolating for L

yields:

$$L = \frac{gT^2}{4\pi^2}$$

The period of the pendulum does not depend upon the mass since the gravitational force is proportional to the mass.

SOLVE   The length of the ropes supporting the swing $L = \dfrac{9.8\ \text{m/s}^2 * (4.1\ \text{s})^2}{4\pi^2} = 4.17$ m.

The period is unchanged when her brother gets on the swing

REFLECT   This length corresponds to a average size park swing. Experience is consistent the result. Imagine pushing a child on a swing, you have to push about every 4 seconds.

71. **ORGANIZE AND PLAN**   If there are N oscillations over a duration Δt the period is time per oscillation or simply $T = \dfrac{\Delta t}{N}$.

Given the period, we obtain the length by assuming $g = 9.8$ m/s^2 inverting the relationship for the period yielding:

$$L = \frac{gT^2}{4\pi^2}$$

SOLVE   Plugging in values: The period is $T = \dfrac{32\ \text{s}}{25\ \text{osc}} = 1.28$ s

The length corresponding to this period is $L = \dfrac{9.8\ \text{m/s}^2 * (1.28\ \text{s})^2}{4\pi^2} = 0.406$ s

REFLECT   The roughly half a meter long pendulum with the 1.28 s period is consistent with experience.

73. **ORGANIZE AND PLAN**   The period of the pendulum with length $L = 1.50$ m yields a period of

$$T = 2\pi\sqrt{\frac{L}{g}} = 2\pi\sqrt{\frac{L}{g}} = 2.46\ \text{s}$$

The angle as a function of time is given in equation 7.13: $\theta = \theta_{max} \cos(\omega t)$

We recall that $\omega = 2\pi/T$ leaving the equation above $\theta = \theta_{max} \cos\left(\dfrac{2\pi t}{T}\right)$.

From $t = 0$ to $t = 10$ s there are $N = \dfrac{10\ \text{s}}{2.46} = 4.07$ periods.

SOLVE   We plot the angle as a function of time below. N.b. the x axis is in units of t/T so $\dfrac{t}{T} = 1$ corresponds to one period.

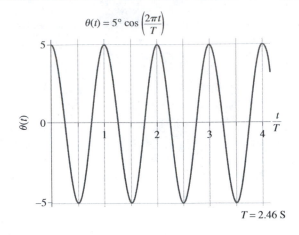

$$\theta(t) = 5° \cos\left(\frac{2\pi t}{T}\right)$$

$$T = 2.46 \text{ S}$$

REFLECT   When plotting functions it is often useful to modify the scale to fit the natural scales in the problem. The natural time scale of an oscillator is the period of oscillation.

75.   **ORGANIZE AND PLAN**   The number of oscillations of a pendulum per day is $N_0 = \frac{sec}{day} * f_0$ where f_0 is the frequency in Hz of the optimally tuned timekeeping pendulum. The fractional error in timekeeping when the pendulum has decreased its frequency to f_1 due to thermal expansion of the pendulum's length is

$$\eta = 1 - \frac{N_1}{N_0}.$$

In terms of the frequencies the fractional error in timekeeping is:

$$\eta = 1 - \frac{f_1}{f_0}$$

The challenge is finding f_1 in terms of the fractional increase in length of the pendulum. We are given the fractional increase of the length $\lambda = 5.0 \times 10^{-5}$.

The frequency is given by $f_1 = 2\pi \sqrt{\frac{L_1}{g}}$. If L_0 is increased by a fractional amount λ then $L_1 = (1 + \lambda)L_0$. Therefore,

$$f_1 = f_0 \sqrt{(1 + \lambda)}$$

It follows then that the fractional error in timekeeping is:

$$\eta = 1 - \sqrt{(1 + \lambda)}$$

The number of seconds lost per day is then:

$$\frac{sec}{day}\eta$$

Preliminary calculation: The number of seconds per day is: 3600 sec/hour* 24 hr/day = 8.64×10^4 sec/day

SOLVE   The fractional error in timekeeping is $\eta = 1 - \sqrt{(1 + 5.0 \times 10^{-5})} = -2.5 \times 10^{-5}$. This yields a loss of 2.16 s/day, roughly 1 minute every 28 days or about 13.1 minutes per year

REFLECT   The fractional change given in the problem corresponds to a change in length of 5 hundredths of a millimeter for a pendulum length of one meter. Thermal fluctuations notwithstanding, it would be challenging to machine a pendulum to this degree of precision in length. Such engineering realities necessitate fine scale calibration.

77. **ORGANIZE AND PLAN** The small amplitude period of a pendulum is $T = 2\pi\sqrt{\dfrac{L}{g}}$

The difference between periods of a pendulum with g_0 and g_1 is:

$$\Delta T = 2\pi\sqrt{L}\left(\frac{1}{\sqrt{g_0}} - \frac{1}{\sqrt{g_1}}\right)$$

SOLVE Plugging in values:

$$\Delta T = 2\pi\sqrt{2.00\ \text{m}}\left(\frac{1}{\sqrt{9.78\ \text{m/s}^2}} - \frac{1}{\sqrt{9.83\ \text{m/s}^2}}\right) = 0.00723\ \text{s}$$

REFLECT This change in time is about 0.25% of the period of the oscillator. To put this into perspective in the time it takes for one to make 393 oscillations the other will have made approximately 394.

79. **ORGANIZE AND PLAN** Criteria for light, critical or heavy damping is shown on table 7.3. The values that must be compared are b^2 and 4 mk.
For the given values $b^2 = 7.02\ \text{kg}^2/\text{s}^2$ and 4 mk = 480 kg^2/s^2
SOLVE From calculated values we note that $b^2 < 4$ mk which corresponds to the lightly damped solution.
REFLECT If set into motion this oscillator will oscillate back and forth around the equilibrium value with ever decreasing amplitude.

81. **ORGANIZE AND PLAN** Criteria for light, critical or heavy damping is shown on table 7.3. The values that must be compared are b^2 and 4 mk.
For the given values $b^2 = 66.4\ \text{kg}^2/\text{s}^2$ and 4 mk = 1890 kg^2/s^2
The oscillation period of the damped motion is obtained as $T_{damped} = 2\pi/\omega_{damped}$ where

$$\omega_{damped} = \sqrt{\frac{k}{m} - \frac{b^2}{4\ m^2}} = \sqrt{\frac{150\ \text{N/m}}{3.15\ \text{kg}} - \frac{66.4\ \text{kg}^2/\text{s}^2}{4(3.15\ \text{kg})^2}} = 6.78\ \text{s}^{-1}$$

The oscillation period of the un-damped motion is obtained by considering the $b = 0$ case $T_{undamped} = 2\pi\sqrt{m/k} = 0.911\ \text{s}$

The attenuation function for the under-damped motion is $e^{\frac{-br}{2m}}$. We need to find the time when the attenuation function equals 1/2 ($t_{1/2}$):

$$\frac{1}{2} = e^{\frac{-bt_{1/2}}{2\ m}}$$

Solving for $t_{1/2}$ yields:

$$t_{1/2} = \frac{2\ m}{b}\ln 2 \text{ ,,}$$

SOLVE The motion is lightly damped because $b^2 < 4$ mk:

$$66.4\ \text{kg}^2/\text{s}^2 < 1890\ \text{kg}^2/\text{s}^2$$

The damped oscillation period $T_{damped} = \dfrac{2\pi}{\omega_{damped}} = \dfrac{2\pi}{6.78\ \text{s}^{-1}} = 0.926\ \text{s}$

The damped oscillation period is 1.7% longer than the undamped oscillation period of 0.911s
The "half-life" of the oscillation is

$$t_{1/2} = \frac{2\ m}{b}\ln 2 = \frac{2*3.15\ \text{kg}}{8.15\ \text{kg/s}}\ln 2 = 0.536\ \text{s}$$

Therefore the amplitude will decrease by half in roughly 1/2 a period.
REFLECT Even when the system is well in the lightly damped regime such that the period is only affected by a small percentage the damping is still significant enough to produce a relatively rapid attenuation of the oscillations.

83.   **ORGANIZE AND PLAN**   We note the given information $t_{1/2} = 12T$. Like the calculation in problem 81 the "half-life" expression in terms of m and b is:

$$t_{1/2} = \frac{2\ m}{b}\ln 2.$$

Solving for b and inserting the given information we find:

$$b = \frac{2\ m}{12 * T}\ln 2$$

We recall the result for the small fractional change in the period in the case when the damping parameter is strong enough to decrease the amplitude of the oscillation by half over half a period. If the damping constant is such that it allows 12 full oscillations before diminishing the amplitude by half we can conclude that the change in the period from the undamped case $\left(T = 2\pi\sqrt{\dfrac{m}{k}} \right)$ is negligible (we can check this for consistency after we obtain the solution). This allows us to plug in the undamped value for the period into the above equation leaving a expression for b in terms of given values m and k:

$$b = \frac{2\sqrt{m * k}}{24\pi}\ln 2$$

SOLVE   Plugging in values:

$$b = \frac{2\sqrt{0.50\ \text{kg} * 12\ \text{N/m}}}{24\pi}\ln 2 = 0.065\ \text{kg/s}$$

REFLECT   The critical simplifying assumption used above is that $T_{\text{damped}} = T_{\text{undamped}}$. If we plug the value obtained for the damping constant we obtain a percent difference between the damped and undamped values is approximately 0.01%, indeed negligible.

As usual, we depend upon the units to ensure that we are not on an obviously wrong track.

85.   **ORGANIZE AND PLAN**   We are given the mass, initial amplitude and the period of the mass-spring system. We proceed in finding the spring constant by assuming that the period is smaller than the half-life and therefore the difference between the damped period and the undamped period is negligible. In this case the spring constant k is given by:

$$k = \frac{4\pi^2 m}{T^2}$$

The expression derived for the "half-life" in problem 82:

$$t_{1/2} = \frac{2\ m}{b}\ln 2.$$

SOLVE   Plugging in values:
The spring constant is

$$k = \frac{4\pi^2 70\ \text{kg}}{\left(3.75\ \text{s} \right)^2} = 197\ \text{N/m}$$

The time it takes for the amplitude to decrease by 1/2 if

$$t_{1/2} = \frac{2 * 70\ \text{kg}}{2.5\ \text{k/s}}\ln 2 = 38.88\ \text{s}$$

REFLECT   Now we need to go back and check that our assumptions produce self-consistent results. The value for $t_{1/2}$ is roughly 10 periods which justifies the approximation that the undamped period is equal to the damped period (as seen in problem 83).

87. **ORGANIZE AND PLAN**   The peak force exerted by a spring on a mass in a SHO is attained when the displacement from equilibrium is at its maximum:

$$F_m = -kA$$

If we know the peak force and the amplitude of the oscillation we can deduce the spring constant in a straight forward manner. $k = \dfrac{F_m}{A}$

Given a known oscillation frequency we can deduce the effective mass as follows:

$$2\pi f = \frac{k}{m}$$

Isolating for mass:

$$m = \frac{k}{4\pi^2 f^2}$$

The wrinkle in this problem is the units. The unit pN $= 10^{-12}$ N while the unit $nm = 10^{-9}$ m.

SOLVE   Plugging in numbers with mks units:

The inferred spring constant is $k = \dfrac{10^{-12} \text{ N}}{15 \times 10^{-9} \text{ m}} = 6.7 \times 10^{-5}$ N/m

The effective mass is

$$m = \frac{6.7 \times 10^{-5} \text{ N/m}}{4\pi^2 (70 \text{ Hz})^2} = 3.4 \times 10^{-10} \text{ kg}$$

REFLECT   The mass of a single hydrogen atom is roughly 2×10^{-27} kg so the effective mass is well beyond the single atom limit. The effective mass is equivalent to approximately 10^{17} hydrogen atoms.

89. **ORGANIZE AND PLAN**   This problem is different in set up from problem 88 but identical in result.

Using the same strategy from problem 88. The potential energy can be written as $U = \dfrac{1}{2} k_{eff} x^2$. In the case of the two identical springs the equilibrium position of the combined system is the same as the equilibrium position would be for a single spring. Therefore, if the mass of the combined system is displaced an amount x the energy is simply the energy stored in a single spring times two. Consequently, like problem 88, $k_{eff} = 2 * k_{single}$ from which follows:

$$T = 2\pi \sqrt{\frac{m}{k_{eff}}} = 2\pi \sqrt{\frac{m}{k}} * \frac{1}{\sqrt{2}} = \frac{1}{\sqrt{2}} T_{single}$$

SOLVE   The oscillation period is $T = 2\pi \sqrt{\dfrac{0.25 \text{ kg}}{2 * 16 \text{ N/m}}} = 0.555$ s

The oscillation period of the mass on a single spring is

$$T_{single} = \sqrt{2} * T = \sqrt{2} * 0.555 \text{ s} = 0.785 \text{ s}$$

REFLECT   As a challenging exercise, calculate the effective spring constant if the springs have different k values and compare to the similar suggested exercise at the end of problem 88.

91. **ORGANIZE AND PLAN**   We first note that twice the oscillation frequency is equivalent to twice the angular frequency since the angular frequency is proportional to the oscillation frequency. Let system 1 have the angular frequency $\omega 1$ and system 2 have the angular frequency $\omega 2$. The problem statement gives $2\omega 1 = \omega 2$

The maximum acceleration is given as an amplitude in equation 7.11: $a_{max} = A\omega^2$

Since both oscillators have the same amplitude we note that $a_{1max} = A\omega_1^2$ and $a_{2max} = A\omega_2^2$.

SOLVE   The frequency of the second system is twice that of the first.

The maximum value of the acceleration of the second system is 4 times that of the first since $a_{2\,max} = A\,\omega_2^2 = A4\,\omega_1^2$

REFLECT   One must resist the temptation to assume everything is directly proportional to everything else. Intuition is built upon the mathematical relationships, let the math guide you and your brain will eventually catch up.

93.   **ORGANIZE AND PLAN**   We address this problem by looking at the acceleration of the mass-spring interface as was done in problem 54. If at any point in the motion of the mass-spring system the acceleration is downward and greater than the acceleration due to gravity the mass will lose contact with the spring. The maximum downward acceleration applied by the spring occurs when the mass is at its apex of motion (just stopped and being accelerated downward). This maximum acceleration is given as the amplitude in equation 7.11: $a_m = A\omega^2$. When $a_m = g$ the system is in a state that results in the mass exerting no force on the spring at the apex of motion. The begins to

occur if the amplitude is increased to $A_c = \dfrac{g}{\omega^2} = \dfrac{gm}{k}$. For values of $A \le A_c$ the mass stays connected to the spring.

SOLVE   The maximum amplitude allowed that keeps the mass in contact with the spring is

$$A_c = \frac{9.8 \text{ m/s}^2 * 0.50 \text{ kg}}{34.0 \text{ N/m}} = 0.144 \text{ m}$$

If you exceed this amplitude just slightly the mass looses the force applied by the spring at the apex of its motion

REFLECT   If the amplitude is greater than A_c the bottom will drop out before the spring reaches its maximum extent. Try this in lab by exciting a mass hooked to a spring at its resonance frequency (causing the amplitude to increase with time).

95.   **ORGANIZE AND PLAN**   We shall assume the atom executes simple harmonic motion. Under this assumption the maximum speed and maximum acceleration are given respectively: $v_m = A\omega$ and $a_m = A\omega^2$. In the problem we are given the oscillation frequency f in THz = 10^{12} Hz. Converting to angular frequency $\omega = 2\pi f$, so, in terms of oscillation frequency: $v_m = A2\pi f$ and $a_m = A4\pi^2 f^2$

We find the energy by noting that total energy is equal to maximum kinetic energy: $E = \dfrac{1}{2}mv_m^2$.

If the atom is a carbon atom (isotope 12) it has a mass of $M_C = 0.012$ kg/mol or

$$m_c = 0.012 \text{ kg/mol} * \frac{1 \text{ mol}}{6.022 \times 10^{23} \text{ atoms}} = 2 \times 10^{-26} \text{ kg/atom}$$

SOLVE   The atoms maximum speed:

$$v_m = 10 \times 10^{-12} \text{ m} * 2 * \pi * 12 \times 10^{12} \text{ Hz} = 750 \text{ m/s}$$

The atoms maximum velocity:

$$a_m = 10 \times 10^{-12} \text{ m} * 4 * \pi^2 * (12 \times 10^{12} \text{ Hz})^2 = 5.6 \times 10^{16} \text{ m/s}^2$$

The energy of the oscillator is:

$$E = \frac{1}{2}mv_m^2 = \frac{1}{2} 2 \times 10^{-26} \text{ kg/atom} * (750 \text{ m/s})^2 = 5.6 \times 10^{-21} \text{ J}$$

REFLECT   Another example of how units must be payed close attention to. In this chapter, most all question come in nice mks form. The rare exception proves the rule that units are important. Losing track of them can have disastrous consequences (think prescription medicine dosing or architectural specifications or in this case nonsense answers for the properties of a carbon atom in a solid).

97.   **ORGANIZE AND PLAN**   We will use conservation of energy to find the initial compression of the spring. Consider the figure below.

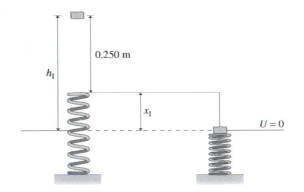

We set the zero of potential energy to be the point at which the spring is fully compressed from the equilibrium value of the spring without the mass. In this case the total energy of the system before the mass is dropped is $E = mgh_i$. When the spring is compressed there is no more gravitational potential energy, all of the energy is stored in the potential energy of the spring $E = \frac{1}{2} k (x_i)^2$. Equating the different forms for the total energy yields:

$$mgh_i = \frac{1}{2} k \left(x_i \right)^2$$

We note, however, from the figure that $h_i = d + x_i$ where $d = 0.250$ m. Subbing in this value for h_i yields the quadratic equation:

$$x_i^2 - \gamma x_i - \gamma d = 0,$$

where $\gamma = \dfrac{2\, mg}{k} = \dfrac{2 * 2.15 \text{ kg} * 9.8 \text{ m/s}^2}{250 \text{ N/m}} = 0.169$ m.

Solving for the roots of this equation yields:

$$x_i = \frac{\gamma}{2} \left(1 \pm \sqrt{1 + \frac{4d}{\gamma}} \right)$$

The negative solution to this problem (where $\pm \rightarrow -$) does not correspond to a physical solution so the displacement from uncompressed state:

$$x_i = \frac{\gamma}{2} \left(1 + \sqrt{1 + \frac{4d}{\gamma}} \right)$$

The amplitude of the oscillation is not this displacement x_i from the spring in the relaxed state. The equilibrium position is determined by force balance on the spring. When the force due to the spring (given by Hooke's law) upward is equal to the downward gravitational force then the spring is in the equilibrium position. In equation form we have $kx_0 = mg$, or $x_0 = mg/k$ which turns our to be equal to $\gamma/2$.

The amplitude of oscillation about the equilibrium position is

$$A = x_i - x_0 = \frac{\gamma}{2} \sqrt{1 + \frac{4d}{\gamma}}$$

SOLVE   The compression of the spring for the given values:

$$x_i = \frac{0.169 \text{ m}}{2} \left(1 + \sqrt{1 + \frac{4 * 0.250 \text{ m}}{0.169 \text{ m}}} \right) = 0.307 \text{ m}$$

The amplitude of the oscillation:

$$A = \frac{0.169 \text{ m}}{2} \sqrt{1 + \frac{4 * 0.250 \text{ m}}{0.169 \text{ m}}} = 0.222 \text{ m}$$

The period of the oscillation:

$$T = 2\pi \sqrt{\frac{m}{k}} = 2\pi \sqrt{\frac{2.15 \text{ kg}}{250 \text{ N/m}}} = 0.583 \text{ s}$$

REFLECT   Aside from units, another method of ruling out obviously wrong symbolic solutions is to imagine limiting situations. You will will notice in the set-up of these problems (all of them) the symbolic solution is always presented. As problems get a little more challenging the functional forms are there for you to check for sanity. For example, observe the form the amplitude of the oscillation as the distance that the mass is dropped goes to zero. We know that this amplitude should be simply the displacement of the equilibrium value. These types of self consistency checks really help when there is no solution in the back of the book.

99.   ORGANIZE AND PLAN   The sum of all forces on the tightrope walker (named Pat) shown in the figure below is

$$F_n = mg - 2F_y$$

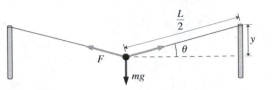

The y-component of the tension force of the rope is $F_y = F\sin\theta$. If we assume the rope does not stretch yet the y value can change (this necessitates distance between the support poles to be flexible which does not matter since that distance will not enter in our calculations.) In this case, the y-component can be written in terms of the distances in the above diagram.

$$F_y = F\frac{y}{L/2}$$

Which yields the net force of:

$$F_n = mg - \frac{4F}{L}y$$

Let us say the equilibrium position (where the total force is zero) of Pat occurs at $y = y_0$. The net force is then:

$$F_n = mg - \frac{4F}{L}y_0 - \frac{4F}{L}d = -\frac{4F}{L}d,$$

where d is the vertical displacement from the equilibrium position. The net force $F_n = -\frac{4F}{L}d$ is in the form of

Hooke's Law $F = -kx$. Consequently, the effective spring constant is $k_{\text{eff}} = \frac{4F}{L}$.

SOLVE   Subbing in the effective spring constant into the expression for the period of a SHO yields:

$$T = 2\pi\sqrt{\frac{m}{k_{\text{eff}}}} = 2\pi\sqrt{\frac{mL}{4F}}$$

REFLECT   Though this is a concrete example of the power of the simple harmonic motion example, all systems near a stable equilibrium point undergo SHM for small deviations from the equilibrium position.

8 ROTATIONAL MOTION

CONCEPTUAL QUESTIONS

1. **SOLVE** The centripetal acceleration is given in Equation 8.13:

$$a_r = \frac{v_t^2}{r} = r\omega^2$$

The tangential velocity, v_t, will be different at different latitudes on Earth, but the angular velocity, ω, will be the same everywhere. The radius, r, in this equation is not the distance to the center of the Earth, but the distance to the Earth's rotational axis. The equator is farther from the axis than a point at higher latitudes ($r_{eq} > r_{45°}$), so the centripetal acceleration will be higher at the equator.

REFLECT A force is needed to supply centripetal acceleration. In this case, it's the Earth's gravity. If the Earth for some reason started spinning much faster, material at the equator would be the first to go flying off.

3. **SOLVE** The information written on a CD is spaced uniformly. To read that information at a uniform rate, the tangential velocity should be the same no matter what region of the CD is being read. From Equation 8.10, the tangential velocity is: $v_t = r\omega$. At the inner edge, the radius is the smallest so the angular velocity has to be the largest to compensate.

REFLECT Information is encoded in bits on a CD, and a laser is used to read this data. The CD spins beneath the laser so that bits move past at a constant rate, as seen in the figure below.

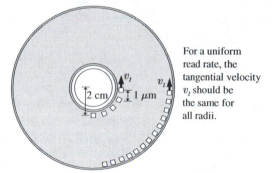

For a uniform read rate, the tangential velocity v_t should be the same for all radii.

Imagine the bits (the squares in the figure) are 1 micron apart and that the read rate is one million bits per second (this is roughly the 1x speed for a CD-ROM drive). This implies that the tangential velocity is 1 m/s (since a micron is one-millionth of a meter). If the inner radius is roughly 2 cm from the center, then the angular velocity is:

$$\omega = \frac{v_t}{r} = \frac{1 \text{ m/s}}{2 \text{ cm}}\left[\frac{100 \text{ cm}}{1 \text{ m}}\right] = 50 \text{ rad/s}\left[\frac{1 \text{ rev}}{2\pi \text{ rad}}\right]\left[\frac{60 \text{ s}}{1 \text{ min}}\right] = 480 \text{ rpm}$$

This roughly matches the 500 rpm maximum rotation rate of the 1x speed. The CD will spin slower as the laser read device moves farther out on the disk.

5. **SOLVE** No, the baseball bat has greater rotational inertia when rotated around the handle end. This is because the bat is heavier on the hitting end than on the handle end. The largest rotational inertia is when the heavy hitting end is the farthest away from the pivot point.

REFLECT   If you try swinging a bat while holding it backwards, it feels lighter. Many baseball players prefer a lighter bat, since they can get the bat speed higher, for a given input energy. But that doesn't mean any of them want to turn their bats backwards!

7.   **SOLVE**   With respect to the center of the wheel, the outer edge is moving in a circle with speed v_{cm}. In particular, the bottom of the wheel is moving backwards at v_{cm}, while the top is moving forwards at v_{cm}. See Figure 8.11. Because the center of the wheel itself is moving forward at v_{cm}, the net velocity of the bottom is zero, whereas the net velocity of the top is $2v_{cm}$.

REFLECT   The fact that the bottom of the wheel is at rest with respect to the ground is important for gripping the road. If the wheel starts to slip, it means the bottom of the wheel is moving with respect to the road, and there's less control of its direction.

9.   **SOLVE**   The Earth's rotational inertia should be less than it is for a uniform sphere, i.e., $I_{\text{Earth}} < \frac{2}{5}MR^2$. This is because the heavier mass elements are at small radii where they contribute less to the rotational inertia.

REFLECT   The Earth is in some sense the opposite of a hollow ball, which has all its mass at large radii where it contributes more to the rotational inertia (see Problem 8.4).

11.   **SOLVE**   The spinning top precesses, meaning its axis rotates around the vertical. When the top's rotation axis is tilted, the center of mass of the top is no longer directly above the bottom point that touches the ground. This induces a torque that aims to rotate the top toward the ground. By the right-hand rule, this torque points horizontally (see Figure 8.26), causing a small change in the angular momentum $\Delta \vec{L}$ that points in the same direction as the torque (Equation 8.24). Because $\Delta \vec{L}$ is parallel to the ground, the angular momentum vector $\vec{L}$ moves in a circle, but it doesn't dip down toward the ground. Since $\vec{L}$ runs along the rotational axis, this means that the top itself doesn't dip toward the ground.

REFLECT   One way to think of this is that the angular momentum is composed of a part that is parallel to the ground (horizontal) and another part that is perpendicular (vertical):

$$\vec{L} = \vec{L}_{\text{horiz}} + \vec{L}_{\text{vert}}$$

The torque only acts on the horizontal part. There's no torque acting on the vertical part, so by conservation of momentum it should remain unchanged, and the top should never dip. In reality, friction will slow down the spinning of the top, causing $\vec{L}_{\text{vert}}$ to decrease and the top to fall.

13.   **SOLVE**   A solid disk has $I = \frac{1}{2}MR^2$, whereas a thin ring with its mass all at the rim has $I = MR^2$ (from Table 8.4). So the wheel in this example is more like a bicycle wheel.

REFLECT   Is it better for a wheel to be more like a disk or more like a thin ring? It depends on the application. A disk has a smaller rotational inertia, so it takes less torque to start and stop than a ring. For a given rotation rate, however, the ring rolls straighter, since it has a larger angular momentum.

MULTIPLE-CHOICE PROBLEMS

15.   **SOLVE**   We're asked to convert one revolution per 24 hours into radians per second.

$$\omega = \frac{1 \text{ rev}}{24 \text{ h}}\left[\frac{2\pi \text{ rad}}{1 \text{ rev}}\right]\left[\frac{1 \text{ h}}{60 \text{ min}}\right]\left[\frac{1 \text{ min}}{60 \text{ sec}}\right] = 7.27\times10^{-5} \text{ rad/s}$$

The correct answer is (d).

REFLECT   That's not terribly fast in comparison to other things that turn, like wheels and CDs. But this is in some sense the speed of the day. It is how fast the sun moves across the sky and half the angular speed of the hour hands on a watch.

17.   **SOLVE**   We'll want to use Equation 8.9 to find the angle displacement, but first we need to find the angular acceleration (Equation 8.6):

$$\alpha = \frac{\Delta\omega}{\Delta t} = \frac{(3.6 \text{ rad/s} - 2.4 \text{ rev/s})}{2.0 \text{ s}} = 0.60 \text{ rad/s}^2$$

Plugging this into Equation 8.9:

$$\Delta\theta = \omega_0 t + \tfrac{1}{2}\alpha t^2 = (2.4 \text{ rad/s})(2.0 \text{ s}) + \tfrac{1}{2}(0.60 \text{ rad/s}^2)(2.0 \text{ s})^2 = 6.0 \text{ rad}$$

The answer is (a).

REFLECT   We could have used Equation 8.10 instead of Equation 8.9 and still have gotten the same answer.

19.   **SOLVE**   During angular acceleration, the angular velocity is increasing, as is the tangential velocity. The centripetal acceleration is proportional to the square of the angular velocity ($a_r = r\omega^2$ from Equation 8.13), so it is increasing as well. Only the tangential acceleration is constant.

The correct answer is (c).

REFLECT   The tangential acceleration is constant at a particular radius, but it increases as one moves farther out on the wheel.

21.   **SOLVE**   The rotational inertia for a uniform ball is $I = \tfrac{2}{5}MR^2$, so plugging in the given values:

$$I = \tfrac{2}{5}(0.046 \text{ kg})(0.0213 \text{ m})^2 = 8.3 \times 10^{-6} \text{ kg} \cdot \text{m}^2$$

The correct answer is (a).

REFLECT   A lot of thought goes into making golf balls. The insides are generally liquid or solid cores. The liquid cores are easier to control but go less far than the solid cores. Can you explain why?

23.   **SOLVE**   Rolling without slipping implies that the bike velocity is equal to the tangential velocity of the wheel: $v_t = r\omega$ (Equation 8.11), so:

$$\omega = \frac{v_{\text{bike}}}{r} = \frac{(40 \text{ km/h})}{(\tfrac{1}{2}0.69 \text{ m})}\left[\frac{1000 \text{ m}}{1 \text{ km}}\right]\left[\frac{1 \text{ h}}{60 \cdot 60 \text{ s}}\right] = 32 \text{ rad/s}$$

The correct answer is (d).

REFLECT   Note that we are given the wheel's diameter, but we need the radius, so we divide the value by 2.

25.   **SOLVE**   To find the angular velocity, we need the angular acceleration. That can be found with a combination of Equations 8.16 and 8.17:

$$\alpha = \frac{\tau}{I} = \frac{rF\sin\theta}{I} = \frac{(0.75 \text{ m})(35 \text{ N})(1)}{(25 \text{ kg} \cdot \text{m}^2)} = 1.05 \text{ rad/s}^2$$

Now applying this to Equation 8.8:

$$\omega = \omega_0 + \alpha t = (1.05 \text{ rad/s}^2)(5.0 \text{ s}) = 5.3 \text{ rad/s}$$

The correct answer is (c).

REFLECT   A force applied tangentially means that it makes an angle of $90°$ with the radius direction, so $\sin\theta = 1$.

27.   **SOLVE**   With no torques operating on the system, we can think of this as a conservation of angular momentum problem:

$$L = I_0\omega_0 = I\omega$$

Before, there is just the flywheel rotating ($I_0 = I_{\text{fly}}$), but when the clutch engages, the rotational inertia increases:

$$I = I_{\text{fly}} + I_{\text{clutch}} = I_{\text{fly}} + \tfrac{1}{2}I_{\text{fly}} = \tfrac{3}{2}I_{\text{fly}}$$

Solving for the final angular velocity:

$$\omega = \frac{I_0}{I}\omega_0 = \left(\frac{I_{\text{fly}}}{\tfrac{3}{2}I_{\text{fly}}}\right)\omega_0 = \tfrac{2}{3}(310 \text{ rad/s}) = 206 \text{ rad/s}$$

The closest answer is (b).

REFLECT   Notice that we never needed to use the given value for the rotational inertia. Doing a little algebra can save time!

29.   **SOLVE**   By the right-hand rule, we let our fingers curl in the direction of the rotation and observe that our thumb points up. By definition, the angular momentum (as well as the angular velocity) points up.

The correct answer is (a).

REFLECT   We may be tempted to think that the angular momentum is tangent to the wheel in the direction of rotation. But this is ill-defined, since the tangent makes a full circle. We need a unique vector, and so it makes sense to choose the direction of the rotational axis.

PROBLEMS

31.   **ORGANIZE AND PLAN**   The distance traveled is just the arc length from Equation 8.1: $s = r\Delta\theta$. The period, T, which is the time it takes to make one revolution, is 9 hours and 50 minutes or in seconds:

$$T = 9 \text{ hr} \left[\frac{60 \text{ min}}{1 \text{ hr}} \right] + 50 \text{ min} = 590 \text{ min} = 35{,}400 \text{ s}$$

Known: $r = 7.14 \times 10^7 \text{ m}$.

SOLVE   Let's first find the angular velocity in radians per second using Equation 8.5:

$$\omega = \frac{2\pi}{T} = \frac{2\pi}{35{,}400 \text{ s}} = 1.8 \times 10^{-4} \text{ rad/s}$$

In this case, the angular displacement is just: $\Delta\theta = \omega \cdot (1 \text{ s}) = 1.8 \times 10^{-4}$ rad. Multiplying by the radius gives the distance traveled.

$$s = r\Delta\theta = (7.14 \times 10^7 \text{ m})(1.8 \times 10^{-4} \text{ rad}) = 1.3 \times 10^4 \text{ m}$$

REFLECT   This says that a point on Jupiter's equator is moving 13 km in one second, or 47,000 km in one hour. Let's compare this to the result from Example 8.1, where we found that a point on the Earth's equator moves 1670 km in one hour. Therefore, a point on Jupiter's equator moves 28 times farther, which makes sense since Jupiter's radius is 11 times bigger than Earth's radius and its angular velocity is 2.4 times faster.

33.   **ORGANIZE AND PLAN**   The *E. Coli.*'s flagellum spins at some 100 rev/s, or more than 600 rad/s. We need only convert the centrifuge spin rate into comparable units.

Known: $\omega = 3000$ rpm.

SOLVE   Converting the given spin rate:

$$\omega = \frac{3000 \text{ rev}}{1 \text{ min}} \left[\frac{1 \text{ min}}{60 \text{ s}} \right] = 50 \text{ rev/s}$$

$$= 50 \text{ rev/s} \left[\frac{2\pi \text{ rad}}{1 \text{ rev}} \right] = 314 \text{ rad/s}$$

REFLECT   Centrifuges spin around extremely fast to separate materials of different mass, but they only spin half as fast as a bacteria's flagellum, which is pretty amazing.

35.   **ORGANIZE AND PLAN**   To find the final angular velocity of the lawnmower after the given time of acceleration, we'll need to use Equation 8.8:

$$\omega = \omega_0 + \alpha t$$

Known: $\omega_0 = 0$ rad/s, $\alpha = 98$ rad/s^2, $t = 2.5$ s.

SOLVE   Plugging the values into Equation 8.8 and expressing the answer in both rad/s and rpm:

$$\omega = \omega_0 + \alpha t = 0 + (98 \text{ rad/s}^2)(2.5 \text{ s}) = 250 \text{ rad/s}$$

$$= 250 \text{ rad/s} \left[\frac{1 \text{ rev}}{2\pi \text{ rad}} \right] \left[\frac{60 \text{ s}}{1 \text{ min}} \right] = 2{,}300 \text{ rpm}$$

REFLECT   Does 2,300 rpm sound reasonable for a lawnmower? A car engine revs around a few 1,000 rpm, so it's conceivable that a lawnmower motor turns at roughly the same rate.

37.   **ORGANIZE AND PLAN**   It takes 24 hours for the Earth to make a full rotation now, but it took nearly half that time 4 billion years ago. We need to find the rotation rates now and back then, so that we can use Equation 8.6 for the average angular acceleration: $\bar{\alpha} = \Delta\omega / \Delta t$.

Known: The period 4 billion years ago: $T_{4bya} = 14$ hrs, and the period now: $T_{now} = 24$ hrs. The time difference: $\Delta t = 14$ billion years.

SOLVE   First, let's determine the angular velocity back then using Equation 8.3:

$$\omega_{4bya} = \frac{2\pi}{T_{4bya}} = \frac{2\pi}{14 \text{ h}} \left[\frac{1 \text{ h}}{60 \cdot 60 \text{ s}} \right] = 1.25 \times 10^{-4} \text{rad/s}$$

Then, the same thing for our current epoch:

$$\omega_{now} = \frac{2\pi}{T_{now}} = \frac{2\pi}{24 \text{ h}} \left[\frac{1 \text{ h}}{60 \cdot 60 \text{ s}} \right] = 7.27 \times 10^{-5} \text{rad/s}$$

The change in angular velocity is therefore:

$$\Delta\omega = \omega_{now} - \omega_{4bya} = 7.27 \times 10^{-5} \text{rad/s} - 1.25 \times 10^{-4} \text{rad/s} = -5.23 \times 10^{-5} \text{rad/s}$$

We can now plug this into Equation 8.6:

$$\bar{\alpha} = \frac{\Delta\omega}{\Delta t} = \frac{(-5.23 \times 10^{-5} \text{rad/s})}{4 \times 10^9 \text{ yr}} \left[\frac{1 \text{yr}}{365 \cdot 24 \cdot 60 \cdot 60 \text{s}} \right] = -4.15 \times 10^{-22} \text{ rad/s}^2$$

REFLECT   The answer is negative as it should be, since the angular velocity has decreased over time. Notice that it is an extremely small deceleration. Thank goodness! We'd rather not have the Earth stop spinning anytime soon.

39.   **ORGANIZE AND PLAN**   There are two stages here: an acceleration stage, followed by a constant spinning stage. The number of revolutions in the first stage can be found with Equation 8.9, since we know the initial angular velocity (zero), the acceleration, and the time. In the second stage, the angular velocity is the final speed attained in the first stage. We can figure out ω using Equation 8.8. To find the number of revolutions, we just multiply by the time (see Equation 8.3).

Known: First stage with acceleration: $\omega_0 = 0$ rad/s, $\alpha = 615$ rad/s^2, $t_1 = 2.10$ s. Second stage without acceleration: $t_2 = 7.50$ s.

SOLVE   Using Equation 8.9 for the first stage:

$$\Delta\theta_1 = \tfrac{1}{2}\alpha t_1^2 = \tfrac{1}{2}(615 \text{ rad/s}^2)(2.10 \text{ s})^2 = 1360 \text{ rad} \left[\frac{1 \text{ rev}}{2\pi \text{ rad}} \right] = 216 \text{ rev}$$

At the end of this stage, the angular velocity will be:

$$\omega = \omega_0 + \alpha t = (615 \text{ rad/s}^2)(2.10 \text{ s}) = 1290 \text{ rad/s}$$

Now using Equation 8.3, we can find the number of revolutions in the second stage:

$$\Delta\theta_2 = \omega t_2 = (1290 \text{ rad/s})(7.50 \text{ s}) = 9680 \text{ rad} \left[\frac{1 \text{ rev}}{2\pi \text{ rad}} \right] = 1540 \text{ rev}$$

The total number of revolutions is then:

$$\Delta\theta_1 + \Delta\theta_2 = 216 \text{ rev} + 1540 \text{ rev} = 1756 \text{ rev}$$

REFLECT   Most of the revolutions occur in the second stage, since the first stage is relatively short. The total number of revolutions is pretty high, but then remember the zzz of the dentist's drill. Ouch!

41.   **ORGANIZE AND PLAN**   We're given the initial and final angular velocity and the time, so we can use Equation 8.8 to find the angular acceleration. Once we have that, we can use either Equation 8.9 or 8.10 to find $\Delta\theta = \theta - \theta_0$. The last part goes back to Equation 8.1, and determining the arc length. We'll have to be sure that the angle is in radians.

Known: $\omega_0 = 9,000$ rpm, $\omega = 5,000$ rpm, $t = 3.50$ s, $r = 9.40$ cm.

SOLVE   (a) Since part (b) asks for the number of revolutions, let's convert the angular velocities into rev/s:

$$\omega_0 = 9000 \text{ rev/min} \left[\frac{1 \text{ min}}{60 \text{ s}} \right] = 150 \text{ rev/s}$$

$$\omega = 5000 \text{ rev/min} \left[\frac{1 \text{ min}}{60 \text{ s}} \right] = 83.3 \text{ rev/s}$$

Plugging these values into Equation 8.8:

$$\alpha = \frac{\omega - \omega_0}{t} = \frac{83.3 \text{ rev/s} - 150 \text{ rev/s}}{3.50 \text{ s}} = -19.1 \text{ rev/s}^2$$

In terms of radians, the answer is : $\alpha = -119$ rad/s^2.

(b) We will use Equation 8.10 (but 8.9 could be used as well):

$$\Delta\theta = \frac{\omega^2 - \omega_0^2}{2\alpha} = \frac{(83.3 \text{ rev/s})^2 - (150 \text{ rev/s})^2}{2(-19.1 \text{ rev/s}^2)} = 407 \text{ rev}$$

(c) Before turning to Equation 8.1, we should convert revolutions into radians:

$$\Delta\theta = 407 \text{ rev} \left[\frac{2\pi \text{ rad}}{1 \text{ rev}} \right] = 2560 \text{ rad}$$

Now we're ready to solve for the arc length:

$$s = r\Delta\theta = (9.40 \text{ cm})(2560 \text{ rad}) = 24,000 \text{ cm}$$

REFLECT   The acceleration is negative, since the centrifuge slows down. Even though it is decelerating, the centrifuge still manages to make 407 revolutions in 3.50 s. A point on the edge covers a distance of 240 m, or roughly two-and-a-half football fields in this short time.

43.   **ORGANIZE AND PLAN**   We have to think carefully about what information is provided. To say that the day is getting longer means that the period (the time to make a revolution) is increasing by 2.3 ms every century. Or to say it in another way, the period 100 years from now will be:

$$T_{100} = 24 \text{ h} + 2.3 \text{ ms} = 86,400.0023 \text{ s}$$

We can use Equation 8.5 to solve for the angular velocity now and in 100 years. Then we can use Equation 8.8 to find the angular acceleration.

Known: $T_{\text{now}} = 24 \text{ h} = 86,400$ s, $t = 100$ y $= 3.15 \times 10^9$ s.

SOLVE   The angular velocity of the Earth now is:

$$\omega_{\text{now}} = \frac{2\pi}{T_{\text{now}}} = \frac{2\pi}{86,400 \text{ s}} = 7.272205217 \times 10^{-5} \text{ rad/s}$$

We are going to need to keep this many digits because, as you will see, there's very little difference between the angular velocity now and what it will be in 100 years:

$$\omega_{100} = \frac{2\pi}{T_{100}} = \frac{2\pi}{86,400.0023 \text{ s}} = 7.272205023 \times 10^{-5} \text{ rad/s}$$

The angular acceleration is the difference in the angular velocities divided by a century's worth of time:

$$\alpha = \frac{\omega_{100} - \omega_{now}}{t} = \frac{-1.94 \times 10^{-12} \text{ rad/s}}{3.15 \times 10^{9} \text{ s}} = -6.14 \times 10^{-22} \text{ rad/s}^2$$

Note: If your calculator doesn't hold enough digits, you can use the following mathematical trick:

$$\frac{1}{1+x} \approx 1 - x \quad \text{for } x \ll 1$$

In this case,

$$\omega_{100} = \frac{2\pi}{24 \text{ h} + 2.3 \text{ ms}} = \frac{2\pi}{24 \text{ h}}\left(\frac{1}{1+x}\right) \approx \frac{2\pi}{24 \text{ h}}(1-x)$$

Where $x = 2.3 \text{ ms}/24 \text{ h} = 2.66 \times 10^{-8}$. Notice how this simplifies the expression for the angular acceleration to:

$$\alpha = \frac{-x\omega_{now}}{t} = \frac{-(2.66 \times 10^{-8})(7.27 \times 10^{-5} \text{ rad/s})}{3.15 \times 10^{9} \text{ s}} = -6.14 \times 10^{-22} \text{ rad/s}^2$$

REFLECT   The Earth's rotation is slowing, so the angular acceleration is negative. But it is an almost imperceptible change.

45.   **ORGANIZE AND PLAN**   We are given the tangential velocity and must find the angular velocity using Equation 8.11: $\omega = v_t/r$.

Known: $v_t = 325 \text{ km/h}$, $r = 18 \text{ m}$.

SOLVE   Solving for the angular velocity in rad/s:

$$\omega = \frac{v_t}{r} = \frac{325 \text{ km/h}}{18 \text{ m}}\left[\frac{1000 \text{ m}}{1 \text{ km}}\right]\left[\frac{1 \text{ h}}{60 \cdot 60 \text{ s}}\right] = 5.02 \text{ rad/s}$$

REFLECT   This is a little less than 1 revolution per second. That sounds about right, although it may not seem as fast as one might imagine. Still, the corresponding wind speed is impressive, simply because the rotation occurs at a substantial radius.

47.   **ORGANIZE AND PLAN**   As we see in the figure below, the weight's acceleration induces a tangential acceleration on the pulley through the tension in the string (i.e., $a_y = a_t$). We can convert this into an angular acceleration with Equation 8.12: $\alpha = a_t/r$. For part (b), the falling weight achieves a velocity, v_y, by the time it reaches the ground. This linear velocity is equal to the tangential velocity of the pulley: $v_y = v_t$. There are different ways to go about this, but we will solve first for the weight's velocity using Equation 2.10: $v_y^2 = v_{y0}^2 + 2a_y\Delta y$, and then use this to find the angular velocity from Equation 8.11: $\omega = v_t/r$.

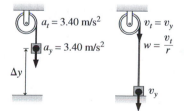

Known: $r = 3.50 \text{ cm}$, $a_y = 3.40 \text{ m/s}^2$, $v_{0y} = 0$, $\Delta y = 1.30 \text{ m}$, where we have defined "down" as the positive y-direction.

SOLVE   (a) The angular acceleration comes directly from the weight's acceleration:

$$\alpha = \frac{a_y}{r} = \frac{3.40 \text{ m/s}^2}{3.50 \text{ cm}}\left[\frac{100 \text{ cm}}{1 \text{ m}}\right] = 97.1 \text{ rad/s}^2$$

(b) The first thing is to solve for the final velocity of the weight:

$$v_y = \sqrt{v_{y0}^2 + 2a_y \Delta y} = \sqrt{2(3.40 \text{ m/s}^2)(1.30 \text{ m})} = 2.97 \text{ m/s}$$

The pulley turns at the same speed, so the angular velocity is:

$$\omega = \frac{v_y}{r} = \frac{2.97 \text{ m/s}}{0.0350 \text{ m}} = 84.9 \text{ rad/s}$$

REFLECT   You may wonder why the weight doesn't fall with the gravitational acceleration of 9.8 m/s^2. The string provides a tension that slows the weight's fall. Notice, too, that we chose the down direction to be positive, but we can choose it to be negative, in which case all the answers would have a negative sign.

49.   **ORGANIZE AND PLAN**   Equation 8.13 gives the relation between tangential velocity and centripetal acceleration: $v_t^2 = ra_r$. For part (b), we'll need to convert the tangential velocity into angular velocity (Equation 8.11) and then solve for angular acceleration using the time it takes to reach this speed (Equation 8.8).

Known: $d = 10.5$ m, $a_r = 5.5$ g, $\omega_0 = 0$, $t = 25$ s.

SOLVE   (a) Solving for the tangential velocity, and realizing that we have to divide the diameter by 2 to obtain the radius:

$$v_t = \sqrt{ra_r} = \sqrt{(\tfrac{1}{2} \cdot 10.5 \text{ m})(5.5 \cdot 9.8 \text{ m/s}^2)} = 16.8 \text{ m/s}$$

(b) First, we'll need to figure out the angular velocity:

$$\omega = \frac{v_t}{r} = \frac{16.8 \text{ m/s}}{\tfrac{1}{2} \cdot 10.5 \text{ m}} = 3.20 \text{ rad/s}$$

Using Equation 8.8 for the angular acceleration:

$$\alpha = \frac{\omega - \omega_0}{t} = \frac{3.20 \text{ rad/s}}{25 \text{ s}} = 0.13 \text{ rad/s}^2$$

REFLECT   The acceleration is positive since the centrifuge is starting from rest and speeding up. The angular velocity is 30 rpm, or one revolution every 2 seconds. That sounds positively stomach turning.

51.   **ORGANIZE AND PLAN**   We can think of the spiral track as a large number (N) of concentric circles, separated by a tiny distance that will call δ_r. The length will be the sum of the circumferences of all the circles. We'll try to be clever in adding up these lengths.

Known: $r_{in} = 2.6$ cm, $r_{out} = 5.7$ cm, $\delta_r = 0.74$ μm, $d_{byte} = 2.3$ μm.

SOLVE   (a) The number of concentric circles is just the distance between the outer and inner edges divided by the separation between successive circles:

$$N = \frac{r_{out} - r_{in}}{\delta_r} = \frac{5.7 \text{ cm} - 2.6 \text{ cm}}{0.74 \text{ μm}} \left[\frac{10^4 \text{ μm}}{1 \text{ cm}} \right] = 42{,}000$$

The circumference of the first (inner) circle is $2\pi r_{in}$, of the 2nd circle is $2\pi(r_{in} + \delta_r)$, of the third circle is $2\pi(r_{in} + 2\delta_r)$, etc. until the final (outer) circle with circumference of $2\pi(r_{in} + N\delta_r) = 2\pi r_{out}$. The total length is the sum of all these circumferences:

$$L = 2\pi r_{in} + 2\pi(r_{in} + \delta_r) + 2\pi(r_{in} + 2\delta_r) + \dots + 2\pi(r_{in} + N\delta_r)$$

Rearranging the terms:

$$L = 2\pi \left[r_{in}(N+1) + \delta_r N \frac{N+1}{2} \right]$$

$$= 2\pi(N+1)\left[r_{in} + \frac{N}{2}\delta_r \right] = 2\pi(N+1)\left[\frac{r_{in} + r_{out}}{2} \right]$$

What this essentially says is that the total length is the number of circles times the circumference of the average radius. Plugging in the given values we get:

$$L = 2\pi(42,001)\left[\frac{2.6\ \text{cm} + 5.7\ \text{cm}}{2}\right] = 1.1 \times 10^6\ \text{cm}$$

This is 11 kilometers.

(b) To find the number of bytes, we divide the length of the track by the byte length:

$$N_{\text{byte}} = \frac{L}{d_{\text{byte}}} = \frac{1.1 \times 10^6\ \text{cm}}{2.3\ \mu\text{m}}\left[\frac{10^4\ \mu\text{m}}{1\ \text{cm}}\right] = 4.8 \times 10^9$$

REFLECT   The DVD has 4.8 billion bytes, which is 4.8 Gigabytes. This is what typical DVDs can carry.

53.   **ORGANIZE AND PLAN**   This problem does not involve revolutions, but movement through a given angle. We want to solve for one downstroke, in which case the top of the stroke is θ_0 and the bottom of the stroke is θ, as shown in the figure below.

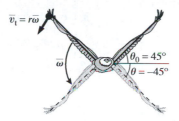

If the eagle flaps 20 times per minute, then it makes a full flap every 3 seconds. A full flap consists of an upstroke and a downstroke, $\Delta t = 1.5\ \text{s}$. This is the time that we can plug into Equation 8.3 to determine the average angular velocity. For the tangential velocity, Equation 8.11 asks for the radius. We are given the wingspan, which is the length of the two wings together. The radius is then just half the wingspan.

Known: $\theta_0 = 45°$, $\theta = -45°$, $\Delta t = 1.5\ \text{s}$, $r = \frac{1}{2} \cdot 2.1\ \text{m} = 1.1\ \text{m}$.

SOLVE   (a) Let's first convert the angles from degrees to radians:

$$\theta_0 = 45°\left[\frac{2\pi\ \text{rad}}{360°}\right] = 0.79\ \text{rad}$$

$$\theta = -45°\left[\frac{2\pi\ \text{rad}}{360°}\right] = -0.79\ \text{rad}$$

We figured out the time it takes to go between these angles above, so the average angular velocity is:

$$\bar{\omega} = \frac{\Delta\theta}{\Delta t} = \frac{(0.79\ \text{rad}) - (-0.79\ \text{rad})}{1.5\ \text{s}} = 1.1\ \text{rad/s}$$

(b) As for the tangential velocity at the tip of the wing:

$$\bar{v}_t = r\bar{\omega} = (1.1\ \text{m})(1.1\ \text{rad/s}) = 1.2\ \text{m/s}$$

REFLECT   The eagle makes a single downstroke in 1.5 s, which seems reasonable. And since its wings are about a meter long each, it makes sense that the tangential velocity is approximately one meter per second.

55.   **ORGANIZE AND PLAN**   The rotational kinetic energy is stated in Equation 8.15: $K = \frac{1}{2}I\omega^2$. We have the rotational inertia, so we only need the angular velocity of the Earth in rad/s. Let's use Equation 8.5 with the period of 24 hours in a day.

Known: $I = 9.72 \times 10^{37}\ \text{kg m}^2$, $T = 24\ \text{h}$.

SOLVE   First, finding the angular velocity:

$$\omega = \frac{2\pi}{T} = \frac{2\pi}{24\ \text{h}}\left[\frac{1\ \text{h}}{60 \cdot 60\ \text{s}}\right] = 7.27 \times 10^{-5}\ \text{rad/s}$$

This is the same answer we got in Problem 15. Plugging this into Equation 8.15:

$$K = \tfrac{1}{2}I\omega^2 = \tfrac{1}{2}(9.72\times10^{37} \text{ kg m}^2)(7.27\times10^{-5} \text{ rad/s})^2 = 2.56\times10^{29} \text{ J}$$

Notice we have cancelled radians from the equation because it is dimensionless. And we have used the fact that:

$$1 \text{ J} = 1 \text{ kg m}^2/\text{s}^2.$$

REFLECT This is a lot of energy, but we expect as much from an entire planet.

57. **ORGANIZE AND PLAN** We're only given the saw's initial rotational kinetic energy (Equation 8.15): $K_0 = \tfrac{1}{2}I\omega_0^2$. The rotation rate drops in half ($\omega = \tfrac{1}{2}\omega_0$), while the rotational inertia, I, does not change.

Known: $K_0 = 44$ J, $\omega = \tfrac{1}{2}\omega_0$.

SOLVE The rotational kinetic energy after the rotation rate drop:

$$K = \tfrac{1}{2}I\omega^2 = \tfrac{1}{2}I(\tfrac{1}{2}\omega_0)^2 = \tfrac{1}{4}(\tfrac{1}{2}I\omega_0^2) = \tfrac{1}{4}K_0 = \tfrac{1}{4}(44 \text{ J}) = 11 \text{ J}$$

This says that the rotational kinetic energy drops to one-fourth its initial value.

Reflect Because the rotation kinetic energy is proportional to the angular velocity squared, any increase or decrease to the angular velocity will be squared in the rotational kinetic energy.

59. **ORGANIZE AND PLAN** From Table 8.4, the rotational inertia of a uniform ball is: $I = \tfrac{2}{5}MR^2$. For the pitched ball, the translational kinetic energy is: $K_{\text{trans}} = \tfrac{1}{2}Mv^2$, and the rotational kinetic energy is: $K_{\text{rot}} = \tfrac{1}{2}I\omega^2$. The spin rate of 20 Hz implies that the ball makes 20 revolutions per second. We'll need to convert this into units of rad/s.

Known: $M = 145$ g, $R = 3.7$ cm, $v = 22$ m/s, $\omega = 20$ rev/s.

SOLVE (a) The rotational inertia of the baseball is:

$$I = \tfrac{2}{5}MR^2 = \tfrac{2}{5}(0.145 \text{ kg})\left(3.7 \text{ cm}\left[\frac{1 \text{ m}}{100 \text{ cm}}\right]\right)^2 = 7.9\times10^{-5} \text{ kg m}^2$$

(b) Let's first put the angular velocity in terms of rad/s:

$$\omega = 20 \text{ rev/s}\left[\frac{2\pi \text{ rad}}{1 \text{ rev}}\right] = 130 \text{ rad/s}$$

The translational and rotational kinetic energy are:

$$K_{\text{trans}} = \tfrac{1}{2}Mv^2 = \tfrac{1}{2}(0.145 \text{ kg})(22 \text{ m/s})^2 = 35 \text{ J}$$
$$K_{\text{rot}} = \tfrac{1}{2}I\omega^2 = \tfrac{1}{2}(7.9\times10^{-5} \text{ kg m}^2)(130 \text{ rad/s})^2 = 0.67 \text{ J}$$

As in previous problems, radians are not kept for the energy units and $1 \text{ J} = 1 \text{ kg m}^2/\text{s}^2$.

REFLECT The answers are in the Joule range, which is about right for human capabilities. The translational kinetic energy is over 50 times the rotational kinetic energy. We can make sense of this by imagining that the pitcher is applying most of his/her force perpendicular to the outer edge of the ball in order to induce translational motion, while much less force is applied tangentially to make the ball rotate.

61. **ORGANIZE AND PLAN** Rolling without slipping means that the center of mass velocity and angular velocity are related by: $v_{\text{cm}} = \omega R$, where the angular velocity is in units of rad/s.

Known: $v_{\text{cm}} = 50$ km/h, $D = 69$ cm.

SOLVE Solving the given relation for the angular velocity:

$$\omega = \frac{v_{\text{cm}}}{R} = \frac{50 \text{ km/h}}{\tfrac{1}{2}\cdot69 \text{ cm}}\left[\frac{100 \text{ cm}}{1 \text{ m}}\frac{1000 \text{ m}}{1 \text{ km}}\right]\left[\frac{1 \text{ h}}{60\cdot60 \text{ s}}\right] = 40 \text{ rad/s}$$

REFLECT The angular velocity is equivalent to 380 rpm. That sounds about right for a bicycle wheel at top speed.

63. **ORGANIZE AND PLAN**   We have to imagine cases where the friction between the tires and the road is not enough to prevent slipping.

SOLVE   (a) In the case when $\omega > v_{cm}/r$, the wheel is spinning faster than the car is moving. This occurs when you accelerate too fast or your car is stuck in snow, and the wheels "spin out." (b) In the case when $\omega < v_{cm}/r$, the wheel is spinning slower than the car is moving. You can imagine this happening when you try braking on ice. The wheels may stop moving but the car continues to skid forward.

REFLECT   Both of these cases are not good, since they mean that the tires are no longer controlling where you are going.

65. **ORGANIZE AND PLAN**   The angular velocity of the wheels is related to the racer's center of mass velocity by: $\omega = v_{cm}/r$, where the angular velocity is in units of rad/s. Notice that we are given the diameter of the wheels, so the radius will be half of this.

Known: $v_{cm} = 140$ km/h, $d = 76.2$ cm.

SOLVE   The angular velocity is:

$$\omega = \frac{v_{cm}}{r} = \frac{140 \text{ km/h}}{\frac{1}{2}(76.2 \text{ cm})}\left[\frac{100 \text{ cm}}{1 \text{ m}} \frac{1000 \text{ m}}{1 \text{ km}}\right]\left[\frac{1 \text{ h}}{60 \cdot 60 \text{ s}}\right] = 102 \text{ rad/s}$$

REFLECT   This is equal to 974 rpm, or roughly 16 revolutions per second. This seems reasonable. Drag racers appear to burn a lot of rubber to accelerate from the start line, so there is a high risk of skidding, i.e., $\omega > v_{cm}/r$.

67. **ORGANIZE AND PLAN**   The translational kinetic energy is: $K_{trans} = \frac{1}{2}Mv_{cm}^2$ and the rotational kinetic energy is: $K_{rot} = \frac{1}{2}I\omega^2$. If we divide by the total kinetic energy, the unknown variables should cancel each other out, as they did in Example 8.10. The rotational inertia for a solid cylinder is: $I = \frac{1}{2}MR^2$ (from Table 8.4), and the angular velocity: $\omega = v_{cm}/R$.

SOLVE   Let's plug in the values for the rotational kinetic energy:

$$K_{rot} = \frac{1}{2}(\frac{1}{2}MR^2)(v_{cm}/R)^2 = \frac{1}{4}Mv_{cm}^2$$

This is exactly half of the translational kinetic energy. The fractions of kinetic energy in translational and rotational motion are:

$$\frac{K_{trans}}{K_{trans} + K_{rot}} = \frac{\frac{1}{2}}{\frac{1}{2} + \frac{1}{4}} = \frac{2}{3}$$

$$\frac{K_{rot}}{K_{trans} + K_{rot}} = \frac{\frac{1}{4}}{\frac{1}{2} + \frac{1}{4}} = \frac{1}{3}$$

Where the term Mv_{cm}^2 was factored out of the equations.

REFLECT   It's interesting to think that these ratios do not depend on mass or radius. Both a little wooden pencil and the big metal drum on a steam roller will roll down an incline with the same ratios in their translational and rotational kinetic energies.

69. **ORGANIZE AND PLAN**   You might be tempted to attack this problem with kinematic equations from Table 8.2. But it will be much easier to derive the answer from the work-energy theorem; see the figure below.

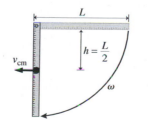

The center of mass of the stick falls a distance equal to half its length: $h = L/2$. This tells us the work done by gravity is mgh, which is equal to the gain in kinetic energy:

$$mgh = K_{trans} + K_{rot}$$

The translational and rotational kinetic energy are as before: $K_{trans} = \frac{1}{2}mv_{cm}^2$ and $K_{rot} = \frac{1}{2}I\omega^2$. The center of mass velocity can be seen from the above figure to be: $v_{cm} = h\omega$, while the rotational inertia for a uniform thin rod held at one end is: $I = \frac{1}{3}mL^2$ (from Table 8.4).

Known: $L = 1$ m.

SOLVE Plugging the kinetic energy values into the work-energy equation:

$$mgh = \frac{1}{2}mv_{cm}^2 + \frac{1}{2}I\omega^2 = \frac{1}{2}m(h\omega)^2 + \frac{1}{2}(\frac{1}{3}mL^2)\omega^2$$

Canceling the mass term and plugging in $h = L/2$, we can solve for the angular velocity:

$$\omega = \sqrt{\frac{\frac{1}{2}gL}{\frac{1}{8}L^2 + \frac{1}{6}L^2}} = \sqrt{\frac{12\,g}{7\,L}} = \sqrt{\frac{12(9.80\text{ m/s}^2)}{7(1\text{ m})}} = 4.1\text{ rad/s}$$

REFLECT The answer implies that center of mass velocity is 2 m/s. If the meter stick simply fell without being held at one end, then the center of mass velocity would be: $v = \sqrt{gL} = 3$ m/s. The velocity is therefore smaller, as it should be, when the stick is held. This is because some of the kinetic energy goes into rotating the stick.

71. **ORGANIZE AND PLAN** As in the previous problem, we can use the work-energy theorem to find the speed of the ball at the bottom of the ramp with length, L, and height, h, where: $h = L\sin\theta$. Once we know this speed, we can use Equation 2.10 to solve for the acceleration.

    SOLVE (a) For a solid sphere rolling without slipping:

    $$mgh = \frac{1}{2}mv^2 + \frac{1}{2}I\omega^2 = \frac{1}{2}mv^2 + \frac{1}{5}mv^2$$

    Where we have used the rotational inertia $I = \frac{2}{5}mR^2$ and $\omega = v/R$. Canceling the mass term and then solving for the velocity:

    $$v = \sqrt{\frac{10}{7}gh}$$

    Plugging this into Equation 2.10 to solve for the acceleration:

    $$a = \frac{v^2 - v_0^2}{2L} = \frac{\left(\sqrt{\frac{10}{7}gh}\right)^2 - 0}{2L} = \frac{5}{7}g\sin\theta$$

    This is slightly smaller than what we would get if the ball slid down the incline without friction and without rolling.

    REFLECT Because some of the kinetic energy is going into rotating the ball, the final velocity at the bottom of the incline will be less. Therefore, the acceleration will be smaller as well.

73. **ORGANIZE AND PLAN** We need to solve for the time that it takes the ball to travel 1.00 m. We don't know the angle of the ramp, which means we don't know the height from which the ball falls. But as in Problem 8.71, we can solve for the acceleration of both the ball and cylinder. We can use Equation 2.9, $\Delta x = v_{x0}t + \frac{1}{2}a_xt^2$, with the ball acceleration to find the time for the ball to travel 1.00 m. Then plug this time into Equation 2.9 using the acceleration of the cylinder to find how far the cylinder travels.

    Known: $L_{ball} = 1.00$ m.

    SOLVE In Problem 8.71, we found that the ball rolls down an incline with the acceleration of:

    $$a_{ball} = \frac{5}{7}g\sin\theta$$

We can do the same derivation for the cylinder. If it falls a height of $h = L\sin\theta$, then its velocity (see Example 8.11) is $v = \sqrt{\frac{4}{3}gh}$. Plugging this into Equation 2.10 to solve for the acceleration:

$$a_{\text{cylinder}} = \frac{v^2 - v_0^2}{2L} = \frac{\left(\sqrt{\frac{4}{3}gh}\right)^2}{2L} = \frac{2}{3}g\sin\theta$$

We've used the fact that the cylinder and the ball both start out at rest $(v_0 = 0)$. We're now ready to solve for the time: plugging the ball acceleration into Equation 2.9:

$$\Delta x = L_{\text{ball}} = \frac{1}{2}a_{\text{ball}}t^2 = \frac{5}{14}g\sin\theta\, t^2$$

$$\Rightarrow\quad t = \sqrt{\frac{14 L_{\text{ball}}}{5 g\sin\theta}}$$

Plugging this time back into the same formula with the cylinder acceleration will give us the distance that the cylinder travels down the ramp:

$$L_{\text{cylinder}} = \frac{1}{2}a_{\text{cylinder}}t^2 = \frac{1}{2}\left(\frac{2}{3}g\sin\theta\right)\left(\sqrt{\frac{14 L_{\text{ball}}}{5 g\sin\theta}}\right)^2 = \frac{14}{15}L_{\text{ball}} = 0.933 \text{ m}$$

(b) Interestingly, the answer does not depend on the inclination angle. This is because the accelerations both scale with $\sin\theta$. In fact, the ratio of the accelerations is equal to the ratio in the distances traveled down the ramp:

$$\frac{a_{\text{cylinder}}}{a_{\text{ball}}} = \frac{L_{\text{cylinder}}}{L_{\text{ball}}} = \frac{14}{15} = 0.933$$

No matter what the inclination, the cylinder will accelerate at 93.3% of the ball's acceleration.

REFLECT Similar relations hold true for any two objects put head-to-head in a "great rolling race."

75. **ORGANIZE AND PLAN** Since the force and the radius are specified, the only variable in the torque equation. $\tau = rF\sin\theta$, is the angle. The sin function has a maximum value of 1 when this angle is 90° (or equivalently when the force is applied in the same direction as the rotational motion).

Known: $r = 35$ cm, $F = 65$ N, $\theta = 90°$.

SOLVE Using 90° as the angle of application, the torque is:

$$\tau = (0.35 \text{ cm})(65 \text{ N})\sin 90° = 23 \text{ N} \cdot \text{m}$$

REFLECT A tool called a torque wrench can actually measure the torque that is being applied to a bolt or screw. They come in different sizes, but a 35-cm-long torque wrench might have a maximum of 70 N·m, so the torque found here seems entirely reasonable.

77. **ORGANIZE AND PLAN** The torque is related to the force by Equation 8.16: $\tau = rF\sin\theta$. We are given the torque and told that the force is applied tangentially $(\sin\theta = 1)$ and at the edge of the flagellum.

Known: $\tau = 400$ pN · nm, $r = 12$ nm, $\sin\theta = 1$.

SOLVE Solving for the force:

$$F = \frac{\tau}{r\sin\theta} = \frac{400 \text{ pN} \cdot \text{nm}}{12 \text{ nm}} = 33 \text{ pN}$$

REFLECT A pico Newton (pN) is one-trillionth (10^{-12}) of a Newton. Not a lot of force, but not much is needed to push an *E. Coli* bacterium around.

79. **ORGANIZE AND PLAN** The children each convey a torque on the seesaw: $\tau = rF\sin\theta$, where the force here is the child's weight. The torque from the child on the left (child 1) would result in a counterclockwise acceleration of the seesaw, so we'll define that as a positive torque. By contrast, the child on the right (child 2) exerts a negative (clockwise) torque. We add these torques together to get the net torque.

The seesaw itself is balanced on its center of mass, so there is no torque from its weight. However, when calculating the angular acceleration from Equation 8.18: $\alpha = \tau_{net}/I$, we'll need to consider the moment of inertia of the children and the seesaw.

Known: $m_1 = 35$ kg, $m_2 = 40$ kg, $L = 3.4$ m, $m_{seesaw} = 25$ kg.

SOLVE   (a) Using the fact that the radius ($r = L/2$) and the angle ($\theta = 90°$) are the same for each child, the torques are :

$$\tau_1 = (\tfrac{1}{2}L)(m_1 g)(1) = (\tfrac{1}{2} \cdot 3.4 \text{ m})(35 \text{ kg})(9.80 \text{ m/s}^2) = 583 \text{ N} \cdot \text{m}$$

$$\tau_2 = -(\tfrac{1}{2}L)(m_2 g)(1) = -(\tfrac{1}{2} \cdot 3.4 \text{ m})(40 \text{ kg})(9.80 \text{ m/s}^2) = -666 \text{ N} \cdot \text{m}$$

We have put a minus sign on the child 2's torque, as described above. The net torque is:

$$\tau_{net} = \tau_1 + \tau_2 = 583 \text{ N} \cdot \text{m} - 666 \text{ N} \cdot \text{m} = -83 \text{ N} \cdot \text{m}$$

(b) The tricky part now is to figure out the rotational inertia for the entire system. The children can each be considered as "particles" at a given radius, so: $I_1 = m_1 r^2$ and $I_2 = m_2 r^2$. The seesaw can be considered as a flat plate (or a thin rod) rotating about its central axis: $I_{seesaw} = \tfrac{1}{12} m_{seesaw} L^2$ (see Table 8.4). The total rotational inertia is the sum of these three components:

$$I_{net} = m_1 r^2 + m_2 r^2 + \tfrac{1}{12} m_{seesaw} L^2$$
$$= (35 \text{ kg})(1.7 \text{ m})^2 + (40 \text{ kg})(1.7 \text{ m})^2 + \tfrac{1}{12}(25 \text{ kg})(3.4 \text{ m})^2 = 240 \text{ kg m}^2$$

The angular acceleration is the net torque divided by the net moment of inertia:

$$\alpha = \frac{\tau_{net}}{I_{net}} = \frac{-83 \text{ N m}}{240 \text{ kg m}^2} = -0.35 \text{ rad/s}^2$$

REFLECT   The acceleration is negative, which by our definition means that the seesaw is turning clockwise. This makes sense, since the heavier child (child 2) will make the seesaw fall on his side.

81.   **ORGANIZE AND PLAN**   The setup is similar to that in the previous problem, except that there are now only two masses and the fulcrum is not at the center. There are three forces to consider: the weights of the two masses (pointing down) and the normal force from the fulcrum, n_f (pointing up). There are two unknowns here: the magnitude of the normal force and its location, x_f. This means that we're going to need to consider both translational and rotational equilibrium. To simplify the net torque equation, we'll choose as our axis the 35-cm-mark, so that the torque will be zero from the first mass.

Known: $x_1 = 35$ cm, $m_1 = 0.20$ kg, $x_2 = 75$ cm, $m_2 = 0.40$ kg.

SOLVE   Defining the up direction as positive, the net force in the system is:

$$F_{net} = n_f - m_1 g - m_2 g = 0$$

And defining counterclockwise as positive, the net torque from the 35-cm mark is:

$$\tau_{net} = \sum rF \sin \theta = r_2 m_2 g - r_f n_f = 0$$

All the angles are 90 degrees, so $\sin \theta = 1$. The radii are defined as the distance from the first mass: $r_2 = 40$ cm and $r_f = x_f - 35$ cm. Solving for n_f in the first equation and substituting it into the second equation gives:

$$r_f = \frac{r_2 m_2 g}{m_1 g + m_2 g} = \frac{(40 \text{ cm})(0.40 \text{ kg})}{(0.20 \text{ kg}) + (0.40 \text{ kg})} = 27 \text{ cm}$$

This means the fulcrum needs to be put at: $x_f = r_f + 35 \text{ cm} = 62 \text{ cm}$.

REFLECT   The answer seems reasonable, since the fulcrum is closer to the heavier mass. Just like on a seesaw, the heavier person needs to sit closer to the fulcrum to balance the two weights.

83.   ORGANIZE AND PLAN   In Example 8.15, there were four forces acting on the ladder: its weight $\vec{w} = m_{ladder}\vec{g}$, the normal force from the wall $\vec{n}_w$, the normal force from the floor $\vec{n}_f$, and static friction $\vec{f}_s$ at the ground. Adding the man means adding a fifth force: his weight $\vec{w} = m_{man}\vec{g}$. We will follow the same steps as in the example, and use the rotational equilibrium, with the net torque calculated from the base of the ladder.

Known: $L = 3.64$ m, $m_{ladder} = 18.2$ kg, $m_{man} = 75$ kg, $\theta = 16°$.

SOLVE   (a) The man is at the midpoint, so the torque from his weight will have the same radius, same angle, and same direction as that of the ladder itself. We choose to calculate the net torque from the base, so that the torques from $\vec{n}_f$ and $\vec{f}_s$ are zero.

$$\tau_{net} = -(\tfrac{1}{2}L)(m_{ladder}g)(\sin 16°) - (\tfrac{1}{2}L)(m_{man}g)(\sin 16°) + (L)(n_w)(\sin 74°)$$

Notice that the angles are different for the weight torques and the wall normal torque (see Figure 8.21). Since the net torque is zero, the normal force from the wall is:

$$n_w = \frac{\tfrac{1}{2}(m_{ladder} + m_{man})(g)(\sin 16°)}{\sin 74°} = (0.14)(18.2 \text{ kg} + 75 \text{ kg})(9.80 \text{ m/s}^2) = 130 \text{ N}$$

(b) The only difference in this case is that the radius for the man's torque is no longer $\tfrac{1}{2}L$, but is now $\tfrac{4}{5}L$. The normal force equation becomes:

$$n_w = \frac{(\tfrac{1}{2}m_{ladder} + \tfrac{4}{5}m_{man})(g)(\sin 16°)}{\sin 74°} = 190 \text{ N}$$

REFLECT   The normal force from the wall is much higher with the man on the ladder than it was without the man in Example 8.15. The normal force becomes higher still as the man climbs up the ladder. This means the friction force (which is equal but opposite to the normal force on the wall, see Figure 8.21). Consequently, the risk that the ladder will slip increases as the man climbs higher.

85.   ORGANIZE AND PLAN   There are four forces in this problem: the weight of the unknown mass $\vec{w} = m\vec{g}$, the upward force from the fulcrum $\vec{n}_f$, the weight of the meter stick $\vec{w} = m_{stick}\vec{g}$, and the string tension $\vec{T}$. If we choose an appropriate axis point, we should be able to solve both parts of this problem with rotational equilibrium. For part (a), we can pick the fulcrum position as our axis, so that its torque is zero. For part (b), we can pick the left end of the meter stick, where the unknown mass hangs.

Known: $m_{stick} = 0.160$ kg, $T = 2.50$ N.

SOLVE   (a) The angles are all 90 degrees, so we will not include the sine terms. From the fulcrum point, the weight from the unknown mass is a positive torque (with radius 0.3 m), whereas the weight of the stick (with radius 0.2m) and the string tension (with radius 0.7 m) are both negative torques. The net torque equation is:

$$\tau_{net} = (0.3 \text{ m}) mg - (0.2 \text{ m}) m_{stick}g - (0.7 \text{ m})T = 0$$

Solving for the unknown mass:

$$m = \frac{(0.2 \text{ m})(0.160 \text{ kg})(9.80 \text{ m/s}^2) + (0.7 \text{ m})(2.50 \text{ N})}{(0.3 \text{ m})(9.80 \text{ m/s}^2)} = 0.702 \text{ kg}$$

Where we have used the fact that $N = \text{kg m/s}^2$.

(b) From the left end of the meter stick, the upward force from the fulcrum is a positive torque (with radius 0.3 m), whereas the weight of the stick (with radius 0.5 m) and the string tension (with radius 1.0 m) are both negative torques. Note how the radii change, as we change the axis point. The net torque in this case is:

$$\tau_{net} = (0.3 \text{ m})n_f - (0.5 \text{ m}) m_{stick}g - (1.0 \text{ m})T = 0$$

Solving for the fulcrum force:

$$n_f = \frac{(0.5 \text{ m})(0.160 \text{ kg})(9.80 \text{ m/s}^2) + (1.0 \text{ m})(2.50 \text{ N})}{(0.3 \text{ m})} = 10.9 \text{ N}$$

REFLECT   We can check if our answers are right by checking that translational equilibrium is satisfied. The net force on the system is:

$$F_{net} = 10.9 \ N - (0.702 \ kg + 0.160 \ kg)(9.80 \ m/s^2) - (2.50 \ N) \approx 0$$

Taking into account rounding errors in the calculations, the answer is zero as it should be.

87.   **ORGANIZE AND PLAN**   The rotational inertia is now $I = Mr^2$ for a particle moving in a circle whose radius is the distance between the Earth and the Sun.

Known: $M = 5.97 \times 10^{24}$ kg, $r = 1.50 \times 10^{11}$ m, $\omega = 1$ rev/year.

SOLVE   Plugging in the given values, the Earth's rotational inertia and angular velocity are:

$$I = (5.97 \times 10^{24} \ kg)(1.50 \times 10^{11} \ m)^2 = 1.34 \times 10^{47} \ kg \ m^2$$

$$\omega = \frac{2\pi}{365 \ d} \left[\frac{1 \ d}{24 \cdot 60 \cdot 60 \ s} \right] = 1.99 \times 10^{-7} \ rad/s$$

Combining these values:

$$L = I\omega = (1.34 \times 10^{47} \ kg \ m^2)(1.99 \times 10^{-7} \ rad/s) = 2.67 \times 10^{40} \ J \ s$$

REFLECT   This is over a million times bigger than the angular momentum due to rotation (calculated in the previous problem). The large difference is because the orbital radius is so much larger than the Earth's radius. You might want to compare these results to the answers in Problems 8.55 and 8.56, where the rotational and orbital energy of the Earth are compared.

89.   **ORGANIZE AND PLAN**   When the disk drops on the turntable, they begin rotating together, so the rotational inertia becomes the sum of the two objects. We are told there are no external forces so the angular momentum before and after is the same.

Known: $\omega_0 = 3.25$ rad/s, $I_{table} = 0.225$ kg m^2, $I_{disk} = 0.104$ kg m^2.

SOLVE   The angular momentum before and after are the same:

$$L = I_{table}\omega_0 = (I_{table} + I_{disk})\omega$$

Solving for the unknown angular velocity:

$$\omega = \frac{I_{table}\omega_0}{I_{table} + I_{disk}} = \frac{(0.225 \ kg \ m^2)(3.35 \ rad/s)}{0.225 \ kg \ m^2 + 0.104 \ kg \ m^2} = 2.29 \ rad/s$$

REFLECT   The rotation speed slows, as we'd expect. Notice how this is similar to the translational case of a moving object colliding and sticking to a stationary object. By conservation of translational momentum, the speed slows down to compensate for the increase in mass.

91.   **ORGANIZE AND PLAN**   In vector form for the angular momentum is: $\vec{L} = I\vec{\omega}$ (Equation 8.23). If the merry-go-round stops, then $\Delta\vec{L} = 0 - \vec{L} = -\vec{L}$, and we can use this to find the torque: $\vec{\tau} = \Delta\vec{L}/\Delta t$ (Equation 8.24).

Known: $I = 35$ kg m^2, $\omega = 1.3$ rad/s, $\Delta t = 10$ s.

SOLVE   (a) The angular momentum points in the same direction as the angular velocity. The merry-go-round rotates clockwise, so the angular velocity either points up or down. Using your right hand, turn your fingers in a clockwise sense, and it should be apparent that your thumb points down. So, the angular momentum points down as well. As for the magnitude:

$$L = I\omega = (35 \ kg \ m^2)(1.3 \ rad/s) = 46 \ J \ s$$

(b) The magnitude of the torque for stopping the merry-go-round is: will point opposite of the angular momentum, so upward. For the magnitude:

$$\tau = \frac{\Delta L}{\Delta t} = \frac{0 - (46 \ J \ s)}{(10 \ s)} = -4.6 \ N \cdot m$$

Note that the common torque unit is the Newton-meter, which is equivalent to the Joule. The negative sign in the answer implies that the torque points in the opposite direction of the angular momentum, which in this case is up.

REFLECT   Since the torque is reducing the angular momentum, it makes sense that it points in the opposite direction.

93.   **ORGANIZE AND PLAN**   The figure below shows the radius and applied force drawn as vectors. We can use this to determine the torque direction.

SOLVE   Start by holding the drawing in front of you. Point the fingers on your right hand in the direction of the radius vector (straight up in this case). Now rotate your fingers toward the direction of the applied force (to the right in this case). Your thumb should be pointing straight ahead, which is into the piece of paper.

REFLECT   Imagine the wrench is turning a screw. The torque is being applied in a clockwise direction, so is the wrench tightening or loosening the screw? Answer: tightening such that the screw goes farther into the paper (some people remember this by "righty-tighty" vs. "lefty-loosey"). The upshot of this is that any torque will point in the direction that an imaginary screw would move in response to that torque.

95.   **ORGANIZE AND PLAN**   In Problem 43, we're told that the day is lengthening by 2.3 ms per century. We calculated that this was equal to a deceleration of: $\alpha = -6.14 \times 10^{-22}$ rad/s^2. We can use this to find the torque: $\vec{\tau} = I\vec{\alpha}$, where the rotational inertia of the Earth was determined in Problem 54 to be: $I = 9.72 \times 10^{37}$ kg m^2.

SOLVE   Picture the Earth rotating on its axis. Looking down from the North Pole, the rotation is in a counter-clockwise direction. This means the angular velocity points up in the direction of north. The rotation rate is slowing, so the acceleration is negative, meaning it points south in the opposite direction of the angular velocity. The torque also points to the south. The magnitude is:

$$\tau = I\alpha = (9.72 \times 10^{37} \text{ kg m}^2)(-6.14 \times 10^{-22} \text{ rad/s}^2) = -5.97 \times 10^{16} \text{ N} \cdot \text{m}$$

REFLECT   This is a humongous torque, but it has to be to have any effect on the entire Earth. Although it's enough to say that the torque points south, we have kept the negative sign just to reiterate that this torque is working opposite to the direction of the Earth's rotation.

97.   **ORGANIZE AND PLAN**   The applied torque can be found from Equation 8.16: $\tau = rF\sin\theta$. The angular acceleration is from Equation 8.17: $\alpha = \tau/I$, in which case we need the formula for the rotational inertia of a disk: $I = \frac{1}{2}mr^2$ (from Table 8.4).

Known: $m = 500$ g, $r = 12.5$ cm, $F = 23.5$ N.

SOLVE   (a) Since the force is applied tangentially, $\sin\theta = 1$ and the torque is: We'll need the rotational inertia for the angular momentum and the torque, so using the formula for a thin rod:

$$\tau = rF\sin\theta = (0.125 \text{ m})(23.5 \text{ N})(1) = 2.94 \text{ N} \cdot \text{m}$$

(b) For the angular acceleration, we insert the formula for the rotational inertia:

$$\alpha = \frac{\tau}{\frac{1}{2}mr^2} = \frac{(2.94 \text{ N} \cdot \text{m})}{\frac{1}{2}(0.500 \text{ kg})(0.125 \text{ m})^2} = 753 \text{ rad/s}^2$$

REFLECT   If there had been friction in the pulley, the net torque would have been reduced ($\tau_{net} = \tau - \tau_{friction}$), and the angular acceleration would have been less as well.

99.   **ORGANIZE AND PLAN**   We're asked for the angle that the saw turns through while stopping, so we could use either Equations 8.9 or 8.10. But in either case, we have to determine the angular acceleration first: $\alpha = \Delta\omega / \Delta t$ (Equation 8.6). For part (b), the kinetic energy is $K = \frac{1}{2}I\omega^2$, given that the saw is a disk with rotational inertia: $I = \frac{1}{2}mr^2$ (from Table 8.4). The torque is readily found from Equation 8.17: $\tau = I\alpha$.

Known: $\Delta t = 5.0$ ms, $\omega_0 = 3500$ rpm, $d = 19.0$ cm, $m = 0.860$ kg.

SOLVE   (a) The blade decelerates from $\omega_0 = 3500$ rpm $\left[2\pi/60\right] = 367$ rad/s to zero in 5.0 ms:

$$\alpha = \frac{\Delta\omega}{\Delta t} = \frac{0-(367\ \text{rad/s})}{5.0\times10^{-3}\ \text{s}} = -7.34\times10^4\ \text{rad/s}^2$$

Let's use Equation 8.10 to find the angle that the blade turns through:

$$\Delta\theta = \frac{\omega^2 - \omega_0^2}{2\alpha} = \frac{0-(367\ \text{rad/s})^2}{2(-7.34\times10^4\ \text{rad/s}^2)} = 0.918\ \text{rad} = 52.6°$$

(b) The rotational inertia of the saw is:

$$I = \frac{1}{2}mr^2 = \frac{1}{2}(0.860\ \text{kg})(\tfrac{1}{2}\cdot0.190\ \text{m})^2 = 3.89\times10^{-3}\ \text{kg m}^2$$

So the change in kinetic energy is just:

$$\Delta K = 0 - \frac{1}{2}I\omega^2 = -\frac{1}{2}(3.89\times10^{-3}\ \text{kg m}^2)(367\ \text{rad/s})^2 = -262\ \text{J}$$

(c) The torque for stopping the blade is:

$$\tau = I\alpha = (3.89\times10^{-3}\ \text{kg m}^2)(-7.34\times10^4\ \text{rad/s}^2) = -285\ \text{N}\cdot\text{m}$$

REFLECT   The blade turns only 0.14 revolutions when shutting off. It's amazing that it can brake so quickly. The required torque is substantial, and negative because it is a stopping torque.

101.   **ORGANIZE AND PLAN**   The lesson from Example 8.11 is that the object with the smallest rotational inertia will win the race. This is because more of its kinetic energy will be put into translational motion, rather than rotational motion. As in Example 8.11, we can calculate the velocity of each object. The fastest will win the race, and from its velocity we can calculate the time when it reaches the bottom, and the distance the others have covered in that time.

Known: $L = 1.50$ m, $\theta = 3.50°$.

SOLVE   (a) In general, the work-energy theorem tells us that:

$$K = \frac{1}{2}mv^2 + \frac{1}{2}I(v/r)^2 = mgh$$

$$\Rightarrow\quad v = \sqrt{\frac{2gL\sin\theta}{1+I/mr^2}}$$

Where we have used the fact that $\omega = v/r$ and $h = L\sin\theta$. By plugging in the rotational inertia for a solid sphere ($I_{\text{sphere}} = \frac{2}{5}mr^2$), solid cylinder ($I_{\text{cylinder}} = \frac{1}{2}mr^2$), hollow cylindrical shell ($I_{\text{shell}} = mr^2$), and hollow ball ($I_{\text{ball}} = \frac{2}{3}mr^2$), we can determine the velocity for each:

$$v_{\text{sphere}} = \sqrt{\tfrac{10}{7}gL\sin\theta} = 1.13\ \text{m/s}$$

$$v_{\text{cylinder}} = \sqrt{\tfrac{4}{3}gL\sin\theta} = 1.09\ \text{m/s}$$

$$v_{\text{shell}} = \sqrt{gL\sin\theta} = 0.947\ \text{m/s}$$

$$v_{\text{ball}} = \sqrt{\tfrac{6}{5}gL\sin\theta} = 1.04\ \text{m/s}$$

The sphere is the fastest, so it will reach the bottom first.

(b) The time it takes for the sphere to reach the bottom:

$$t = \frac{v_{\text{sphere}}}{L} = \frac{1.13\ \text{m/s}}{1.50\ \text{m}} = 0.753\ \text{s}$$

In this much time the other objects travel smaller distances:

$$x_{cylinder} = v_{cylinder}t = (1.09 \text{ m/s})(0.753 \text{ s}) = 0.821 \text{ m}$$

$$x_{shell} = v_{shell}t = (0.947 \text{ m/s})(0.753 \text{ s}) = 0.713 \text{ m}$$

$$x_{ball} = v_{ball}t = (1.04 \text{ m/s})(0.753 \text{ s}) = 0.783 \text{ m}$$

REFLECT   From fastest to slowest, we have: solid sphere, solid cylinder, hollow ball, hollow cylindrical shell. This is also the ranking of smallest to largest rotational inertia for the same mass and radius. The solid sphere has the least percentage of its mass at large radii, while the hollow cylindrical shell has the largest percentage of its mass at large radii (in fact all of its mass is at the outer edge of the object). This means more of its kinetic energy is in rotational motion, and less is in translational motion down the ramp.

103.   **ORGANIZE AND PLAN**   We have the initial and final velocity of the ball, and the distance over which it comes to rest, so we can figure the constant acceleration with Equation 2.10. Since the ball was moving horizontally, it strikes the raised arm in a tangential direction. So we can relate the acceleration of the ball to the angular acceleration of the arm through Equation 8.12: $a_t = r\alpha$. The torque is then: $\tau = I\alpha$ (Equation 8.17). But what is I in this case? It's the thing being acted on, which is the ball. Its rotational inertia is: $I = mr^2$, where r is the radius of the arm.

Known: $m = 0.145 \text{ kg}$, $v_{x0} = 42.5 \text{ m/s}$, $\Delta x = 5.00 \text{ cm}$, $r = 63.5 \text{ cm}$.

SOLVE   Using the info on how the ball stops, we can determine the acceleration:

$$a_x = \frac{v^2 - v_0^2}{2\Delta x} = \frac{(0)^2 - (42.5 \text{ m/s})^2}{2(0.0500 \text{ m})} = -1.81 \times 10^4 \text{ m/s}^2$$

As argued above, this deceleration can be related to the arm's angular acceleration: $\alpha = a_x / r$. Plugging this into the torque equation with the rotational inertia of the ball moving in a circle at the end of the player's arm:

$$\tau = I\alpha = mra_x = (0.145 \text{ kg})(0.635 \text{ m})(-1.81 \times 10^4 \text{ m/s}^2) = -1670 \text{ N} \cdot \text{m}$$

REFLECT   Notice that the same equation for the torque can be derived by using Equation 8.16 ($\tau = rF\sin 90°$), with the force on the ball: $F = ma_x$.

105.   **ORGANIZE AND PLAN**   We'll be comparing the periods of the physical pendulum of a meterstick ($T = 2\pi\sqrt{2L/3g}$) and the simple pendulum of a mass on a string ($T = 2\pi\sqrt{L/g}$).

SOLVE   (a) The period for the meter stick is:

$$T = 2\pi\sqrt{\frac{2L}{3g}} = 2\pi\sqrt{\frac{2(1.00 \text{ m})}{3(9.80 \text{ m/s}^2)}} = 1.64 \text{ s}$$

(b) We want to solve for the length of a simple pendulum that gives the same period:

$$L = g\left(\frac{T}{2\pi}\right)^2 = (9.80 \text{ m/s}^2)\left(\frac{1.64 \text{ s}}{2\pi}\right)^2 = 0.667 \text{ m}$$

REFLECT   The simple pendulum is shorter because we need to reduce its rotational inertia. The meter stick has a relatively small rotational inertia for a given length, so it accelerates faster (think of a small car accelerating faster than a big one). To speed up the simple pendulum, we shorten its length.

107.   **ORGANIZE AND PLAN**   We will be following the general derivation of Problem 104, but allowing for a more general set of parameters.

SOLVE (a) The torque due to gravity is applied at the center of mass: $\tau = -(d)(Mg)\sin\theta$, where we have included a negative sign because the torque tries to pull the pendulum to zero angle. The acceleration due to this torque is:

$$\alpha = \frac{\tau}{I} = -\left(\frac{dMg}{I}\right)\sin\theta \approx -\omega^2\theta$$

In the last step we take the small angler approximation and define $\omega = \sqrt{dMg/I}$, like in Problem 8.104. The period is then

$$T = \frac{2\pi}{\omega} = 2\pi\sqrt{\frac{I}{dMg}} = 2\pi\sqrt{\frac{d}{g}}\sqrt{\frac{I}{Md^2}}$$

(b) For the uniform stick, $d = \frac{1}{2}L$ and $I = \frac{1}{3}ML^2$, so $T = 2\pi\sqrt{2\,L/3\,g}$. For the simple pendulum, $d = L$ and $I = ML^2$, so $T = 2\pi\sqrt{L/g}$.

REFLECT This derivation makes clear how the period of a pendulum varies with the rotational inertia. The smaller the rotational inertia the shorter the period, i.e., the faster the pendulum oscillates back and forth.

9

GRAVITATION

CONCEPTUAL QUESTIONS

1. **ORGANIZE AND PLAN** Review section 9.1, especially the part that discusses the acceleration due to gravity and its relation to latitude.

   SOLVE Since the Earth rotates, objects on its surface must experience a centripetal acceleration [Eq. 1]

   $$a = \frac{v(\theta)^2}{R_E}$$

   to remain on the surface of the Earth. In Eq. 1, v is a function of the latitude θ. Near the poles, the circumference of the Earth is smaller so the distance covered during one rotation of the Earth is less than at the equator. Thus, the speed in Eq. 1 is smaller at the poles than at the equator. Near the equator, a greater fraction of an object's acceleration due to gravity is spent simply providing the necessary centripetal acceleration, leaving less gravitational acceleration available to contribute to an object's weight. This reduces the apparent value of g at the equator relative to the poles.

   The centripetal acceleration effect also results in the Earth's radius RE being slightly larger at the equator than at the poles. If you think of the Earth as a giant centrifuge, the greatest centrifugal "force" is at the equator. This concentrates more mass near the equator, which by Newton's law of gravity (see Eq. 9.1) will increase the force due to gravity, and thus the acceleration g due to gravity is also increased (by Newton's second law).

   But we're not finished. The fact that the Earth's radius is slightly larger at the equator than at the poles also contributes to a decrease in the force due to gravity, since the denominator of Eq. 9.1 is greater at the equator than at the poles. And again by Newton's second law this reduces the acceleration g due to gravity at the equator relative to the poles.

   Combining these effects leads to an overall decrease in the acceleration g due to gravity at the equator relative to the poles.

   REFLECT What would happen if Earth's gravitational field were magically turned off? By Newton's first law, objects would continue with their given tangential velocity and simply fly off of the Earth's surface. Can you find the analytical form of the velocity in Eq. 1 as a function of latitude?

3. **ORGANIZE AND PLAN** Eq. 5.5 ($W_g = -mg\Delta y$) expresses the work done by gravity. For the current problem, Δy is the change in the satellite's altitude above the Earth. Recall that for a force to do work on an object, the force must *not* be directed perpendicular to the object's displacement. If the force is perpendicular to the displacement, no work is done.

   SOLVE If the satellite travels in an elliptical orbit, its distance from the Earth varies as it travels (see, e.g., Figure 9.10). Thus, $\Delta y \neq 0$, so the Earth does work on the satellite.

   For a circular orbit, no work is done by the Earth on the satellite. This is because the satellite is always moving in a direction perpendicular to the direction in which gravity is pulling; i.e., tangential to its circular orbit.

   REFLECT Gravity is a conservative force, so the total work done in a single orbit is zero because the satellite regains its initial position. Thus, in a single orbit, all positive work is canceled out by an equal amount of negative work.

5. **ORGANIZE AND PLAN**   Escape speed is given by Eq. 9.6

$$v_{esc} = \sqrt{\frac{2GM_E}{R_E}}$$

and was derived using the principle of conservation of energy. Consider what happens if a non-conservative force such as drag acts on the projectile.

SOLVE   Drag from air resistance is a force that acts counter to the velocity (see Chapter 4) and that will do negative work on the projectile (see Chapter 5). Thus more work must be done to overcome this loss of energy (work) due to drag. The initial potential energy of the projectile cannot be increased, since it is position-dependent. Therefore, the initial kinetic energy must be increased by increasing the initial velocity. The escape velocity would therefore increase.

REFLECT   Is it possible to increase the mass of the projectile instead of its initial speed to generate the needed increase in initial kinetic energy? Why would this not work?

7. **ORGANIZE AND PLAN**   The Earth spins eastward on its axis once every 24 hours. This equates to a speed of 465 m/s at the equator.

SOLVE   Imagine that the rocket were launched from the bed of a truck traveling at 465 m/s. Would it be best to launch the rocket in the same direction as the truck's velocity, or in the opposite direction? To take advantage of the kinetic energy of the truck, it is best to launch in the same direction as the truck's velocity. The same is true when launching from a rotating planet. It is thus best to put the rotational speed of the Earth to use to give the rocket an extra boost, and thus to launch eastward.

Since the rotational speed of the Earth's surface is greatest at the equator and goes to zero at the poles, it is best to launch the rocket from as close as possible to the equator. Thus Florida is a better choice than Alaska.

REFLECT   An analogous situation is launching a rock using a slingshot (a la David and Goliath). Imagine fitting a small rocket engine to your rock and being able to launch it in any direction desired. The rock's launch speed would be greatest if you launched it in the direction in which the slingshot was turning.

9. **ORGANIZE AND PLAN**   Review the derivation of the escape velocity (Eq. 9.6), as well as Fig, 9.15 describing Newton's thought experiment about orbiting objects.

SOLVE   The derivation of the escape velocity (Eq. 9.6) is based on conservation of mechanical energy in the limit that the final position of the projectile is much, much greater than the radius of the Earth. Thus, the Earth may be treated as if all the mass were concentrated at a point at the center of the Earth, and the angle at which the projectile is launched makes no difference (provided the projectile trajectory is aimed above the horizon).

REFLECT   Every time you throw a ball you put it in orbit about the Earth. Of course, the ball strikes the surface of the Earth before it can complete a revolution. If you impart the ball with a greater and greater initial velocity, eventually it will orbit the Earth (like orbit A in Fig. 9.15). The orbital path will be elliptic (with a circle being a special case), but the ball will not escape from the Earth. If the initial velocity of the ball is greater than the escape velocity, the orbit will become parabolic and will not return to the Earth. The ball has escaped!

Consider dropping a ball so that it falls with no horizontal velocity component. May we consider this ball in orbit about the Earth?

11. **ORGANIZE AND PLAN**   To travel from the Earth to the Moon, the spacecraft must escape the Earth's gravitational field. On the return trip, the spacecraft must escape the Moon's gravitational field. Use the escape velocity therefore to provide a rough approximation of the relative amounts of fuel needed for each trip. Use the data from Appendix E to find the escape velocities.

Known: $M_E = 5.97 \times 10^{24}$ kg, $M_M = 7.35 \times 10^{22}$ kg, $R_E = 6.37 \times 10^6$ m, $R_M = 1.74 \times 10^6$ m.

SOLVE   The escape velocity is given by Eq. 9.6, reproduced here.

$$v_{esc} = \sqrt{\frac{2GM}{R}}$$

Since the kinetic energy is proportional to the velocity squared, the ratio of kinetic energy needed to escape the Earth to that needed to escape the Moon is [Eq. 1]

$$\left(\frac{v_{esc}^{E}}{v_{esc}^{M}}\right)^{2} = \frac{M_{E}R_{M}}{R_{E}M_{M}} = \frac{(5.97\times10^{24}\text{ kg})(1.74\times10^{6}\text{ m})}{(6.37\times10^{6}\text{ m})(7.35\times10^{22}\text{ kg})} = 22.19$$

Thus we estimate that an order of magnitude more fuel is needed for the Earth-Moon trip than for the return trip.

REFLECT   What are some of the many factors neglected in making this estimation?

13.   **ORGANIZE AND PLAN**   Review the derivation of escape speed (Eq. 9.6). Note that the escape speed is the speed needed to attain an infinite orbital radius.

SOLVE   The total mechanical energy of an orbiting object is [Eq. 1]

$$E = \frac{1}{2}mv^{2} - \frac{GMm}{r}$$

To attain an orbit of infinite radius, both terms in Eq. 1 must tend to zero. The kinetic term because the minimum speed at infinite orbit is zero, and the potential term because $r \to \infty$. Since energy is conserved, the total initial energy of the object (at the Earth's surface in this example) must also be zero, or [Eq. 2]

$$\frac{1}{2}mv_{esc}^{2} = \frac{GM_{E}m}{R_{E}} \therefore$$

Eq. 2 does not depend on the mass of the orbiting object, since it appears on both sides of the equation. Thus, escape speed is independent of mass.

REFLECT   The escape speed does not depend on the mass of the object being put into orbit, but it does depend on the mass of the object being escaped from!

15.   **ORGANIZE AND PLAN**   Review Kepler's third law.

SOLVE   From Kepler's third law [Eq. 1],

$$\frac{a^{3}}{T^{2}} = C$$

the orbital period of an object depends only on the semimajor axis a of the orbit and the value of C. For circular orbits, the value of C is given by Eq. 9.4, reproduced here.

$$C_{circular} = \frac{GM}{4\pi^{2}}$$

It turns out that for elliptical orbits, the value of C is the same [Eq. 2]

$$C_{elliptical} = C_{circular} = \frac{GM}{4\pi^{2}}$$

Thus, the orbital period of a planet orbiting a star does not depend on the mass of the planet, only the mass of the star and the semimajor axis of the orbit.

REFLECT   Another way to write Kepler's third law is [Eq. 3] $a^{3} \propto T^{2}$.

MULTIPLE-CHOICE PROBLEMS

17.   **ORGANIZE AND PLAN**   Use Newton's law of gravitation (Eq. 9.1) and Newton's second law ($\vec{F}_{net} = m\vec{a}$) from chapter 4 to calculate the acceleration due to gravity on the surface of Mars.

Known: $M_{Mars} = 6.42\times10^{23}$ kg, $R_{Mars} = 3.37\times10^{6}$ m.

SOLVE   Newton's law of gravitation applied to Mars gives [Eq. 1]

$$F = \frac{GM_{Mars}m}{R_{Mars}^{2}}$$

Applying Newton's second law to find the acceleration gives [Eq. 2]

$$F_{net} = ma = \frac{GM_{Mars}m}{R_{Mars}^2}$$

$$a = \frac{GM_{Mars}m}{R_{Mars}^2} = \frac{(6.67 \times 10^{-11} \text{ N} \cdot \text{m}^2/\text{kg}^2)(6.42 \times 10^{23} \text{ kg})}{(3.37 \times 10^6 \text{ m})^2} = 3.77 \text{ m/s}^2$$

where the acceleration is directed toward the center of the planet. Thus, the response is (a).

REFLECT   Can you convince yourself that the units in the quotient in Eq. 2 divide out properly to give m/s²?

19.   **ORGANIZE AND PLAN**   Use Kepler's law for circular orbits (Eq. 9.4) to calculate the mass of the planet. Be careful to convert all quantities to SI units.

Known: m = 150 kg, $R = 7.1 \times 10^6$ m, $T = 4$ h.

SOLVE   Kepler's law for circular orbits is [Eq. 1]

$$\frac{R^3}{T^2} = \frac{GM}{4\pi^2}$$

where R is the satellite's orbital radius, T is its orbital period, M is the planet's mass, and G is Newton's universal gravitation constant. Solving for the planet's mass M gives [Eq. 2]

$$M = \frac{R^3}{T^2}\frac{4\pi^2}{G} = \frac{(7.1 \times 10^6 \text{ m})^3}{(4 \text{ h})^2}\left(\frac{1 \text{ h}}{60 \text{ min}}\right)^2\left(\frac{1 \text{ min}}{60 \text{ s}}\right)^2\frac{4\pi^2}{(6.67 \times 10^{-11} \text{ N} \cdot \text{m}^2/\text{kg}^2)} = 1.0 \times 10^{24} \text{ kg}$$

which is response (d).

REFLECT   This planet is about 6 times less massive than the Earth, but slightly more massive than Mars. Note that the mass of the satellite did not figure in the calculation.

21.   **ORGANIZE AND PLAN**   This problem is the same as the previous, except that we must solve for the new orbital radius given the ratio of the periods. Therefore, use the same strategy as for problem 20.

Known: $a_0 = 7,000$ km, $T = 2 T_0$.

SOLVE   Using Eq. 4 of problem 20 gives [Eq. 1]

$$\frac{a_0^3}{a^3} = \frac{T_0^2}{T^2}$$

$$a = a_0\sqrt[3]{\frac{T^2}{T_0^2}} = (7,000 \text{ km})\sqrt[3]{4} = 11,100 \text{ km}$$

which is response (b).

REFLECT   Note that we did not have to express the initial radius in SI units, since the result is simply a multiple of the initial radius.

23.   **ORGANIZE AND PLAN**   This problem is the reverse of problem 22. It can be solved using the same strategy, but the initial and final positions are reversed. This initial height of the ball $h = 250$ km above the surface of the Earth, so from the center of the Earth [Eq. 1], $R_0 = R_E + h$.

Known: $v_0 = 0$, $h = 250$ km, $R = R_E$.

SOLVE   By conservation of energy, equate the initial and final mechanical energy. Use Eq. (1) to express the initial radial position of the ball [Eq. 2]

$$K_0 + U_0 = K + U$$

$$\frac{1}{2}mv_0^2 - \frac{GM_E m}{R_0} = \frac{1}{2}mv^2 - \frac{GM_E m}{R}$$

$$-\frac{GM_E}{R_E + h} = \frac{1}{2}v^2 - \frac{GM_E}{R_E}$$

$$v = \sqrt{2GM_E\left(\frac{1}{R_E} - \frac{1}{R_E + h}\right)}$$

$$= \sqrt{2\left(6.67\times10^{-11}\,\text{N·m}^2/\text{kg}^2\right)\left(5.97\times10^{24}\,\text{kg}\right)\left(\frac{1}{\left(6.37\times10^6\,\text{m}\right)} - \frac{1}{\left(6.37\times10^6\,\text{m}\right)+\left(0.25\times10^6\,\text{m}\right)}\right)}$$

$$= 2200\ \text{m/s}$$

which is response (b).

REFLECT   Notice that the ball's speed, 2,200 m/s, is slightly less than the initial speed of the rocket in Problem 22. This is reasonable since the rocket climbs to a slightly higher altitude (300 km) than the altitude from which the ball was dropped.

25.   **ORGANIZE AND PLAN**   Consider Kepler's first law, which states that the orbit of each planet is an ellipse, with the Sun at one focus, and Fig. 9.10, which defines some terms for ellipses.

SOLVE   Kepler's first law states that the Sun is at one focus of each of the elliptical planetary orbits. From Fig. 9.10, we see that the closest approach of an object to the focus of its elliptical orbit is called the perihelion, which is response (b).

REFLECT   The Greek word "peri" means about, near, enclosing, and the Greek word helion refers to the Sun.

27.   **ORGANIZE AND PLAN**   Use Eq. 9.4 to find the period of the satellite. If the satellite is at a height h above the surface of the Earth, then the orbital radius is [Eq. 1] $R = R_E + h$. Converting h to SI units gives [Eq. 2]

$$h = 4{,}000\ \text{km}\frac{10^3\ \text{m}}{\text{km}} = 4\times10^6\ \text{m}$$

Known: $h = 4\times10^6$ m, $M = M_E$

SOLVE   Use Eq. 9.4 and Eqs. (1) and (2) to solve for the period T [Eq. 3].

$$\frac{R^3}{T^2} = \frac{GM}{4\pi^2}$$

$$T = 2\pi\sqrt{\frac{\left(R_E + h\right)^3}{GM_E}}$$

$$= 2\pi\sqrt{\frac{\left(6.37\times10^6\ \text{m} + 4\times10^6\ \text{m}\right)^3}{\left(6.67\times10^{-11}\ \text{N·kg}^2/\text{m}^2\right)\left(5.97\times10^{24}\ \text{kg}\right)}}$$

$$= 1.05\times10^4\ \text{s}$$

Converting seconds to hours gives [Eq. 4]

$$1.05\times10^4\ \text{s}\frac{1\ \text{min}}{60\ \text{s}}\frac{1\ \text{h}}{60\ \text{min}} = 2.92\ \text{h}$$

which is response (b).

REFLECT   Is it possible for this satellite to have a geosynchronous orbit?

PROBLEMS

29.   **ORGANIZE AND PLAN**   Use Newton's law of gravitation (Eq. 9.1) to find the approximate attractive force due to gravity between the two football players.

Known: $r = 25.0$ m, $m_1 = m_2 = 115$ kg.

SOLVE   Newton's law of gravitation gives [Eq. 1]

$$F = \frac{Gm_1m_2}{r^2} = \frac{\left(6.67\times10^{-11}\ \text{N}\cdot\text{kg}^2/\text{m}^2\right)\left(115\ \text{kg}\right)^2}{\left(25.0\ \text{m}\right)^2} = 1.41\times10^{-9}\ \text{N} = 1.41\ \text{nN}$$

where the symbol "nN" represents nano-newtons.

REFLECT   Note that this is an approximate calculation since the football players are not far enough apart relative to their size to be considered point particles. If they were, say, 25 km apart, the calculation would be much more accurate.

31.   **ORGANIZE AND PLAN**   Use Newton's law of gravitation to calculate the forces between the stars, and between the Sun and the Earth.

Known: $m_1 = 2.3\times10^{30}$ kg, $m_2 = 6.8\times10^{30}$ kg, $r = 8.8\times10^{11}$ m.

SOLVE   The force between the two stars is [Eq. 1]

$$F = \frac{Gm_1m_2}{r^2}$$

$$F = \frac{\left(6.67\times10^{-11}\ \text{N}\cdot\text{kg}^2/\text{m}^2\right)\left(2.3\times10^{30}\ \text{kg}\right)\left(6.8\times10^{30}\ \text{kg}\right)}{\left(8.8\times10^{11}\ \text{m}\right)^2} = 1.35\times10^{27}\ \text{N}$$

The force between the Sun and the Earth is [Eq. 2]

$$F = \frac{Gm_1m_2}{r^2}$$

$$F = \frac{\left(6.67\times10^{-11}\ \text{N}\cdot\text{kg}^2/\text{m}^2\right)\left(5.97\times10^{24}\ \text{kg}\right)\left(1.99\times10^{30}\ \text{kg}\right)}{\left(150\times10^6\ \text{m}\right)^2} = 3.52\times10^{28}\ \text{N}$$

which is approximately 26 times more force than between the binary stars.

REFLECT   The Earth is much less massive than the stars, yet the force between the Earth and Sun is an order of magnitude greater than between the binary stars. This is because the binary stars are farther apart, and the distance between the two objects is squared, increasing the effect this factor plays.

33.   **ORGANIZE AND PLAN**   Newton's law of gravitation works for protons and electrons as well as for heavenly bodies. The mass of the proton is [Eq. 1] $m_p = 1.67\times10^{-27}$ kg, and of the electron is [Eq. 2] $m_{e^-} = 9.11\times10^{-31}$ kg

Known: $r = 5.29\times10^{-11}$ m.

SOLVE   Using Newton's law of gravitation, the force between the electron and proton is [Eq. 3]

$$F = \frac{Gm_p m_{e^-}}{r^2}$$

$$F = \frac{\left(6.67\times10^{-11}\ \text{N}\cdot\text{kg}^2/\text{m}^2\right)\left(1.67\times10^{-27}\ \text{kg}\right)\left(9.11\times10^{-31}\ \text{kg}\right)}{\left(5.29\times10^{-11}\ \text{m}\right)^2} = 3.63\times10^{-47}\ \text{N}$$

REFLECT   The electromagnetic force is some 10^{36} times stronger than the force due to gravity. This means the electromagnetic force between two protons a light-year apart would be approximately the same as the gravitational force between two protons a centimeter apart.

35.   **ORGANIZE AND PLAN**   Combine Newton's law of gravity with his second law, $\vec{F} = m\vec{a}$, to find the acceleration at the surface of Saturn and Jupiter.

SOLVE   Using the data from Appendix E, the acceleration due to gravity on Saturn is [Eq. 1]

$$F = \frac{GM_{Sat}m}{R_{Sat}^2} = ma$$

$$a = \frac{GM_{Sat}}{R_{Sat}^2} = \frac{\left(6.67\times10^{-11}\ \text{N}\cdot\text{kg}^2/\text{m}^2\right)\left(569\times10^{24}\ \text{kg}\right)}{\left(56.8\times10^6\ \text{m}\right)^2} = 11.8/s^2$$

The acceleration due to gravity on Jupiter is [Eq. 2]

$$F = \frac{GM_{Jup}m}{R_{Jup}^2} = ma$$

$$a = \frac{GM_{Jup}}{R_{Jup}^2} = \frac{\left(6.67\times10^{-11}\ \text{N}\cdot\text{kg}^2/\text{m}^2\right)\left(1.90\times10^{27}\ \text{kg}\right)}{\left(69.1\times10^6\ \text{m}\right)^2} = 26.5\ \text{m/s}^2$$

REFLECT   The acceleration due to gravity on Saturn is slightly greater than on Earth. On Jupiter, the acceleration due to gravity is almost three times greater than on earth.

37.   **ORGANIZE AND PLAN**   Use Kepler's third law applied to circular orbits (Eq. 9.4). Convert all units to SI units in the calculation.
Known: $R = 23{,}500$ km, $T = 1.26$ days.
SOLVE   Applying Eq. 9.4 gives [Eq. 1]

$$\frac{R^3}{T^2} = \frac{GM}{4\pi^2}$$

$$M = 4\pi^2\frac{R^3}{GT^2} = 4\pi^2\frac{\left(23{,}500\ \text{km}\right)^3}{\left(6.67\times10^{-11}\ \text{N}\cdot\text{m}^2/\text{kg}^2\right)\left(1.26\ \text{days}\right)^2}\left(\frac{10^3\ \text{m}}{\text{km}}\right)^3\left(\frac{\text{day}}{24\ \text{h}}\right)^2\left(\frac{\text{h}}{60\ \text{min}}\right)^2\left(\frac{\text{min}}{60\ \text{s}}\right)^2 = 6.48\times10^{23}\ \text{kg}$$

This is within 1% of the value given in Appendix E.

REFLECT   What factors are neglected in this calculation that might lead to the difference in this result compared with Appendix E? The problem states that the orbit is "nearly" circular, so that a more accurate calculation would use the semimajor axis a of the elliptical orbit instead of a radius R of a circular orbit.

39.   **ORGANIZE AND PLAN**   To use Newton's law of gravitation (Eq. 9.1), the distance between the centers of the balls is needed. This is [Eq. 1]

$$r = \frac{D_1}{2} + \frac{D_2}{2} + d$$

where D_i is the diameter of ball i and d is the separation between the surfaces of the balls. Also needed is the mass of the balls, which is [Eq. 2]

$$m_i = \rho V_i = \rho\frac{4}{3}\pi\left(\frac{D_i}{2}\right)^3$$

where $i = 1,2$.
Known: $D_1 = 1.0$ cm, $D_2 = 3.4$ cm, $d = 0.50$ cm, $\rho = 11{,}350\ \text{kg/m}^3$, $l = 15.0$ cm.

SOLVE   (a) Using Newton's law of gravitation, Eq. (1) and Eq. (2), the magnitude of the attractive force between the small and large balls is [Eq. 3]

$$F = \frac{Gm_1m_2}{r^2} = \frac{G\left(\rho\frac{4}{3}\pi\left(\frac{D_1}{2}\right)^3\right)\left(\rho\frac{4}{3}\pi\left(\frac{D_2}{2}\right)^3\right)}{\left(\frac{D_1}{2} + \frac{D_2}{2} + d\right)^2}$$

$$= \frac{16}{144}\pi^2\frac{\left(6.67\times10^{-9}\ \text{N}\cdot\text{m}^2/\text{kg}^2\right)\left(11{,}350\ \text{kg/m}^3\right)^2\left(0.010\ \text{m}\right)^3\left(0.034\ \text{m}\right)^3}{\left(\frac{0.01\ \text{m}}{2} + \frac{0.034\ \text{m}}{2} + 0.50\ \text{m}\right)^2} = 3.86\times10^{-9}\ \text{N}$$

(b) The torque on the wire is given by Eq. 8.14, with $\theta = 90°$. Since there are two small balls (one at each end of the rod), and the forces on them are both contributing to a torque in the same direction, the total torque will be the sum of the individual torques, or [Eq. 4]

$$\tau = 2\left(\frac{l}{2} + \frac{D_1}{2}\right)F\sin(\theta) = (0.15\text{ m} + 0.010\text{ m})(3.86\times10^{-9}\text{ N}) = 6.17\times10^{-10}\text{ N}\cdot\text{m}$$

REFLECT   Notice that the distance used to calculate the torque is from the center of the ball to the wire at the center of the rod.

41.   **ORGANIZE AND PLAN**   Use Newton's law of gravitation to find the height h above sea level at which the acceleration is equal to g. The distance R between the Earth's center and the object being accelerated is [Eq. 1] $R = R_E + h$.

Known: $g = 9.8$ m/s^2

SOLVE   The acceleration of an object with mass m at a distance R from the center of the Earth may be found from Newton's second law and his law of gravitation [Eq. 2].

$$F = ma$$

$$\frac{GM_E m}{R^2} = ma$$

$$a = \frac{GM_E}{(R_E + h)^2}$$

Solving for h gives [Eq. 3]

$$h = \sqrt{\frac{GM_E}{a}} - R_E = \sqrt{\frac{(6.67\times10^{-11}\text{ N}\cdot\text{m}^2/\text{kg}^2)(5.97\times10^{24}\text{ kg})}{a}} - 6.37\times10^6\text{ m}$$

(a) Substituting $g(1-0.001)$ for a in Eq. 3 gives $h = 7.56$ km.
(b) Substituting $g(1-0.01)$ for a in Eq. 3 gives $h = 36.5$ km.
(c) Substituting $g(1-0.1)$ for a in Eq. 3 gives $h = 349$ km.

REFLECT   Notice that as $a \to 0$ in Eq. 3, $h \to \infty$.

43.   **ORGANIZE AND PLAN**   Refer to Fig. 9.8 (a) for the definition of the major (a) and minor (b) axes of an ellipse. Use Eq. 9.3 to find the eccentricity e. We are told that [Eq. 1] $a = 2b$.

SOLVE   Using Eq. 9.3 and Eq. (1), the eccentricity is [Eq. 2]

$$e = \sqrt{1 - \left(\frac{b}{a}\right)^2} = \sqrt{1 - \left(\frac{1}{2}\right)^2} = 0.866$$

REFLECT   The eccentricity is thus intermediate between that of the ellipses shown in Fig. 9.9.

45.   The distance of the Sun from the geometric center of Mars' orbit is given by ([Eq. 1]; see Eq. 9.3) $c = ea$, where a is the semimajor axis and e is the eccentricity of the orbit. Use the data from Appendix E to find the distance c.

Known: $e = 0.093$, $a = 228\times10^6$ km (Appendix E).

SOLVE   Inserting the numerical values into Eq. (1) gives [Eq. 2]

$$c = 0.093(228\times10^6\text{ km}) = 21.2\times10^6\text{ km}$$

as the distance from the Sun to the geometric center of the orbit of Mars.

REFLECT   This is a rather short distance on the scale of the solar system.

47.   **ORGANIZE AND PLAN**   Use the data from Appendix E to calculate the constant C for the given planetary orbits. Convert all units to SI units.

Known: $a_{\text{Merc}} = 57.9\times10^6$ km, $a_{\text{Ven}} = 108\times10^6$ km, $a_{\text{Jup}} = 779\times10^6$ km, $T_{\text{Merc}} = 88.0$ days, $T_{\text{Ven}} = 225$ days, $T_{\text{Jup}} = 4.33\times10^3$ days.

SOLVE   From the data for Mercury, the constant C is [Eq. 1]

$$C = \frac{a_{Merc}^3}{T_{Merc}^2} = \frac{(57.6 \times 10^6 \text{ km})^3}{(88.0 \text{ days})^2} \left(\frac{10^3 \text{ m}}{1 \text{ km}}\right)^3 \left(\frac{1 \text{ day}}{24 \text{ h}}\right)^2 \left(\frac{1 \text{ h}}{60 \text{ min}}\right)^2 \left(\frac{1 \text{ min}}{60 \text{ s}}\right)^2$$

$$= 3.31 \times 10^{18} \text{ m}^3/\text{s}^2$$

From the data for Venus, the constant C is [Eq. 2]

$$C = \frac{a_{Ven}^3}{T_{Ven}^2} = \frac{(108 \times 10^6 \text{ km})^3}{(225 \text{ days})^2} \left(\frac{10^3 \text{ m}}{1 \text{ km}}\right)^3 \left(\frac{1 \text{ day}}{24 \text{ h}}\right)^2 \left(\frac{1 \text{ h}}{60 \text{ min}}\right)^2 \left(\frac{1 \text{ min}}{60 \text{ s}}\right)^2$$

$$= 3.33 \times 10^{18} \text{ m}^3/\text{s}^2$$

From the data for Jupiter, the constant C is [Eq. 3]

$$C = \frac{a_{Jup}^3}{T_{Jup}^2} = \frac{(779 \times 10^6 \text{ km})^3}{(4.33 \times 10^3 \text{ days})^2} \left(\frac{10^3 \text{ m}}{1 \text{ km}}\right)^3 \left(\frac{1 \text{ day}}{24 \text{ h}}\right)^2 \left(\frac{1 \text{ h}}{60 \text{ min}}\right)^2 \left(\frac{1 \text{ min}}{60 \text{ s}}\right)^2$$

$$= 3.37 \times 10^{18} \text{ m}^3/\text{s}^2$$

REFLECT   Within rounding errors, these results agree.

49.   **ORGANIZE AND PLAN**   Use Kepler's third law applied to circular orbits (Eq. 9.4). The radii of the satellites' orbits are related by [Eq. 1] $R_A = 2R_B$

SOLVE   Applying Kepler's third law to the orbit of satellites A and B, and using Eq. (1), gives [Eq. 2]

$$\frac{R_B^3}{T_B^2} = \frac{GM}{4\pi^2} = \frac{R_A^3}{T_A^2}$$

$$\frac{T_B}{T_A} = \sqrt{\frac{R_B^3}{R_A^3}} = \sqrt{\left(\frac{1}{2}\right)^3} = \sqrt{1/8} \approx 0.35$$

REFLECT   Reducing the radius by a factor of 2 results in a factor of ~3 reduction in the period.

51.   **ORGANIZE AND PLAN**   Use Kepler's third law applied to circular orbits (Eq. 9.4). If the satellite is at a height $h = R_E$ above the surface of the Earth, then its distance from the *center* of the Earth is [Eq. 1]

$$R = R_E + h$$
$$R = 2 R_E$$

where R_E is the radius of the Earth.

SOLVE   (a) For a circular orbit of radius R, Kepler's third law gives the orbital period as [Eq. 2]

$$\frac{R^3}{T^2} = \frac{GM_E}{4\pi^2}$$

$$T = 2\pi \sqrt{\frac{8R_E^3}{GM_E}} = 2\pi \sqrt{\frac{8(6.37 \times 10^6 \text{ m})^3}{(6.67 \times 10^{11} \text{ N} \cdot \text{m}^2/\text{kg}^2)(5.97 \times 10^{24} \text{ kg})}} = 1.43 \times 10^4 \text{ s}$$

where the second line is obtained using Eq. (1).

(b) If the period is doubled, the new height will be [Eq. 3]

$$2T = 2\pi \sqrt{\frac{(R_E + h_{new})^3}{GM_E}}$$

$$h_{new} = \sqrt[3]{\frac{T^2 GM_E}{\pi^2}} - R_E = (2R_E)\sqrt[3]{4} - R_E = 1.39 \times 10^7 \text{ m}$$

REFLECT   Doubling the period results in an increase in the orbital radius of approximately 10%, because the radius is cubed in Kepler's law and the period is only squared.

53.   **ORGANIZE AND PLAN**   The elliptical orbit of the spacecraft will have a perihelion distance equal to Earth's orbital radius, and an aphelion distance equal to Jupiter's orbital radius [Eq. 1],

$$d_p = r_E$$
$$d_a = r_J$$

Using Fig. 9.11, we can find the semimajor axis of the spacecraft's orbit [Eq. 2].

$$2a = d_p + d_a$$

$$a = \frac{1}{2}\left(d_p + d_a\right)$$

The travel time from the Earth to Jupiter is half the period of the orbit, which can be found using Kepler's third law.

Known: $r_E = 150 \times 10^9$ m, $r_J = 778 \times 10^9$ m (Appendix E).

SOLVE   The period of the spacecraft's orbit is [Eq. 3]

$$T = \sqrt{\frac{a^3}{C}} = \sqrt{\frac{\left(r_E + r_J\right)^3}{8C}} = \sqrt{\frac{\left(150 \times 10^9 \text{ m} + 778 \times 10^9 \text{ m}\right)^3}{8\left(3.36 \times 10^{18} \text{ m}^3/\text{s}^2\right)}} = 1.72 \times 10^8 \text{ s} = 1,996 \text{ days}$$

so half the period is approximately 1,000 days.

REFLECT   The travel time is approximately three Earth years, but somewhat less than ¼ of a year for Jupiter.

55.   **ORGANIZE AND PLAN**   Apply Eq. 9.5 to the Sun-Jupiter and Sun-Saturn systems.

Known: $M_S = 1.99 \times 10^{30}$ kg, $M_J = 1.90 \times 10^{27}$ kg, $M_{Sat} = 569 \times 10^{24}$ kg, $r_{S-J} = 778 \times 10^9$ m, $r_{Sat-S} = 1.43 \times 10^{12}$ m (Appendix E).

SOLVE   (a) Using Eq. 9.5, the potential energy of the Sun-Jupiter system is [Eq. 1]

$$U_{S-J} = -\frac{GM_S M_J}{r_{S-J}} = -\frac{\left(6.67 \times 10^{-11} \text{ N} \cdot \text{m}^2/\text{kg}^2\right)\left(1.99 \times 10^{30} \text{ kg}\right)\left(1.90 \times 10^{27} \text{ kg}\right)}{778 \times 10^9 \text{ m}} = -3.24 \times 10^{35} \text{ J}$$

(b) For the Sun-Saturn system it is [Eq. 2]

$$U_{S-Sat} = -\frac{GM_S M_{Sat}}{r_{S-Sat}} = -\frac{\left(6.67 \times 10^{-11} \text{ N} \cdot \text{m}^2/\text{kg}^2\right)\left(1.99 \times 10^{30} \text{ kg}\right)\left(569 \times 10^{24} \text{ kg}\right)}{1.43 \times 10^{12} \text{ m}} = -5.28 \times 10^{34} \text{ J}$$

REFLECT   The potential energy of the Sun-Jupiter system is 1 order of magnitude more than that of the Sun-Saturn system.

57.   **ORGANIZE AND PLAN**   Use conservation of mechanical energy to find the speed of the ball when it strikes the ground. Since the initial velocity with respect to the Earth is zero, the initial total energy is [Eq. 1]

$$E_0 = -\frac{GM_E m_{ball}}{r_0}$$

where r_0 is the distance from the center of the earth, and is given by [Eq. 2] $r_0 = R_E + h$, where h is the height from which the ball is dropped.

The final energy will be [Eq. 3]

$$E_f = \frac{1}{2}m_{ball}v^2 - \frac{GM_E m_{ball}}{R_E}$$

Known: $m_{ball} = 1$ kg, $M_E = 5.97 \times 10^{24}$ kg, $R_E = 6.37 \times 10^6$ m (Appendix E).

SOLVE   (a) Using conservation of mechanical energy, equate Eqs. (1) and (3), then solve for the final velocity v. For a height h of 10 km, this gives [Eq. 4]

$$-\frac{GM_E m_{ball}}{R_E + h} = \frac{1}{2}m_{ball}v^2 - \frac{GM_E m_{ball}}{R_E}$$

$$v = \sqrt{2GM_E\left(\frac{1}{R_E} - \frac{1}{R_E + h}\right)}$$

$$= \sqrt{2\left(6.67 \times 10^{-11} \text{ N} \cdot \text{m}^2/\text{kg}^2\right)\left(5.97 \times 10^{24} \text{ kg}\right)\left(\frac{1}{6.37 \times 10^6 \text{ m}} - \frac{1}{6.37 \times 10^6 \text{ m} + 10^4 \text{ m}}\right)}$$

$$= 443 \text{ m/s}$$

(b) For a height h of 1,000 km, this gives [Eq. 5]

$$v = \sqrt{2(6.67\times10^{-11}\ \text{N}\cdot\text{m}^2/\text{kg}^2)(5.97\times10^{24}\ \text{kg})\left(\frac{1}{6.37\times10^6\ \text{m}} - \frac{1}{6.37\times10^6\ \text{m}+10^6\ \text{m}}\right)} = 4{,}119\ \text{m/s}$$

(c) For a height h of 10^7 m, this gives [Eq. 6]

$$v = \sqrt{2(6.67\times10^{-11}\ \text{N}\cdot\text{m}^2/\text{kg}^2)(5.97\times10^{24}\ \text{kg})\left(\frac{1}{6.37\times10^6\ \text{m}} - \frac{1}{6.37\times10^6\ \text{m}+10^7\ \text{m}}\right)} = 8{,}739\ \text{m/s}$$

REFLECT At 10^7 m above the surface, the ball is nearing the escape speed of the earth. At what height would the ball achieve escape speed upon striking the Earth? If we drop the ball from beyond this distance, will it still be attracted towards the Earth by gravity?

59. **ORGANIZE AND PLAN** The perihelion and aphelion distances for the Earth are (from Ex. 9.7) [Eq. 1]

$$d_p = 1.47\times10^{11}\ \text{m}$$
$$d_a = 1.52\times10^{11}\ \text{m}$$

The change in the potential energy is simply [Eq. 2] $\Delta U = U_p - U_a$, where the potential energy U is given by Eq. 9.5. By conservation of energy, the change in the kinetic energy [Eq. 3],

$$\Delta K = K_p - K_a = (1/2)M_E\left(v_p^2 - v_a^2\right)$$

must cancel the change in the potential energy [Eq. 4],

$$\Delta U + \Delta K = 0.$$

Known: $M_E = 5.97\times10^{24}$ kg, $M_S = 1.99\times10^{30}$ kg (Appendix E).

SOLVE Using Eq. 2 and the known quantities, the change in the Earth's potential energy is [Eq. 6]

$$\Delta U = -GM_E M_S\left(\frac{1}{d_p} - \frac{1}{d_a}\right)$$

$$= -(6.67\times10^{-11}\ \text{N}\cdot\text{m}^2/\text{kg}^2)(1.99\times10^{30}\ \text{kg})(5.97\times10^{24}\ \text{kg})\left(\frac{1}{1.47\times10^{11}\ \text{m}} - \frac{1}{1.52\times10^{11}\ \text{m}}\right)$$

$$= -1.77\times10^{32}\ \text{J}$$

Conservation of angular momentum demands that [Eq. 7]

$$d_a v_a = d_p v_p$$
$$v_a = \frac{d_p}{d_a}v_p$$

To find the change in the velocity between the apogee and perigee, use Eqs. (3), (4), and Eq. 7:

$$\Delta U = -\Delta K = -(1/2)M_E\left(v_p^2 - v_a^2\right) = -(1/2)M_E\left[1 - \frac{d_p^2}{d_a^2}\right]v_p^2$$

$$v_p = \sqrt{\frac{2\Delta U}{M_E\left[\dfrac{d_p^2}{d_a^2} - 1\right]}}$$

$$v_p - v_a = \left(1 - \frac{d_p}{d_a}\right)\sqrt{\frac{2\Delta U}{M_E\left[\dfrac{d_p^2}{d_a^2} - 1\right]}} = 996\ \text{m/s}$$

REFLECT 996 m/s corresponds to about 2200 mph – and this is just the *difference* in the speed of the Earth between the apogee and perigee.

61.   **ORGANIZE AND PLAN**   Use Eq. 9.7 to calculate the total energy of the satellite. The radius of the satellite's orbit is [Eq. 1] $r = R_E + h$, where h is the height of the satellite above the surface of the earth.

Known: $m = 500$ kg, $h = 1500$ km, $M_E = 5.97 \times 10^{24}$ kg, $R_E = 6.37 \times 10^6$ m (Appendix E).

SOLVE   Using Eq. (1) in Eq. 9.7 gives [Eq. 2]

$$E = -\frac{GM_E m}{2(R_E + h)} = -\frac{(6.67 \times 10^{-11}\ \text{N} \cdot \text{m}^2/\text{kg}^2)(5.97 \times 10^{24}\ \text{kg})(500\ \text{kg})}{2(6.37 \times 10^6\ \text{m} + 1.5 \times 10^6\ \text{m})} = 2.53 \times 10^{10}\ \text{J}$$

REFLECT   Notice that the height was converted to SI units to be consistent with the other quantities in the calculation.

63.   **ORGANIZE AND PLAN**   If the rocket is dropped a great distance from the Moon's surface the initial potential energy is zero ($r \to \infty$ in Eq. 9.5). The initial speed is also zero, making the initial total mechanical energy zero. The final speed can be found using conservation of mechanical energy.

Known: $M_M = 7.35 \times 10^{22}$ kg, $R_M = 1.74 \times 10^6$ m (Appendix E).

SOLVE   Equating the total initial and final mechanical energies gives [Eq. 1]

$$0 = -\frac{GM_M m}{R_M} + \frac{1}{2}mv^2$$

$$v = \sqrt{\frac{2GM_M}{R_M}} = \sqrt{\frac{2(6.67 \times 10^{-11}\ \text{N} \cdot \text{m}^2/\text{kg}^2)(7.35 \times 10^{22}\ \text{kg})}{1.74 \times 10^6\ \text{m}}} = 2374\ \text{m/s}$$

REFLECT   The speed with which the object strikes the Moon is much greater than the collision speed of the two asteroids in Problem 62.

65.   **ORGANIZE AND PLAN**   Use Kepler's third law applied to circular orbits (Eq. 9.4) to find the orbital radius.

Known: $T = 48$ hours, $M_E = 5.97 \times 10^{24}$ kg (Appendix E).

SOLVE   Using Kepler's third law and solving for the orbital radius R gives [Eq. 1]

$$\frac{R^3}{T^2} = \frac{GM_E}{4\pi^2}$$

$$R = \sqrt[3]{\frac{GM_E T^2}{4\pi^2}} = \sqrt[3]{\frac{(6.67 \times 10^{-11}\ \text{N} \cdot \text{m}^2/\text{kg}^2)(5.97 \times 10^{24}\ \text{kg})(48\ \text{h})^2}{4\pi^2} \left(\frac{60\ \text{min}}{1\ \text{h}}\right)^2 \left(\frac{60\ \text{s}}{1\ \text{min}}\right)^2}$$

$$= 6.70 \times 10^7\ \text{m}$$

REFLECT   This is some 60,000 km above the Earth's surface.

67.   **ORGANIZE AND PLAN**   The perigee and apogee (d_p and d_a) are analogous to the perihelion and aphelion for orbits around the Sun. From Example 9.7, these are related to the semimajor axis of the orbit by [Eq. 1]

$$d_p + d_a = 2a$$

$$a = \frac{1}{2}(d_p + d_a)$$

Since the problem states the perigee and apogee relative to the Earth's surface, the corresponding quantities relative to the center of the Earth are [Eq. 2]

$$d_p = R_E + h_p$$

$$d_a = R_E + h_a$$

For orbits about the Earth, the constant C in Kepler's third law is derived in Ex. 9.9, with the result reproduced here [Eq. 3]: $C = 1.02 \times 10^{13}$ m^3/s^2. With this information, use Kepler's third law to find the period of the satellite.

Known: $h_p = 200$ km, $h_a = 1600$ km, $M_E = 5.97 \times 10^{24}$ kg (Appendix E).

SOLVE Using Kepler's third law and Eqs. (1), (2), and (3) the orbital period is [Eq. 4]

$$\frac{a^3}{T^2} = C$$

$$T = \sqrt{\left(\frac{1}{2}\right)^3 \frac{\left(2R_E + h_p + h_a\right)^3}{C}} = \sqrt{\left(\frac{1}{8}\right) \frac{\left(2\left(6.37 \times 10^6 \text{ m}\right) + 2 \times 10^5 \text{ m} + 1.6 \times 10^6 \text{ m}\right)^3}{1.02 \times 10^{13} \text{ m}^3/\text{s}^2}}$$

$$= 6.14 \times 10^3 \text{ s}$$

REFLECT The orbital period is about 4 hours and 15 minutes.

69. **ORGANIZE AND PLAN** The escape speed from the surface of Callisto and of a neutron star can be found using Eq. 9.6, replacing the Earth's mass and radius with the mass and radius of Callisto and the neutron star.
Known: $M_{Cal} = 1.07 \times 10^{23}$ kg, $R_{Cal} = 2.40 \times 10^6$ m, $M_{ns} = 1.99 \times 10^{30}$ kg, $R_{ns} = 6000$ m.

SOLVE (a) The escape speed from the surface of Callisto is [Eq. 1]

$$v_{esc} = \sqrt{\frac{2GM_{Cal}}{R_{Cal}}} = \sqrt{\frac{2\left(6.67 \times 10^{-11} \text{ N} \cdot \text{m}^2/\text{kg}^2\right)\left(1.07 \times 10^{23} \text{ kg}\right)}{2.40 \times 10^6 \text{ m}}} = 2.44 \text{ km/s}$$

(b) From the surface of the neutron star, the escape speed is [Eq. 2]

$$v_{esc} = \sqrt{\frac{2GM_{ns}}{R_{ns}}} = \sqrt{\frac{2\left(6.67 \times 10^{-11} \text{ N} \cdot \text{m}^2/\text{kg}^2\right)\left(1.99 \times 10^{30} \text{ kg}\right)}{6000 \text{ m}}} = 2.10 \times 10^5 \text{ km/s}$$

REFLECT Due to the large mass and small size of the neutron star, the escape speed from its surface is 100,000 times more than for Callisto, and approximately 20,000 more than for the Earth.

71. **ORGANIZE AND PLAN** By definition, a lunosynchronous satellite has a period equal to the period of rotation of the Moon. Use this and Kepler's third law applied to circular orbits (Eq. 9.4) to find the height of the lunosynchronous satellite. The lunar rotation period is the same as the lunar orbital period (which is why we always see the same side of the moon).
Known: $T = 27.3$ days

SOLVE Using Kepler's third law applied to circular orbits and the known quantities gives [Eq. 1]

$$\frac{R^3}{T^2} = \frac{GM_{Moon}}{4\pi^2}$$

$$R = \sqrt[3]{\frac{GM_{Moon}T^2}{4\pi^2}}$$

$$= \sqrt[3]{\frac{\left(6.67 \times 10^{-11} \text{ N} \cdot \text{m}^2/\text{kg}^2\right)\left(7.35 \times 10^{22} \text{ kg}\right)\left(27.3 \text{ day}\right)^2}{4\pi^2}\left(\frac{24 \text{ h}}{1 \text{ day}}\right)^2 \left(\frac{60 \text{ min}}{1 \text{ h}}\right)^2 \left(\frac{60 \text{ s}}{1 \text{ min}}\right)^2}$$

$$= 8.84 \times 10^7 \text{ m}$$

This is about 50 times the radius of the Moon.

REFLECT The orbital radius of a geosynchronous satellite is about 24 times the radius of the Earth, so the lunosynchronous orbit is larger relative to the radius of its central body (i.e., the Moon).

73. **ORGANIZE AND PLAN** In a binary star system each star orbits about the center of mass of the system, which is given by Eq. 6.19. Assuming a Cartesian coordinate system in which the origin is located at the center of one of the stars, the center of mass for the binary system is [Eq. 1]

$$X_{cm} = \frac{0M + dM}{2M} = \frac{d}{2}$$

where M is the mass of each star and d is there separation. Thus, the distance to each star from the center of mass is $d/2$, so [Eq. 2] $R = d/2$.

Each star orbits about the center of mass as if the total mass of the system were concentrated there. Thus, the mass of the central body is [Eq. 3] $M_{cm} = M + M = 2M$.

To find the orbital period of the stars, use Kepler's third law applied to circular orbits (Eq. 9.4).

SOLVE   (a) Using Kepler's third law applied to circular motion, along with Eqs. (2) and (3), the period of each star is [Eq. 4]

$$\frac{R^3}{T^2} = \frac{GM_{cm}}{4\pi^2}$$

$$T = 2\pi\sqrt{\frac{(d/2)^3}{G(2M)}} = \frac{\pi}{2}\sqrt{\frac{d^3}{GM}}$$

(b) If one star is much, much smaller than the other, then the center of mass (Eq. (1)) becomes [Eq. 5]

$$X_{cm} = \frac{0m + dM}{M + m} \approx d; \; m \ll M$$

assuming the origin of the coordinate system is at the center of the small star. The center of mass is located essentially at the center of the large star, so the orbital radius of the small star is simply d. Furthermore, the equivalent mass located at the center of mass is [Eq. 6; see Eq. (3)] $M_{cm} = m + M \approx M; \; m \ll M$. Inserting these results into Kepler's third law for circular orbits yields [Eq. 7]

$$\frac{R^3}{T^2} = \frac{GM_{cm}}{4\pi^2}$$

$$T = 2\pi\sqrt{\frac{d^3}{GM}}$$

which is 4 times as long as the period for the equal-mass binary star system.

REFLECT   What if the origin (in Eq. (5)) is placed at the center of the large star, instead of the small one? Then Eq. (5) would become [Eq. 8]

$$X_{cm} = \frac{0M + dm}{M + m} = \frac{dm}{M}d$$

However, the radius of orbit for the small star would then be [Eq. 9]

$$R = d - \frac{dm}{M} \approx d; \; m \ll M$$

so the result is the same.

75.   **ORGANIZE AND PLAN**   To find the kinetic energy of the Moon, the orbital speed is needed. This is given by the circumference of the orbit divided by the orbital period, or [Eq. 1]

$$v = \frac{2\pi R_M}{T_M}$$

The potential energy is given by Eq. 9.5, and the total mechanical energy is simply the sum of the potential and kinetic energies. Convert all units to SI units for the calculation.

Known: $M_M = 7.35 \times 10^{22}$ kg, $R_M = 3.85 \times 10^5$ km, $T_M = 27.3$ days, $M_E = 5.97 \times 10^{24}$ kg (Appendix E).

SOLVE   (a) Using Eq. (1), the kinetic energy of the Moon is [Eq. 2]

$$K = \frac{1}{2}M_M v^2 = \frac{1}{2}M_{Moon}\left(\frac{2\pi R_M}{T_M}\right)^2$$

$$= \frac{1}{2}(7.35 \times 10^{22} \text{ kg})\left(\frac{2\pi(3.85 \times 10^8 \text{ m})}{27.3 \text{ days}}\right)^2\left(\frac{1 \text{ day}}{24 \text{ h}}\right)^2\left(\frac{1 \text{ h}}{60 \text{ min}}\right)^2\left(\frac{1 \text{ min}}{60 \text{ s}}\right)^2 = 3.87 \times 10^{28} \text{ J}$$

(b) The potential energy of the Moon is ([Eq. 3]; see Eq. 9.5)

$$U = -\frac{GM_M M_E}{R_M} = -\frac{(6.67 \times 10^{-11} \text{ N} \cdot \text{m}^2/\text{kg}^2)(7.35 \times 10^{22} \text{ kg})(5.97 \times 10^{24} \text{ kg})}{3.85 \times 10^8 \text{ m}} = -7.60 \times 10^{28} \text{ J}$$

(c) The total mechanical energy is the sum of the kinetic and potential energies, or [Eq. 4]

$$E = K + U = -3.74 \times 10^{28} \text{ J}$$

REFLECT   The Moon's potential energy has about twice the magnitude of its kinetic energy.

77. **ORGANIZE AND PLAN** Express the total mechanical energy as the sum of the potential and kinetic energy. Use Eq. 9.5 to express the potential energy, with the only unknown being the mass. Use Kepler's third law applied to circular orbits (Eq. 9.4) to find the orbital period, and then calculate the orbital velocity by dividing the circumference of the orbit by the period. From this, express the kinetic energy of the satellite, with the only unknown being the mass. Solve for the mass, and then use the known mass to calculate the kinetic energy of the satellite.

Note that the radius of the orbit is [Eq. 1] $R = R_E + h$, where h is the height of the orbit above the surface of the Earth.

Known: $h = 10^6$ m, $R_E = 6.37 \times 10^6$ m, $M_E = 5.97 \times 10^{24}$ kg.

SOLVE (a) From Kepler's third law applied to circular orbits [Eq. 2],

$$\frac{R^3}{T^2} = \frac{GM_E}{4\pi^2}$$

$$T = 2\pi\sqrt{\frac{R^3}{GM_E}}$$

The velocity of the satellite is the circumference divided by the period, or [Eq. 3]

$$v = \frac{2\pi R}{T} = \sqrt{\frac{GM_E}{R}}$$

$$= \sqrt{\frac{\left(6.67\times10^{-11}\ \text{N}\cdot\text{m}^2/\text{kg}^2\right)\left(5.97\times10^{24}\ \text{kg}\right)}{\left(6.37\times10^6\ \text{m}\right)+\left(1.00\times10^6\ \text{m}\right)}} = 7.35\times10^3\ \text{m/s}$$

(b) The mass of the satellite may be found from the expression for the total energy [Eq. 4].

$$E = U + K = -\frac{GM_E m}{R} + \frac{1}{2}mv^2$$

$$m = \frac{E}{-\dfrac{GM_E}{R} + \dfrac{1}{2}v^2} = \frac{-4.0\times10^{10}\ \text{J}}{-\dfrac{\left(6.67\times10^{-11}\ \text{N}\cdot\text{m}^2/\text{kg}^2\right)\left(5.97\times10^{24}\ \text{kg}\right)}{\left(6.37\times10^6\ \text{m}\right)+\left(1.00\times10^6\ \text{m}\right)} + \dfrac{1}{2}\left(7.35\times10^3\ \text{m/s}\right)^2} = 1.48\times10^3\ \text{kg}$$

(c) Knowing the mass, the kinetic energy is [Eq. 5]

$$K = \frac{1}{2}mv^2 = \frac{1}{2}\left(1.48\times10^3\ \text{kg}\right)\left(7.35\times10^3\ \text{m/s}\right)^2 = 4.00\times10^{10}\ \text{J}$$

REFLECT The potential energy is thus -8.00×10^{10} J.

79. **ORGANIZE AND PLAN** Use Kepler's third law applied to circular orbits (Eq. 9.4) to find the orbital period. The orbital radius is [Eq. 1] $R = R_M + h$, where $h = 60$ miles.

Known: 1 mi = 1609 m, $M_M = 7.35 \times 10^{22}$ kg, $R_M = 1.74 \times 10^6$ m (Appendix E).

SOLVE The orbital period is [Eq. 2]

$$\frac{R^3}{T^2} = \frac{GM_M}{4\pi^2}$$

$$T = 2\pi\sqrt{\frac{\left(R_{\text{Moon}} + h\right)^3}{GM_M}} = 2\pi\sqrt{\frac{\left(1.74\times10^6\ \text{m} + 60\ \text{mi}\dfrac{1609\ \text{m}}{1\ \text{mi}}\right)^3}{\left(6.67\times10^{-11}\ \text{N}\cdot\text{m}^2/\text{kg}^2\right)\left(7.35\times10^{22}\ \text{kg}\right)}} = 7.06\times10^3\ \text{s}$$

REFLECT This is a slightly longer period than found in Problem 78, which means that the astronauts' orbit was slightly higher than the highest mountain on the Moon.

81.   **ORGANIZE AND PLAN**   Use Newton's law of gravitation (Eq. 9.1) and Newton's second law ($F = ma$) to find the acceleration due to the gravity of the black hole. Convert all quantities to SI units for the calculation.
Known: $m = 9.9 \times 10^{30}$ kg, $R = 2 \times 10^4$ m.
SOLVE   Using Newton's second law and his law of gravitation gives [Eq. 1]

$$F = ma = \frac{GMm}{R^2}$$

$$a = \frac{GM}{R^2} = \frac{\left(6.67 \times 10^{-11} \text{ N} \cdot \text{m}^2/\text{kg}^2\right)\left(9.9 \times 10^{30} \text{ kg}\right)}{\left(2 \times 10^4 \text{ m}\right)^2} = 1.65 \times 10^{12} \text{ m/s}^2$$

REFLECT   How long would it take to reach the center of the black hole, assuming zero initial velocity (and that the classical physics remains pertinent inside a black hole)? About 15 ms.

83.   **ORGANIZE AND PLAN**   The distance to use for the water droplet on the side of the Earth nearest the Moon is [Eq. 1] $R_{\text{near}} = R_{E-M} - R_E$, where R_{E-M} is the Earth-Moon center-to-center distance, and R_E is the radius of the Earth. For the droplet on the side of the Earth farthest from the Moon, the distance is [Eq. 2] $R_{\text{near}} = R_{E-M} + R_E$.
Use Newton's law of gravitation and Newton's second law to find the acceleration of the water droplets.
Known: $M_M = 7.35 \times 10^{22}$ kg, $R_E = 6.37 \times 10^6$ m, $R_{E-M} = 3.84 \times 10^8$ m.
SOLVE   (a) Using Newton's second law, his law of gravitation, and Eq. (1) the acceleration of a water droplet (or any other object) on the side of the Earth nearest the Moon is [Eq. 3]

$$F = ma = \frac{GM_M m}{R_{\text{near}}^2}$$

$$a = \frac{GM_M}{\left(R_{E-M} - R_E\right)^2} = \frac{\left(6.67 \times 10^{-11} \text{ N} \cdot \text{m}^2/\text{kg}^2\right)\left(7.35 \times 10^{22} \text{ kg}\right)}{\left(3.84 \times 10^8 \text{ m} - 6.37 \times 10^6 \text{ m}\right)^2} = 3.44 \times 10^{-5} \text{ m/s}^2$$

(b) The analogous calculation for the far side of the Earth is [Eq. 4]

$$F = ma = \frac{GM_M m}{R_{\text{far}}^2}$$

$$a = \frac{GM_M}{\left(R_{E-M} + R_E\right)^2} = \frac{\left(6.67 \times 10^{-11} \text{ N} \cdot \text{m}^2/\text{kg}^2\right)\left(7.35 \times 10^{22} \text{ kg}\right)}{\left(3.84 \times 10^8 \text{ m} + 6.37 \times 10^6 \text{ m}\right)^2} = 3.22 \times 10^{-5} \text{ m/s}^2$$

(c) The difference between these two accelerations is 2.2×10^{-6} m/s^2.
REFLECT   How far will a drop of water move due to this force in a 12-hour period [Eq. 5]?

$$d = \frac{1}{2} at^2 = \frac{1}{2}\left(2.2 \times 10^{-6} \text{ m/s}^2\right)\left(12 \text{ h}\right)^2 \left(\frac{60 \text{ min}}{1 \text{ h}}\right)^2 \left(\frac{60 \text{ s}}{1 \text{ min}}\right)^2 \approx 2 \text{ km!}$$

Of course, this assumes no other forces acting in opposition.

85.   **ORGANIZE AND PLAN**   Repeat the calculation of problem 83, but use the asteroid-Saturn distance R_{a-S} instead of R_{E-M}, the mass of the Saturn instead of the mass of the Moon, and the radius of the asteroid R_a instead of the radius of the Earth.
Known: $M_S = 569 \times 10^{24}$ kg, $R_{a-S} = 75 \times 10^6$ m, $R_a = 5.00 \times 10^5$ m.
(a) An object on the near side of the asteroid will experience an acceleration of [Eq. 1]

$$F = ma = \frac{GM_S m}{R_{\text{near}}^2}$$

$$a = \frac{GM_S}{\left(R_{a-S} - R_a\right)^2} = \frac{\left(6.67 \times 10^{-11} \text{ N} \cdot \text{m}^2/\text{kg}^2\right)\left(569 \times 10^{24} \text{ kg}\right)}{\left(75 \times 10^6 \text{ m} - 5.00 \times 10^5 \text{ m}\right)^2} = 6.84 \text{ m/s}^2$$

(b) An object on the far side of the asteroid will experience an acceleration of [Eq. 2]

$$F = ma = \frac{GM_S m}{R_{\text{far}}^2}$$

$$a = \frac{GM_M}{\left(R_{a-S} + R_a\right)^2} = \frac{\left(6.67 \times 10^{-11} \text{ N} \cdot \text{m}^2/\text{kg}^2\right)\left(569 \times 10^{24} \text{ kg}\right)}{\left(75 \times 10^6 \text{ m} + 5.00 \times 10^5 \text{ m}\right)^2} = 6.66 \text{ m/s}^2$$

(c) The difference is 0.18 m/s^2

REFLECT   The difference in acceleration between each side of the asteroid due to Saturn is almost four orders of magnitude greater than that for the Earth due to the Moon.

87.   **ORGANIZE AND PLAN**   Use Newton's law of gravitation as the force in his second law to find the acceleration on the surface of each planet. Express the acceleration as a multiple of g.

SOLVE   [Eq. 1]

$$(a) \quad a = \frac{GM_E}{R_E^2} = g$$

$$(b) \quad a = \frac{G(2M_E)}{(2R_E)^2} = 0.5g$$

$$(c) \quad a = \frac{G(0.5M_E)}{(0.5R_E)^2} = 2g$$

$$(d) \quad a = \frac{G(1.8M_E)}{(1.5R_E)^2} = 0.8g$$

$$(e) \quad a = \frac{G(0.75M_E)}{(0.90R_E)^2} = 0.9g$$

Thus, the gravitational acceleration at the surface of the planets in increasing order is (b), (d), (e), (a), (c).

REFLECT   Notice that there was no need to calculate the value of the acceleration for any of the planets – knowing the ratio with respect to g sufficed.

89.   **ORGANIZE AND PLAN**   Use the kinematic equation 2.13 to relate the height jumped to the initial velocity. Notice that the final velocity, v_f (at the peak of the jump) is zero. Since the takeoff conditions are the same on Mars as on Earth, assume the initial velocities are the same.

For the second part, the horizontal velocity on Earth and on Mars is the same, so find ratio of the time spent in the air is the same as the ratio of the horizontal distance traveled.

Known: $v_f = 0$, $h_E = 2.34$ m, $l_E = 8.59$ m, $g_E = 9.81$ m/s^2, $g_M = 3.74$ m/s^2 (Appendix E).

SOLVE   From Eq. 2.13, the initial velocity is related to the heights jumped by [Eq. 1]

$$h_M = \frac{v_0^2}{2g_M}$$

$$h_E = \frac{v_0^2}{2g_E}$$

where the subscripts M and E refer to Mars and Earth, respectively. Taking the ratio gives [Eq. 2]

$$h_M = h_E \frac{g_E}{g_M} = (2.34 \text{ m})\frac{9.81}{3.74} = 6.14 \text{ m}$$

From Eqs. 2.11 and 2.12, the time spent in the air for the long jump is [Eq. 3]

$$t_M = \sqrt{\frac{2h_M}{g_M}}$$

$$t_E = \sqrt{\frac{2h_E}{g_E}}$$

where $h_{E,M}$ now represent the maximum height reached during the long jump. Taking the ratio, and using Eq. (2), gives [Eq. 4]

$$t_M = t_E \sqrt{\frac{2h_M}{g_M}\frac{g_E}{2h_E}} = t_E \frac{g_E}{g_M}$$

Since the horizontal velocity is the same on Earth and on Mars, the distance traveled in the long jump will scale up by the same ratio as the time, or [Eq. 5]

$$l_M = l_E \frac{g_E}{g_M} = (8.59 \text{ m}) \frac{9.81}{3.74} = 22.5 \text{ m}$$

REFLECT   The athletes' performances would be slightly improved if air resistance is considered, since it is less on Mars than on Earth.

91.   From conservation of angular momentum, or Kepler's second law, we know that [Eq. 1] $d_p v_p = d_a v_a$, which can be used to find the distance at the perihelion. From Ex. 9.7, we know that [Eq. 2]

$$d_p + d_a = 2a$$
$$d_a = (1+e)a$$

which can be used to find the eccentricity *e*.
Known: $d_a = 6.99 \times 10^{10}$ m, $v_a = 38.8$ km/s, $v_p = 59.0$ km/s
SOLVE   (a) The distance at the perihelion is [Eq. 3]

$$d_p = d_a \frac{v_a}{v_p} = (6.99 \times 10^{10} \text{ m}) \frac{38.8}{59.0} = 4.60 \times 10^{10} \text{ m}$$

(b) The eccentricity may be found by combining Eqs. (2) to get [Eq. 4]

$$e = \frac{2d_a}{d_p + d_a} - 1 = \frac{2(6.99)}{6.99 + 4.60} - 1 = 0.206$$

which is within 0.3% of the value reported in Appendix E.
REFLECT   The eccentricity of Mercury's orbit is an order of magnitude larger than that of Earth's orbit. What do you think the seasons would be like if Earth had an orbit with an eccentricity similar to Mercury's?

93.   **ORGANIZE AND PLAN**   Recall that, for circular orbits, the orbital speed *v* is related to the orbital period *T* by [Eq. 1]

$$v = \frac{2\pi R}{T}$$

where *R* is the radius of the orbit. Since the problem asked for the difference in height *h* above the surface of the Earth, recall that this is related to the orbital radius *R* by [Eq. 2] $R = R_E + h$. Use Kepler's third law applied to circular orbits (Eq. 9.4) to express the orbital height in terms of the orbital speed, and then take the difference between the heights for the different speeds and periods.
Known: $R_E = 6.37 \times 10^6$ m, $h = 5.5 \times 10^6$ m, $v_+ = 1.1v$, $T_- = 0.9T$.
SOLVE   (a) Using Eqs. (1) and (2) and Kepler's third law applied to circular orbits yields the orbital height *h* in terms of the orbital speed *v* [Eq. 3].

$$\frac{R^3}{T^2} = \frac{GM_E}{4\pi^2}$$

$$R^3 = \frac{GM_E}{4\pi^2} \left(\frac{2\pi R}{v} \right)^2$$

$$R = \frac{GM_E}{v^2}$$

$$R_E + h = \frac{GM_E}{v^2}$$

$$h = \frac{GM_E}{v^2} - R_E$$

The difference in height for an orbital speed increase of 10% is [Eq. 4]

$$h - h_+ = \frac{GM_E}{v^2} - R_E - \left(\frac{GM_E}{v_+^2} - R_E\right) = \frac{GM_E}{v^2}\left(1 - \frac{1}{1.1^2}\right)$$

$$= \left(R_E + h\right)\left(1 - \frac{1}{1.1^2}\right) = \left(1 - \frac{1}{1.1^2}\right)(6.37 + 5.5) \times 10^6 \text{ m} = 2.05 \times 10^6 \text{ m}$$

Thus the new altitude of the satellite is 5.50 Mm – 2.05 Mm = 3.45 Mm above the surface of the Earth.

(b) Again use Kepler's third law applied to circular orbits, but this time solve for the orbital radius R in terms of the orbital period T. Use Eq. (2) to express the height h in terms of the orbital period, then take the difference between the two heights [Eq. 5].

$$\frac{R^3}{T^2} = \frac{GM_E}{4\pi^2}$$

$$R = \sqrt[3]{\frac{GM_E T^2}{4\pi^2}}$$

$$h = \sqrt[3]{\frac{GM_E T^2}{4\pi^2}} - R_E$$

The difference in orbital height for a 10% reduction in the period T is [Eq. 6]

$$h - h_- = \sqrt[3]{\frac{GM_E T^2}{4\pi^2}} - R_E - \left(\sqrt[3]{\frac{GM_E T_-^2}{4\pi^2}} - R_E\right) = \sqrt[3]{\frac{GM_E T^2}{4\pi^2}}\left(1 - 0.9^{2/3}\right)$$

$$= \left(R_E + h\right)\left(1 - 0.9^{2/3}\right) = (6.37 + 5.50)\left(1 - 0.9^{2/3}\right) \times 10^6 \text{ m} = 0.805 \times 10^6 \text{ m}$$

Thus the new altitude of the satellite is 5.500 Mm – 0.805 Mm = 4.695 Mm above the surface of the Earth.

REFLECT   A 10% increase in the orbital speed results in a greater drop in altitude for the satellite than a 10% decrease in the orbital period.

95.   **ORGANIZE AND PLAN**   Use Kepler's law applied to circular orbits (Eq. 9.4) to find the difference in orbital radii of the Moon for the two different orbital periods, T_0 and T_f. The two periods are related by [Eq. 1]
$T_f = T_0 + 35 \times 10^{-3} \text{ s}$.
Convert the Moon's orbital period, given in Appendix E, from days to seconds [Eq. 2].

$$\left(27.3 \text{ days}\right)\left(\frac{24 \text{ h}}{1 \text{ day}}\right)\left(\frac{60 \text{ min}}{1 \text{ h}}\right)\left(\frac{60 \text{ s}}{1 \text{ min}}\right) = 2358720 \text{ s}$$

Known: $T_0 = 2358720 \text{ s}$, $M_E = 5.97 \times 10^{24} \text{ kg}$.

SOLVE   From Kepler's third law, the orbital radius of the Moon is [Eq. 3]

$$\frac{R^3}{T^2} = \frac{GM_E}{4\pi^2}$$

$$R = \sqrt[3]{\frac{GM_E T^2}{4\pi^2}}$$

The difference between the radii for the periods T_0 and T_f is [Eq. 4]

$$R_f - R_0 = \sqrt[3]{\frac{GM_E T_f^2}{4\pi^2}} - \sqrt[3]{\frac{GM_E T_0^2}{4\pi^2}} = \sqrt[3]{\frac{GM_E}{4\pi^2}}\left(T_f^{2/3} - T_0^{2/3}\right)$$

$$= \sqrt[3]{\frac{(6.67 \times 10^{-11} \text{ N} \cdot \text{m}^2/\text{kg}^2)(5.97 \times 10^{24} \text{ kg})}{4\pi^2}}\left(2358720.035^{2/3} - 2358720^{2/3}\right)(\text{s})^{2/3} = 3.79 \text{ m}$$

Thus the Moon is moving away from the Earth at a rate of almost 4 meters/year.

REFLECT   As a fraction of its orbital radius, the Moon is moving away from the Earth at a rate of about one hundred millionth (10^{-8}) of its orbital radius per year.

97.   **ORGANIZE AND PLAN**   Use Eq. 9.6, substituting the mass and radius of the asteroid for that of the Earth. Recall that the mass of the asteroid is given by [Eq. 1]

$$M_a = V\rho = \frac{4}{3}\pi R^3 \rho$$

Use the kinematic equation 2.13 to find the initial velocity of the jump, which will be the same on the asteroid as it is on Earth.

Known: $h = 0.55$ m, $\rho = 4000$ kg/m^3 .

SOLVE   Using Eq. 2.13, the initial velocity v_0 is [Eq. 2]

$$v_f^2 = v_0^2 - 2gh$$

$$v_0 = \sqrt{2gh}$$

because the final velocity (at the peak of the jump) is zero.

Using Eq. 9.6 and Eq. (1) yields [Eq. 3]

$$v_{esc} = \sqrt{\frac{2GM_a}{R_a}}$$

$$= R_a\sqrt{\frac{8\pi G\rho}{3}}$$

$$R_a = v_{esc}\sqrt{\frac{3}{8\pi G\rho}}$$

Inserting the initial velocity of the jump yields [Eq. 4]

$$R_a = \sqrt{\frac{3gh}{4\pi G\rho}} = \sqrt{\frac{3(9.8 \text{ m/s}^2)(0.55 \text{ m})}{4\pi(6.67\times10^{-11} \text{ N}\cdot\text{m}^2/\text{kg}^2)(4000 \text{ kg/m}^3)}} = 2196 \text{ m}$$

REFLECT   Thus you could escape from an asteroid approximately 4 km in diameter.

99.   **ORGANIZE AND PLAN**   Use Eq. 9.6 to relate the escape velocity to the radius of the object escaping from, substituting the mass and radius of the Sun for that of the Earth.

Known: $M_S = 1.99 \times 10^{30}$ kg, $c = 3 \times 10^8$ m/s, $M_G = 10^{11} M_S$.

SOLVE   If the escape speed is to be the speed of light c, the Sun's radius must be [Eq. 1]

$$v_{esc} = c = \sqrt{\frac{2GM_S}{R_S}}$$

$$R_S = \frac{2GM_S}{c^2} = \frac{2(6.67\times10^{-11} \text{ N}\cdot\text{m}^2/\text{kg}^2)(1.99\times10^{30} \text{ kg})}{(3\times10^8 \text{ m/s})^2} = 2950 \text{ m}$$

(b) For an object with the mass M_G of the galaxy, we have [Eq. 2]

$$R_S = \frac{2GM_G}{c^2} = \frac{2(6.67\times10^{-11} \text{ N}\cdot\text{m}^2/\text{kg}^2)(1.99\times10^{30} \text{ kg})(10^{11})}{(3\times10^8 \text{ m/s})^2} = 2.95\times10^{14} \text{ m}$$

REFLECT   Thus, if Newtonian gravitation held, the galactic black hole would have a radius larger than the size of the solar system.

SOLIDS AND FLUIDS

CONCEPTUAL QUESTIONS

1. **SOLVE** The molecules in a liquid are close to each other, but the molecules in a gas are far apart. More molecules in a given volume means that liquids should be significantly denser than gases.

REFLECT Liquids pool at the bottom of a container. Gases expand to fill a container.

3. **SOLVE** A vacuum pump is actually using the atmospheric pressure to hold up the water column. For the 10 m tall water column, the difference between vacuum and the atmospheric pressure equals the weight of the water column divided by the cross-sectional area of the water column.

REFLECT To pump water to a height of 15 m you must apply a pressure difference that is larger than atmospheric pressure. There are several ways of doing this, for example by using a rotating turbine or by placing water in the U-shaped pipe with a liquid denser than water in one side of the U.

5. **SOLVE** The steel ship displaces an amount of water that weighs as much as the ship. The buoyant force from the displaced water balances the gravitational force on the ship, and the ship floats.

Another way of thinking about it is that it's not just the density of the steel that matters, but the density of the air inside the ship. The air inside lowers the average density of the ship far below the density of steel. If the ship were to take in water, the density of the ship goes up. If it takes in enough water, the ship will sink.

REFLECT The same argument applies to balloons. The skin of the balloon is much denser than air, but the total weight of the balloon can equal (or even be less than) the weight of the air displaced by the balloon. A lead zeppelin could actually fly!

7. **SOLVE** To a first approximation, no it doesn't. The buoyant force equals the density of the liquid times the submerged volume times g. When the submarine is fully submerged, none of these quantities change.

REFLECT For an extremely precise calculation, you would have to consider that water is not 100% incompressible. Its density increases slightly the deeper you go. This would increase the buoyant force. On the other hand, the submarine will also get compressed ever so slightly the deeper it goes. That would decrease the buoyant force. However, both of these effects are extremely small and do not play a role in most practical situations.

9. **SOLVE** The air inside the car is heavier than the helium balloon, so the air is "thrown back," creating a forward buoyant force on the balloon. The balloon moves forward.

REFLECT In an accelerating reference frame, the buoyant force is always in the direction of the acceleration.

11. **SOLVE** A gas is a non-ideal fluid when its compressibility plays role, for example in a gas flow near or above the speed of sound.

A glass of water spilled out on top of a table is a non-ideal fluid because the velocity at each point in the fluid changes with time.

The air around the wing tips of an aircraft in flight is a non-ideal fluid because it flows in a vortex that is rotational. Molasses poured out of a bottle is a non-ideal fluid because it's viscous.

REFLECT There are other ways in which a fluid can be non-ideal. For example, fluids can be magnetized, meaning their motions are affected by external and internal magnetic fields.

13.   **Solve**   According to Bernoulli's principle for a flow with no elevation changes, if the flow speed increases in one part of a flow, the pressure decreases, and vice versa. Consequently, when you blow across the top of the paper, you lower the air pressure above the paper compared to the air pressure below it. This results in a net force that is directed upwards, which makes the paper rise.

    Reflect   When the paper rises, the airflow is directed at a slight downward angle by Newton's third law. This downward airflow is what makes airplanes fly.

MULTIPLE-CHOICE PROBLEMS

15.   **Solve**   The volume of the sphere is:

$$V = \frac{m}{\rho} = \frac{(10 \text{ kg})}{(2700 \text{ kg/m}^3)} = 3.7 \times 10^{-3} \text{ m}^3$$

From the formula for the volume of sphere we can then calculate its radius:

$$V = \tfrac{4\pi}{3} r^3$$

$$r = \sqrt[3]{\frac{3V}{4\pi}} = 9.6 \text{ cm}$$

The correct answer is (d).

    Reflect   We could probably have gotten the correct answer with a simple estimate. A 10-cm cube of aluminum has a volume of 1 L and would weigh 2.7 kg. A sphere with a 10 cm radius contains approximate four 10-cm cubes, or approximate 10 kg.

17.   **Solve**   The stress is force divided by the cross-sectional area the force is acting on:

$$\frac{F}{A} = \frac{F}{\frac{\pi}{4} d^2} = \frac{(650 \text{ N})}{\frac{\pi}{4}(1.0 \text{ cm})^2} = 8.3 \times 10^6 \text{ N/m}^2$$

The correct answer is (b).

    Reflect   This is another value we should have been able to estimate well enough to get the correct answer. There are 10^4 cm^2 in 1 m^2 and $\frac{\pi}{4}$ is a little less than 1, so the correct answer in SI units must be a bit more than 6.5×10^6.

19.   **Solve**   The pressure difference between atmospheric pressure and the pressure at a depth of 1.5 km is:

$$\Delta P = \rho_{\text{fluid}} g h = (1000 \text{ kg/m}^3)(9.80 \text{ m/s}^2)(1.5 \text{ km}) = 1.5 \times 10^7 \text{ Pa}$$

The correct answer is (c).

    Reflect   You can easily figure out the correct answer by doing the multiplication in your head.

21.   **Solve**   The maximum depth of the submarine is:

$$h = \frac{P}{\rho_{\text{fluid}} g} = \frac{(4.2 \text{ MPa})}{(1000 \text{ kg/m}^3)(9.80 \text{ m/s}^2)} = 430 \text{ m}$$

The correct answer is (d).

    Reflect   We have used the formula for pressure as a function of height (or depth) in a fluid column and rewritten it to express the depth as a function of pressure.

23.   **Solve**   The volume flow rate is the flow speed times the cross-sectional area. Given the flow rate and the cross-sectional area (or in this case the diameter of a circular cross-section) we can calculate the flow speed:

$$v = \frac{Q}{A} = \frac{Q}{\frac{\pi}{4} d^2} = \frac{(5.2 \text{ L/min})}{\frac{\pi}{4}(1.5 \text{ cm})^2} = 0.49 \text{ m/s}$$

The correct answer is (c).

    Reflect   Always be careful to do the unit conversions correctly.

PROBLEMS

25. **ORGANIZE AND PLAN** From the density given in Table 10.1 we can calculate the mass of 1 L of liquid water. With the mass known we can calculate the volume of gas (vapor).

Known: $V_l = 1\,\text{L};\, \rho_l = 1000\,\text{kg/m}^3;\, \rho_g = 0.804\,\text{kg/m}^3.$

SOLVE The mass of water is:

$$m = V_l\rho_l = (1\,\text{L})(1000\,\text{kg/m}^3) = 1\,\text{kg}$$

The volume of this mass of water in the gas state is:

$$V_g = \frac{m}{\rho_g} = \frac{(1\,\text{kg})}{(0.804\,\text{kg/m}^3)} = 1\,\text{m}^3$$

REFLECT The volume expands by approximately 3 orders of magnitude.

27. **ORGANIZE AND PLAN** Two objects with the same size and shape have the same volume. If the volume is the same, the mass ratio equals the density ratio. We can get the densities from Table 10.1.

Known: $\rho_{Pb} = 11{,}300\,\text{kg/m}^3;\, \rho_U = 19{,}100\,\text{kg/m}^3.$

SOLVE The mass of the uranium bullet divided by the mass of the lead bullet is:

$$\frac{m_U}{m_{Pb}} = \frac{\rho_U}{\rho_{Pb}} = \frac{(19{,}100\,\text{kg/m}^3)}{(11{,}300\,\text{kg/m}^3)} = 1.6903$$

The uranium bullet is approximately 1.7 times heavier than a lead bullet of the same size and shape.

REFLECT There are elements that are even heavier than uranium. Why are those elements not used as armor-penetrating bullets?

29. **ORGANIZE AND PLAN** The person's goal is to change his volume so that it equals 65 L (that is the volume of 65 kg of water). If we first calculate the volume of the person with 2.4 L in his lungs, the amount of air he has to let out is the difference between this volume and 65 L.

Known: $m = 65\,\text{kg};\, \rho_{2.4\,\text{L}} = 990\,\text{kg/m}^3;\, \rho_{\text{water}} = 1000\,\text{kg/m}^3.$

SOLVE The person's goal is to change his volume so it equals:

$$V = \frac{m}{\rho_{\text{water}}} = \frac{(65\,\text{kg})}{(1000\,\text{kg/m}^3)} = 65\,\text{L}$$

The person's initial volume equals:

$$V_{2.4\,\text{L}} = \frac{m}{\rho_{2.4\,\text{L}}} = \frac{(65\,\text{kg})}{(990\,\text{kg/m}^3)} = 66\,\text{L}$$

The amount of air to let out is:

$$\Delta V = V_{2.4\,\text{L}} - V = (66\,\text{L}) - (65\,\text{L}) = 0.66\,\text{L}$$

REFLECT The starting volume of air in the person's lungs did not enter into the calculation, but we should of course check that he has enough air to let out to achieve the density change. He can let out 0.66 L of air because this number is smaller than what he started with (2.4 L).

31. **ORGANIZE AND PLAN** The force compressing the block is the gravitational force on the mass of the man. From the force we can calculate the compression using Equation 10.1 if we know the Young's modulus of concrete. We can find the Young's modulus in Table 10.2.

Known: $L = 28\,\text{cm};\, m = 95\,\text{kg};\, Y = 3 \times 10^{10}\,\text{N/m}^2.$

SOLVE Equation 10.1 relates stress and strain:

$$\frac{F}{A} = Y\frac{\Delta L}{L}$$

Rewrite this equation to calculate the compression, keeping in mind that the cross-sectional area A of a cube equals L^2 :

$$\Delta L = \frac{F}{AY}L = \frac{mg}{L^2Y}L = \frac{mg}{LY} = \frac{(95\ \text{kg})(9.80\ \text{m/s}^2)}{(28\ \text{cm})(3 \times 10^{10}\ \text{N/m}^2)} = 1.1 \times 10^{-7}\ \text{m}$$

REFLECT   This is an example of a problem where it is difficult know if the numerical answer makes sense or not, because we can't tell by our own senses how much a block of concrete compresses as we climb up on it. It is always good to double-check all the calculations in cases like these.

33.   **ORGANIZE AND PLAN**   The required force can be calculated from Equation 10.1 if we know the Young's modulus of steel. We can find the Young's modulus in Table 10.2. The mass to hang on the rod is the required force divided by g.
Known: $L = 1.5\ \text{m}$; $d = 1.2\ \text{mm}$; $\Delta L = 0.50\ \text{mm}$; $Y = 20 \times 10^{10}\ \text{N/m}^2$.
SOLVE   Calculate the required force from Equation 10.1:

$$\frac{F}{A} = Y\frac{\Delta L}{L}$$

$$F = AY\frac{\Delta L}{L} = \frac{\pi}{4}d^2Y\frac{\Delta L}{L} = \frac{\pi}{4}(1.2\ \text{mm})^2(20 \times 10^{10}\ \text{N/m}^2)\frac{(0.50\ \text{mm})}{(1.5\ \text{m})} = 75\ \text{N}$$

The mass to stretch the steel rod 0.50 mm is:

$$m = \frac{F}{g} = \frac{(75\ \text{N})}{(9.80\ \text{m/s}^2)} = 7.7\ \text{kg}$$

REFLECT   Equation 10.1 tells us that if we had placed the mass on top of the rod instead, we would have compressed the rod by the same amount, 0.50 mm. While this is mathematically correct, it would be difficult to do this in practice because the rod is very thin and would likely bend rather than compress.

35.   **ORGANIZE AND PLAN**   The stretch can be calculated from Equation 10.1 if we know the Young's modulus of steel. We can find the Young's modulus in Table 10.2.
Known: $L = 73\ \text{cm}$; $d = 0.15\ \text{mm}$; $F = 15\ \text{N}$; $Y = 20 \times 10^{10}\ \text{N/m}^2$.
SOLVE   Calculate the stretch from Equation 10.1:

$$\frac{F}{A} = Y\frac{\Delta L}{L}$$

$$\Delta L = \frac{F}{AY}L = \frac{F}{\frac{\pi}{4}d^2Y}L = \frac{(15\ \text{N})}{\frac{\pi}{4}(0.15\ \text{mm})^2(20 \times 10^{10}\ \text{N/m}^2)}(73\ \text{cm}) = 0.31\ \text{cm}$$

REFLECT   We have assumed that the material in the guitar string is not stressed beyond its elastic limit. The stretch is about 0.42% of the original length of the guitar string. This should be within the elastic limit.

37.   **ORGANIZE AND PLAN**   The gravitational force on the mass of a column of liquid causes a pressure difference between the top of the liquid and the bottom, as described by Equation 10.4. In a barometer, the top of the fluid is at zero pressure ($P_0 = 0$) because the liquid is contained in an evacuated tube. Other than this we only need to know the density of the liquid to calculate the height of the liquid column in a barometer.
Known: $P = 1\ \text{atm}$; $P_0 = 0$; $\rho = 1000\ \text{kg/m}^3$.
SOLVE   Calculate the height of the water column in a barometer from Equation 10.1:

$$P = P_0 + \rho gh$$

$$h = \frac{P - P_0}{\rho g} = \frac{(1\ \text{atm}) - (0)}{(1000\ \text{kg/m}^3)(9.80\ \text{m/s}^2)} = 10\ \text{m}$$

REFLECT   A water barometer is not practical if you follow the standard design of a mercury barometer. However, by building a barometer in a U-shape where the two columns have different cross-sectional area, you can build a water barometer similar to the hydraulic lift in Figure 10.7.

39.   **ORGANIZE AND PLAN**   The pressure difference between sea level ($P_0 = 1$ atm) and the ocean trench is given by Equation 10.4. The fractional volume change due to compression forces can be calculated from Equation 10.2, where the force per unit area is the pressure. We also need to know the bulk modulus of steel, and that is listed in Table 10.2.

Known: $h = 5.75$ km; $P_0 = 1$ atm; $\rho = 1000$ kg/m³; $B = 16 \times 10^{10}$ N/m².

SOLVE   (a) The pressure at a depth of 5.75 km is:

$$P = P_0 + \rho g h = (1 \text{ atm}) + (1000 \text{ kg/m}^3)(9.80 \text{ m/s}^2)(5.75 \text{ km}) = 56.5 \text{ MPa}$$

(b) The fractional volume change of the steel spoon is given by Equation 10.2:

$$\frac{F}{A} = -B\frac{\Delta V}{V}$$

$$P = -B\frac{\Delta V}{V}$$

$$\frac{\Delta V}{V} = -\frac{P}{B} = -\frac{(5.65 \times 10^7 \text{ N/m}^2)}{(16 \times 10^{10} \text{ N/m}^2)} = -3.53 \times 10^{-4} \text{ m}$$

REFLECT   The spoon shrinks due to the compression forces.

41.   **ORGANIZE AND PLAN**   The pressure difference between sea level ($P_0 = 1$ atm) and an ocean depth is given by Equation 10.4. We can rewrite this equation to solve for a depth given a pressure.

Known: $P = 3$ atm; $P_0 = 1$ atm; $\rho = 1000$ kg/m³.

SOLVE   Rewrite Equation 10.4 to solve for the depth:

$$P = P_0 + \rho g h$$

$$h = \frac{P - P_0}{\rho g} = \frac{(3 \text{ atm}) - (1 \text{ atm})}{(1000 \text{ kg/m}^3)(9.80 \text{ m/s}^2)} = 2 \times 10^1 \text{ m}$$

REFLECT   When diving, the pressure increases by about 1 atm for every 10 m depth.

43.   **ORGANIZE AND PLAN**   The fractional volume change due to compression forces is given by Equation 10.2. Since pressure is force per unit area we can use this equation (and the bulk modulus of steel from Table 10.2) to calculate the required pressure to compress the steel ball. The pressure at a certain depth is given by Equation 10.4, which we will rewrite to solve for the depth.

Known: $\Delta V/V = 0.015\%$; $P_0 = 1$ atm; $\rho = 1000$ kg/m³; $B = 16 \times 10^{10}$ N/m².

SOLVE   The required pressure to compress the steel ball is:

$$P = \frac{F}{A} = -B\frac{\Delta V}{V} = -(16 \times 10^{10} \text{ N/m}^2)(-0.015\%) = 2.4 \times 10^7 \text{ N/m}^2$$

The ocean depth with this pressure is:

$$P = P_0 + \rho g h$$

$$h = \frac{P - P_0}{\rho g} = \frac{(2.4 \times 10^7 \text{ N/m}^2) - (1 \text{ atm})}{(1000 \text{ kg/m}^3)(9.80 \text{ m/s}^2)} = 2.4 \text{ km}$$

REFLECT   The problem text didn't say whether the original volume of the steel ball was measured in vacuum or in atmospheric pressure. Does that change the numerical answer?

45.   **ORGANIZE AND PLAN**   The diastolic pressure is a gauge pressure, i.e., relative to the pressure of the surrounding atmosphere. A gauge pressure equals the height of a column of liquid times the density of the liquid times g. Since we know the pressure we can calculate the height.

Known: $\Delta P = 70$ mm Hg; $\rho = 1060$ kg/m³.

SOLVE   The diastolic pressure should equal the pressure difference ΔP over a column with height h:

$$\Delta P = \rho g h$$

Solve for h:

$$h = \frac{\Delta P}{\rho g} = \frac{(70 \text{ mm Hg})}{(1060 \text{ kg/m}^3)(9.80 \text{ m/s}^2)} = \frac{(70 \text{ mm Hg})\left(\dfrac{101325 \text{ Pa}}{760 \text{ mm Hg}}\right)}{(1060 \text{ kg/m}^3)(9.80 \text{ m/s}^2)} = 90 \text{ cm}$$

REFLECT   The ratio between 90 cm and 70 mm equals the ratio between the density of mercury and the density of blood.

47.   **ORGANIZE AND PLAN**   The pressure on each piston is the air pressure plus the applied force on that piston divided by the piston area. The pressures on the two pistons are equal when the system is in equilibrium. Because the pistons are at the same height, the air pressure is the same on both pistons.
Known: $A_1 = 0.50 \text{ m}^2$; $A_2 = 5.60 \text{ m}^2$; $F_1 = 2.0 \text{ kN}$.
SOLVE   The system is in equilibrium when:

$$\frac{F_1}{A_1} = \frac{F_2}{A_2}$$

This means that the larger piston can support a force:

$$F_2 = \frac{A_2}{A_1}F_1 = \frac{(5.60 \text{ m}^2)}{(0.50 \text{ m}^2)}(2.0 \text{ kN}) = 22 \text{ kN}$$

i.e., it can support a mass:

$$m_2 = \frac{F_2}{g} = \frac{(22 \text{ kN})}{(9.80 \text{ m/s}^2)} = 2.3 \times 10^3 \text{ kg}$$

REFLECT   The first equation in our solution can be rewritten:

$$\frac{F_1}{F_2} = \frac{A_1}{A_2}$$

49.   **ORGANIZE AND PLAN**   The buoyant force is given by Archimedes's principle, Equation 10.5. We can calculate the buoyant force since we know that the density of water is 1000 kg/m^3.
Known: $V = 185 \text{ m}^3$; $\rho_{\text{fluid}} = 1000 \text{ kg/m}^3$.
SOLVE   Insert the known values into Equation 10.5 to calculate the buoyant force:

$$F_B = \rho_{\text{fluid}}gV = (1000 \text{ kg/m}^3)(9.80 \text{ m/s}^2)(185 \text{ m}^3) = 1.81 \times 10^6 \text{ N}$$

REFLECT   This is a large force. It's difficult to submerge a submarine!

51.   **ORGANIZE AND PLAN**   To calculate the buoyant force using Equation 10.5 we need to know the parachutist's volume, which is his mass divided by his density. We also need to know the density of air: 1.28 kg/m^3.
Known: $m = 70 \text{ kg}$; $\rho = 1050 \text{ kg/m}^3$; $\rho_{\text{fluid}} = 1.28 \text{ kg/m}^3$.
SOLVE   First we calculate the volume of the parachutist:

$$V = \frac{m}{\rho} = \frac{(70 \text{ kg})}{(1050 \text{ kg/m}^3)} = 0.067 \text{ m}^3$$

Then we use Equation 10.5 to calculate the buoyant force:

$$F_B = \rho_{\text{fluid}}gV = (1.28 \text{ kg/m}^3)(9.80 \text{ m/s}^2)(0.067 \text{ m}^3) = 0.84 \text{ N}$$

The weight of the parachutist is:

$$W = mg = (70 \text{ kg})(9.80 \text{ m/s}^2) = 6.9 \times 10^2 \text{ N}$$

The weight of the parachutist is obviously much larger than the buoyant force from the air. To compare the two we can calculate their ratio:

$$\frac{W}{F_B} = \frac{(6.9 \times 10^2 \text{ N})}{(0.84 \text{ N})} = 8.2 \times 10^2$$

REFLECT   The ratio we calculated equals $\rho / \rho_{\text{fluid}}$.

53.   **ORGANIZE AND PLAN**   Since the iceberg is not accelerating, the net force acting on the iceberg must be zero. The net force is the sum of the gravitational force and the buoyant force. That means the buoyant force is equal in magnitude to the gravitational force but in the opposite direction. Once we know the buoyant force we can use Archimedes's principle to calculate the volume of water displaced by the iceberg. Dividing this volume with the total volume of the iceberg we get the fraction of the iceberg's volume that is below the water line.

Known: $m = 6500$ kg; $\rho_{ice} = 931$ kg/m³; $\rho_{fluid} = 1030$ kg/m³.

SOLVE   (a) The buoyant force is:

$$F_B = -F_g = -(-mg) = mg = (6500 \text{ kg})(9.80 \text{ m/s}^2) = 63.7 \text{ kN}$$

(b) The volume of displaced water is calculated from Equation 10.5:

$$F_B = \rho_{fluid} g V$$

$$V = \frac{F_B}{\rho_{fluid} g} = \frac{(63.7 \text{ kN})}{(1030 \text{ kg/m}^3)(9.80 \text{ m/s}^2)} = 6.311 \text{ m}^3$$

(c) The fraction of the iceberg's volume that is below that water line equals the volume of displaced water divided by the total volume of the iceberg:

$$\frac{V}{\dfrac{m}{\rho_{ice}}} = \frac{\rho_{ice} V}{m} = \frac{(931 \text{ kg/m}^3)(6.311 \text{ m}^3)}{(6500 \text{ kg})} = 90.4\%$$

REFLECT   The fraction equals $\rho_{ice} / \rho_{fluid}$.

55.   **ORGANIZE AND PLAN**   In Problem 10.53 we learned that the fraction of a floating object that is below that water line is $\rho_{solid} / \rho_{fluid}$. From this we can calculate the density of the wood ball.

Known: $\rho_{fluid} = 1000$ kg/m³; $\rho_{solid} / \rho_{fluid} = \frac{1}{2}$.

SOLVE   The answer is:

$$\rho_{solid} = \frac{\rho_{fluid}}{2} = \frac{(1000 \text{ kg/m}^3)}{2} = 500.0 \text{ kg/m}^3$$

REFLECT   In this problem (and in Problems 10.53 and 10.54) we have neglected the buoyant force from the air. This is reasonable because the buoyant force from air is 3 orders of magnitude smaller than the other forces. However, it would be prudent to round off our calculated density of the wood ball to 500 kg/m³.

57.   **ORGANIZE AND PLAN**   The scale measures the net downward force acting on the scale. The net force in this case is the gravitational force acting on the person reduced by the buoyant force acting on the person. Since we know what the scale reads and the person's mass, we can calculate the buoyant force. From the buoyant force we can calculate the person's volume and then his density.

Known: $m = 69.5$ kg; $F_{net} = -22.0$ N; $\rho_{fluid} = 1000$ kg/m³.

SOLVE   The net force on the scale is $F_{net} = F_g + F_B$ where the gravitational force is:

$$F_g = -mg = -(69.5 \text{ kg})(9.80 \text{ m/s}^2) = -681 \text{ N}$$

and the buoyant force is:

$$F_B = \rho_{fluid} g V$$

We can solve for the person's volume:

$$F_{net} = F_g + \rho_{fluid} g V$$

$$V = \frac{F_{net} - F_g}{\rho_{fluid} g} = \frac{(-22.0 \text{ N}) - (-681 \text{ N})}{(1000 \text{ kg/m}^3)(9.80 \text{ m/s}^2)} = 6.73 \times 10^{-2} \text{ m}^3$$

Finally, from the person's volume and mass we can calculate the person's density:

$$\rho = \frac{m}{V} = \frac{(69.5 \text{ kg})}{(6.73 \times 10^{-2} \text{ m}^3)} = 1.03 \times 10^3 \text{ kg/m}^3$$

REFLECT   The net force must be in the same direction as the gravitational force; otherwise the person wouldn't be able to sit on the scale but would float to the surface!

59.   ORGANIZE AND PLAN   The scale measures the force F_{net} acting on the scale, giving the measurement as an apparent mass $m_a = -F_{net}/g$. The net force in this case is the gravitational force reduced by the buoyant force.

Known: $d = 5.0$ cm; $\rho = 2700$ kg/m³; $\rho_{air} = 1.28$ kg/m³; $\rho_{water} = 1000$ kg/m³.

SOLVE   (a) The gravitational force is:

$$F_g = -mg = -\rho V g = -\rho d^3 g = -(2700 \text{ kg/m}^3)(5.0 \text{ cm})^3 (9.80 \text{ m/s}^2) = -3.3 \text{ N}$$

The buoyant force in air is calculated from Equation 10.5:

$$F_B = \rho_{air} g V = \rho_{air} g d^3 = (1.28 \text{ kg/m}^3)(9.80 \text{ m/s}^2)(5.0 \text{ cm})^3 = 1.5 \text{ mN}$$

In air, the scale reads:

$$m_a = \frac{-F_{net}}{g} = \frac{-(F_g + F_B)}{g} = \frac{-((-3.3 \text{ N}) + (1.5 \text{ mN}))}{(9.80 \text{ m/s}^2)} = 0.34 \text{ kg}$$

(b) The buoyant force in water is:

$$F_B = \rho_{water} g V = \rho_{water} g d^3 = (1000 \text{ kg/m}^3)(9.80 \text{ m/s}^2)(5.0 \text{ cm})^3 = 1.2 \text{ N}$$

In water, the scale reads:

$$m_a = \frac{-F_{net}}{g} = \frac{-(F_g + F_B)}{g} = \frac{-((-3.3 \text{ N}) + (1.2 \text{ N}))}{(9.80 \text{ m/s}^2)} = 0.21 \text{ kg}$$

REFLECT   As you can see, the buoyant force from air is very small, typically 3 orders of magnitude smaller than the gravitational force on a solid. Only for unusual materials with very low densities does the buoyant force from air matter (see Problem 10.62 for an example).

61.   ORGANIZE AND PLAN   We are only asked to estimate the percentage body fat, not calculate it precisely. A reasonable estimate would be to place the woman's density on a linear interpolation between the density of 100% fat and the density of 0%.

Known: $\rho_{fat} = 900$ kg/m³; $\rho_{nonfat} = 1100$ kg/m³; $\rho = 1060$ kg/m³.

SOLVE   The estimate of the woman's body fat is:

$$\frac{\rho - \rho_{nonfat}}{\rho_{fat} - \rho_{nonfat}} \times 100\% = \frac{(1060 \text{ kg/m}^3) - (1100 \text{ kg/m}^3)}{(900 \text{ kg/m}^3) - (1100 \text{ kg/m}^3)} \times 100\% = 20\%$$

REFLECT   A precise calculation is possible. Begin by dividing the woman's mass with her volume:

$$\rho = \frac{m}{V} = \frac{m_{fat} + m_{nonfat}}{V_{fat} + V_{nonfat}} = \frac{m_{fat} + m_{nonfat}}{\dfrac{m_{fat}}{\rho_{fat}} + \dfrac{m_{nonfat}}{\rho_{nonfat}}}$$

Divide this equation with ρ_{nonfat} on both sides and define a variable $x = m_{fat}/m_{nonfat}$ to rewrite the equation:

$$\frac{\rho}{\rho_{nonfat}} = \frac{x + 1}{\dfrac{\rho_{nonfat}}{\rho_{fat}} x + 1}$$

This equation can be solved for x. The solution is:

$$x = \frac{1 - \dfrac{\rho}{\rho_{nonfat}}}{\dfrac{\rho}{\rho_{fat}} - 1} = \frac{1 - \dfrac{(1060 \text{ kg/m}^3)}{(1100 \text{ kg/m}^3)}}{\dfrac{(1060 \text{ kg/m}^3)}{(900 \text{ kg/m}^3)} - 1} = 0.205$$

We can now calculate the precise value for the woman's percentage body fat:

$$\frac{m_{fat}}{m} = \frac{m_{fat}}{m_{fat} + m_{nonfat}} = \frac{1}{1 + \dfrac{m_{nonfat}}{m_{fat}}} = \frac{1}{1 + \dfrac{1}{x}} = \frac{1}{1 + \dfrac{1}{0.205}} = 0.170 = 17\%$$

Our simple estimate of 20% was reasonably close to the precise value of 17%.

63. **ORGANIZE AND PLAN**   The acceleration is given by Newton's second law and is the net upward force divided by the submarine's mass. The net upward force equals the displaced weight of 1.5 m³ of sea water. We can figure out the submarine's mass because we know that prior to "blowing the tank" the buoyant force and the gravitational force were in balance.

Known: $V = 135 \text{ m}^3$; $V_{fluid} = 1.5 \text{ m}^3$; $\rho_{fluid} = 1030 \text{ kg/m}^3$.

SOLVE   We can calculate the mass of the submarine from knowing that initially the apparent weight was zero:

$$0 = w_a = mg - F_B = mg - \rho_{fluid}gV$$
$$m = \rho_{fluid}V = (1030 \text{ kg/m}^3)(135 \text{ m}^3) = 1.39 \times 10^3 \text{ kg}$$

The net upward force on the submarine after "blowing the tank" is:

$$a = \frac{F_{net}}{m} = \frac{\rho_{fluid}gV_{fluid}}{m} = \frac{(1030 \text{ kg/m}^3)(9.80 \text{ m/s}^2)(1.5 \text{ m}^3)}{(1.39 \times 10^3 \text{ kg})} = 0.11 \text{ m/s}^2$$

REFLECT   If the submarine had been in a fresh water lake instead of sea water, would the acceleration have been any different?

65. **ORGANIZE AND PLAN**   The volume flow rate is the flow speed times the cross-sectional area.

Known: $d = 2.75$ cm; $v = 0.450$ m/s.

SOLVE   The volume flow rate is:

$$Q = Av = \tfrac{\pi}{4}d^2v = \tfrac{\pi}{4}(2.75 \text{ cm})^2(0.450 \text{ m/s}) = 2.67 \times 10^{-4} \text{ m}^3/\text{s}$$

REFLECT   Depending on the application, it may be more useful to express the volume flow rate in units of liters per second. In this case $Q = 0.267$ L/s.

67. **ORGANIZE AND PLAN**   The volume flow rate is the same through the hose as it is through the nozzle, and it equals the flow speed times the cross-sectional area.

Known: $d_a = 2.25$ cm; $v_a = 0.320$ m/s; $d_b = 0.30$ cm.

SOLVE   (a) The volume flow rate is:

$$Q = A_a v_a = \tfrac{\pi}{4}d_a^2 v_a = \tfrac{\pi}{4}(2.25 \text{ cm})^2(0.320 \text{ m/s}) = 1.27 \times 10^{-4} \text{ m}^3/\text{s}$$

(b) The flow speed through the nozzle is:

$$v_b = \frac{Q}{A_b} = \frac{Q}{\tfrac{\pi}{4}d_b^2} = \frac{(1.27 \times 10^{-4} \text{ m}^3/\text{s})}{\tfrac{\pi}{4}(0.30 \text{ cm})^2} = 18.0 \text{ m/s}$$

REFLECT   Because the flow rate is constant, the flow speed in the nozzle can also be expressed as:

$$v_b = v_a\frac{A_a}{A_b} = v_a\frac{d_a^2}{d_b^2}$$

69. **ORGANIZE AND PLAN**   Because the total cross-sectional area of the two small pipes is smaller than the cross-sectional area of the large pipe, the flow will fill both of the small pipes, meaning half the volume flow rate goes through each of the small pipes after the large pipe branches.

Known: $Q_1 = 1.20 \times 10^{-4}$ m³/s; $d_1 = 2.0$ cm; $d_2 = 1.0$ cm.

SOLVE   The volume flow rate in each of the smaller pipes is:

$$Q_2 = \frac{Q_1}{2} = \frac{(1.20 \times 10^{-4} \text{ m}^3/\text{s})}{2} = 6.00 \times 10^{-5} \text{ m}^3/\text{s}$$

REFLECT If a single small pipe had branched into two large pipes, the exact geometry of the branch would become important, because all the water could continue in just one of the big pipes, or some fraction could go in one pipe with the rest in the other pipe.

71. **ORGANIZE AND PLAN** Bernoulli's equation relates fluid pressure to flow speed and elevation. We know all the needed quantities to calculate gauge pressure at point B.

Known: $v_A = 1.55$ m/s; $p_A = 180$ kPa; $y_B - y_A = 7.50$ m; $v_B = 1.75$ m/s; $\rho = 900$ kg/m³.

SOLVE Bernoulli's equation from this problem is:

$$\left(P_0 + P_A\right) + \tfrac{1}{2}\rho v_A^2 + \rho g y_A = \left(P_0 + P_B\right) + \tfrac{1}{2}\rho v_B^2 + \rho g y_B$$

Here, P_0 is the surrounding ambient pressure and can be dropped since it occurs on both sides of the equation. If we rewrite the equation we can calculate the gauge pressure at B:

$$P_B = P_A - \rho\left(\tfrac{1}{2}\left(v_B^2 - \tfrac{1}{2}v_A^2\right) + g\left(y_B - y_A\right)\right)$$
$$= \left(180 \text{ kPa}\right) - \left(900 \text{ kg/m}^3\right)\left(\tfrac{1}{2}\left(\left(1.75 \text{ m/s}\right)^2 - \left(1.55 \text{ m/s}\right)^2\right) + \left(9.80 \text{ m/s}^2\right)\left(7.50 \text{ m}\right)\right)$$
$$= 114 \text{ kPa}$$

REFLECT In this case, the difference in elevation made a much larger impact on the pressure difference than what the speed difference made. During design it is often important to identify what aspects of a problem matter and which can be ignored.

73. **ORGANIZE AND PLAN** We will use Bernoulli's equation to relate the fluid pressure, the flow speed, and the elevation difference between the top of the container (subscript 1) and the location of the small hole (subscript 2). The pressures on both sides of the equation are the same (equal to the atmospheric pressure) and cancel. If the diameter of the hole is much smaller than the diameter of the container, we know from the continuity equation that the flow speed over the cross-sectional area of the container is extremely small, so we will neglect it (set it to approximately zero).

Known: $y_1 - y_2 = 0.75$ m; $v_1 = 0$; $\rho = 1000$ kg/m³.

SOLVE Bernoulli's equation with the pressures cancelling each other and $v_1 = 0$ is:

$$\rho g y_1 = \tfrac{1}{2}\rho v_2^2 + \rho g y_2$$

Rewrite this equation to calculate the flow speed through the hole:

$$g y_1 = \tfrac{1}{2}v_2^2 + g y_2$$
$$v_2 = \sqrt{2g\left(y_1 - y_2\right)} = \sqrt{2\left(9.80 \text{ m/s}^2\right)\left(0.75 \text{ m}\right)} = 3.8 \text{ m/s}$$

REFLECT The density of the fluid didn't matter. We would get the same result for any type of fluid as long as we can neglect the viscosity (see Section 10.6).

75. **ORGANIZE AND PLAN** From Bernoulli's equation we see that the difference in wind speed between outdoors and indoors (where the wind speed is zero) creates a fluid pressure on the window. If we multiply this pressure with the area we get the force. We will use subscript 1 for indoors, subscript 2 for outdoors.

Known: $v_2 = 90$ km/h; $A = 4.5$ m²; $\rho = 1.28$ kg/m³.

SOLVE The elevation is the same on either side of the window, and the indoors wind speed $v_1 = 0$. With the terms that equal zero removed, Bernoulli's equation is:

$$P_1 = P_2 + \tfrac{1}{2}\rho v_2^2$$

We see that the faster the wind blows, the lower the outdoors pressure P_2 must be for the right-hand side of the equation to equal P_1. Consequently, the direction of the force on the window must be outwards. The magnitude of this force is the difference between the indoors and outdoors pressures, multiplied by the surface area of the window:

$$F_{\text{net}} = \left(P_1 - P_2\right)A = \tfrac{1}{2}\rho v_2^2 A = \tfrac{1}{2}\left(1.28 \text{ kg/m}^3\right)\left(90 \text{ km/h}\right)^2\left(4.5 \text{ m}^2\right) = 1.8 \text{ kN}$$

REFLECT Do you think a window could break just from the wind blowing past it?

77.   **ORGANIZE AND PLAN**   We can use Poiseulle's law to calculate what change in radius or diameter reduces the volume flow rate by 10%, i.e., reduces the volume flow rate to 90% of the original value. We will use subscript 0 for the initial quantities and subscript 1 for quantities after the blood flow has been reduced.

Known: $Q_1/Q_0 = 0.90$.

SOLVE   Poiseulle's law before the blow flow has been reduced is:

$$Q_0 = \frac{\pi R_0^4 \Delta P}{8\eta L}$$

The pressure difference ΔP, the viscosity η, and the length L of the artery remains constant as we change the radius of the artery. Poiseulle's law after the blow flow has been reduced is:

$$Q_1 = \frac{\pi R_1^4 \Delta P}{8\eta L}$$

If we divide the second equation with the first we get:

$$\frac{Q_1}{Q_0} = \frac{R_1^4}{R_0^4}$$

Which we can solve for the ratio between the artery radius after and before the flow change:

$$\frac{R_1}{R_0} = \left(\frac{Q_1}{Q_0}\right)^{1/4} = 0.90^{1/4} = 0.97$$

Consequently, the radius (or the diameter) of an arterial wall has to decrease by 3% to reduce the blood flow rate by 10%.

REFLECT   The technique demonstrated here, dividing an equation of "after-values" with the same equation of "before-values" is often useful to determine how fractional changes in one quantity varies with fractional changes in another quantity.

79.   **ORGANIZE AND PLAN**   We can use Poiseulle's law to calculate the pressure difference, but we need to know the viscosity of water at 50°C. Table 10.3 lists the viscosities for water at 20°C and 100°C. We can interpolate between these two values to get a good estimate for the viscosity of water at 50°C.

Known:

$T_{50} = 50°C; L = 2.50 \text{ km}; d = 10 \text{ cm}; Q = 12 \text{ L/min}; \eta_{20} = 1.0 \times 10^{-3} \text{ Pa} \cdot \text{s}; T_{20} = 20°C; \eta_{20} = 2.8 \times 10^{-4} \text{ Pa} \cdot \text{s}; T_{100} = 100°C.$

SOLVE   First we estimate the viscosity of water at 50° C by interpolation:

$$\eta_{50} = (\eta_{100} - \eta_{20})\frac{T_{50} - T_{20}}{T_{100} - T_{20}} + \eta_{20} =$$

$$= \left((2.8 \times 10^{-4} \text{ Pa} \cdot \text{s}) - (1.0 \times 10^{-3} \text{ Pa} \cdot \text{s})\right)\frac{(50°C) - (20°C)}{(100°C) - (20°C)} + (1.0 \times 10^{-3} \text{ Pa} \cdot \text{s}) = 7.3 \times 10^{-4} \text{ Pa} \cdot \text{s}$$

Rewrite Poiseulle's law to calculate the pressure difference between the ends of the pipe:

$$Q = \frac{\pi R^4 \Delta P}{8\eta_{50}L}$$

$$\Delta P = \frac{8\eta_{50}LQ}{\pi R^4} = \frac{128\eta_{50}LQ}{\pi d^4} = \frac{128(7.3 \times 10^{-4} \text{ Pa} \cdot \text{s})(2.50 \text{ km})(12 \text{ L/min})}{\pi(10 \text{ cm})^4} = 1.5 \times 10^2 \text{ Pa}$$

REFLECT   This is a small pressure difference that a well-engineered pumping station should have no problem with.

81.   **ORGANIZE AND PLAN**   The surface area in contact with the bed is the normal force divided by the pressure. The normal force is equal in magnitude to the gravitational force.

Known: $m = 130 \text{ kg}; P = 4.7 \text{ kPa}.$

SOLVE   The normal force equals:

$$N = mg = (130 \text{ kg})(9.80 \text{ m/s}^2) = 1.27 \text{ kN}$$

The normal force is distributed over the surface area in contact with the bed, creating the given pressure. The surface area is:

$$A = \frac{N}{P} = \frac{(1.27\ \text{kN})}{(4.7\ \text{kPa})} = 0.27\ \text{m}^2$$

REFLECT   We have assumed that the bodies don't sink into an only partially filled water bed, so that the surface area normal is approximately vertical everywhere.

83.   **ORGANIZE AND PLAN**   The mass is the volume of the barrel multiplied with the density of oil. From the formula for the volume of a circular cylinder we can calculate the diameter. The pressure difference is the density times the height times g.
Known: $V = 42$ gallons; $h = 32$ inches; $\rho = 900\ \text{kg/m}^3$.
SOLVE   (a) The mass of the oil in the barrel is:

$$m = V\rho = (42\ \text{gallons})(900\ \text{kg/m}^3) = (42\ \text{gallons})(3.8 \times 10^{-3}\ \text{m}^3/\text{gallon})(900\ \text{kg/m}^3) = 1.4 \times 10^2\ \text{kg}$$

(b) The inside diameter of the barrel is:

$$V = \tfrac{\pi}{4}d^2 h$$

$$d = \sqrt{\frac{4V}{\pi h}} = \sqrt{\frac{4(42\ \text{gallons})}{\pi(32\ \text{inches})}} = \sqrt{\frac{4(42\ \text{gallons})(3.8\ \text{L/gallon})}{\pi(32\ \text{inches})(2.54\ \text{cm/inch})}} = 0.50\ \text{m}$$

(c) The pressure difference between the top and the bottom of the barrel is:

$$\Delta P = \rho g h = (900\ \text{kg/m}^3)(9.80\ \text{m/s}^2)(32\ \text{inches}) = (900\ \text{kg/m}^3)(9.80\ \text{m/s}^2)(32\ \text{inches})(2.54\ \text{cm/inch}) = 7.2\ \text{kPa}.$$

REFLECT   All these calculations are straightforward, but it is important to do the unit conversions correctly. Particularly it is important to be aware that there are several different "gallons" used around the world!

85.   **ORGANIZE AND PLAN**   The required force is the difference in pressure multiplied by the surface area.
Known: $A = (90\ \text{cm})(50\ \text{cm}) = 0.45\ \text{m}^2$; $P_1 = 0.75$ atm; $P_2 = 0.25$ atm.
SOLVE   The required force is:

$$F = (P_1 - P_2)A = ((0.75\ \text{atm}) - (0.25\ \text{atm}))(0.45\ \text{m}^2) = 23\ \text{kN}$$

REFLECT   Not even the world's strongest human could pull this escape window inward. (At least not under normal circumstances; faced with a life-threatening situation, some humans have been documented to exhibit extraordinary strength. However, even in such a situation it is difficult to imagine anyone being able to make this pull.)

87.   **ORGANIZE AND PLAN**   The force must equal the blood pressure multiplied by the cross-sectional area of the plunger. The speed of the plunger is the volume flow rate divided by the cross-sectional area of the plunger. The speed of the emerging fluid is the volume flow rate divided by the cross-sectional area of the needle.
Known: $d_{\text{plunger}} = 1.2$ cm; $P = 130$ mm Hg; $Q = 1.5$ mL/s; $d_{\text{needle}} = 220\ \mu\text{m}$.
SOLVE   (a) The health provider must push on the plunger with a force:

$$F = PA = P\tfrac{\pi}{4}d_{\text{plunger}}^2 = (130\ \text{mm Hg})\tfrac{\pi}{4}(1.2\ \text{cm})^2 = 2.0\ \text{N}$$

(b) The plunger is moving at a speed:

$$v = \frac{Q}{A} = \frac{Q}{\tfrac{\pi}{4}d_{\text{plunger}}^2} = \frac{(1.5\ \text{mL/s})}{\tfrac{\pi}{4}(1.2\ \text{cm})^2} = 1.3\ \text{cm/s}$$

(c) The fluid emerges from the needle at a speed:

$$v = \frac{Q}{A} = \frac{Q}{\tfrac{\pi}{4}d_{\text{needle}}^2} = \frac{(1.5\ \text{mL/s})}{\tfrac{\pi}{4}(220\ \mu\text{m})^2} = 39\ \text{m/s}$$

REFLECT   These are fairly straightforward calculations, but it is important to pay attention to all the various unit conversions and make sure the answers seem reasonable.

89.   **ORGANIZE AND PLAN**   To lift the payload, the buoyant force on the balloon must be equal to or larger than the combined gravitational forces on the payload and the helium in the balloon.

Known: $m_{payload} = 340$ kg; $\rho_{helium} = 0.179$ kg/m³; $\rho_{air} = 1.28$ kg/m³.

SOLVE   (a) The total mass of payload plus balloon is:

$$m = m_{payload} + m_{helium} = m_{payload} + \rho_{helium} V$$

where V is the volume of the balloon. The buoyant force is:

$$F_B = \rho_{air} g V$$

To lift the payload, the net upward force on the payload must be positive:

$$F_{net} = F_g + F_B = -mg + \rho_{air} g V \geq 0$$

This gives us an equation we can solve for the balloon's volume:

$$-mg + \rho_{air} g V \geq 0$$
$$-m + \rho_{air} V \geq 0$$
$$-m_{payload} - \rho_{helium} V + \rho_{air} V \geq 0$$
$$(\rho_{air} - \rho_{helium}) V \geq m_{payload}$$
$$V \geq \frac{m_{payload}}{\rho_{air} - \rho_{helium}} = \frac{(340 \text{ kg})}{(1.28 \text{ kg/m}^3) - (0.179 \text{ kg/m}^3)} = 309 \text{ m}^3$$

If we multiply this volume by the density of helium we get the amount of helium required:

$$m_{helium} = \rho_{helium} V = (0.179 \text{ kg/m}^3)(309 \text{ m}^3) = 55.3 \text{ kg}$$

(b) If the balloon is spherical, the volume of the balloon is related to its diameter by:

$$V = \frac{\pi d^3}{6}$$

from which we can calculate balloon's diameter:

$$d = \sqrt[3]{\frac{6V}{\pi}} = \sqrt[3]{\frac{6(309 \text{ m}^3)}{\pi}} = 8.39 \text{ m}$$

REFLECT   If you want to double the payload, you would have to double the amount of helium. In other words, there is a fixed ratio between the mass of the payload and the required amount of helium:

$$\frac{m_{payload}}{m_{helium}} = \frac{(340 \text{ kg})}{(55.3 \text{ kg})} = 6.15$$

91.   **ORGANIZE AND PLAN**   We can calculate the new gauge pressure from Bernoulli's equation, which relates pressure to flow speeds (and also to elevation changes, but there's no elevation change in this problem). We need to know the density of blood, which is listed in Table 10.1.

Known: $P_1 = 120$ mm Hg; $v_1 = 35$ cm/s; $v_2 = 1.4$ m/s; $\rho = 1060$ kg/m³.

SOLVE   Bernoulli's equation (without the terms that apply to elevation changes) is:

$$P_1 + \tfrac{1}{2} \rho v_1^2 = P_2 + \tfrac{1}{2} \rho v_2^2$$

We can rewrite this to calculate the new gauge pressure P_2 in the aorta with plaque build-up:

$$P_2 = P_1 + \tfrac{1}{2} \rho (v_1^2 - v_2^2) = (120 \text{ mm Hg}) + \tfrac{1}{2}(1060 \text{ kg/m}^3)\left((35 \text{ cm/s})^2 - (1.4 \text{ m/s})^2\right) = 113 \text{ mm Hg}$$

REFLECT   This result may seem surprising at first: shouldn't a person with so much plaque build-up have an increased blood pressure? Does the answer depend on if we are looking at the body's largest blood vessel, the aorta, or at smaller blood vessels?

93.   **ORGANIZE AND PLAN**   The apparent weight of the crown under water is the gravitational force reduced by the buoyant force from the displaced water. If the crown is 90% gold and 10% silver by weight, it is as if we are weighing two different objects simultaneously, one made of gold and one silver.

Known: $W = 25.0$ N; $\rho_{Au} = 19{,}300$ kg/m³; $\rho_{Ag} = 10{,}500$ kg/m³; $\rho_{H_2O} = 1000$ kg/m³.

SOLVE   (a) If the crown is pure gold, its mass is:

$$m = \frac{W}{g} = \frac{(25.0 \text{ N})}{(9.80 \text{ m/s}^2)} = 2.55 \text{ kg}$$

Its volume is:

$$V = \frac{m}{\rho_{Au}} = \frac{(2.55 \text{ kg})}{(19,300 \text{ kg/m}^3)} = 0.132 \text{ L}$$

The buoyant force of this crown under water is:

$$F_B = \rho_{\text{fluid}} g V = (1000 \text{ kg/m}^3)(9.80 \text{ m/s}^2)(0.132 \text{ L}) = 1.30 \text{ N}$$

The scale would read:

$$W_a = W - F_B = (25.0 \text{ N}) - (1.30 \text{ N}) = 23.7 \text{ N}$$

(b) If the crown is 90% gold and 10% silver by weight, it is as if we are weighing two different objects. The first object is pure gold weighing $W_{Au} = 0.9 \times W = 0.9 \times (25.0 \text{ N}) = 22.5 \text{ N}$ and the second object is pure silver weighing $W_{Ag} = 0.1 \times W = 0.1 \times (25.0 \text{ N}) = 2.50 \text{ N}$. The scale will read the combined apparent weight of these two objects, so all we need to do to get the final answer is repeat part (a) for these two objects and then add the apparent weights together.

The mass of the gold is:

$$m_{Au} = \frac{W_{Au}}{g} = \frac{(22.5 \text{ N})}{(9.80 \text{ m/s}^2)} = 2.30 \text{ kg}$$

Its volume is:

$$V_{Au} = \frac{m_{Au}}{\rho_{Au}} = \frac{(2.30 \text{ kg})}{(19,300 \text{ kg/m}^3)} = 0.119 \text{ L}$$

The buoyant force on the gold is:

$$F_{B,Au} = \rho_{\text{fluid}} g V_{Au} = (1000 \text{ kg/m}^3)(9.80 \text{ m/s}^2)(0.119 \text{ L}) = 1.17 \text{ N}$$

The apparent weight of the gold is:

$$W_{a,Au} = W_{Au} - F_{B,Au} = (22.5 \text{ N}) - (1.17 \text{ N}) = 21.3 \text{ N}$$

The mass of the silver is:

$$m_{Ag} = \frac{W_{Ag}}{g} = \frac{(2.50 \text{ N})}{(9.80 \text{ m/s}^2)} = 0.255 \text{ kg}$$

Its volume is:

$$V_{Ag} = \frac{m_{Ag}}{\rho_{Ag}} = \frac{(0.255 \text{ kg})}{(10,500 \text{ kg/m}^3)} = 0.0243 \text{ L}$$

The buoyant force on the silver is:

$$F_{B,Ag} = \rho_{\text{fluid}} g V_{Ag} = (1000 \text{ kg/m}^3)(9.80 \text{ m/s}^2)(0.0243 \text{ L}) = 0.238 \text{ N}$$

The apparent weight of the silver is:

$$W_{a,Ag} = W_{Ag} - F_{B,Ag} = (2.50 \text{ N}) - (0.238 \text{ N}) = 2.26 \text{ N}$$

The scale would read the combined apparent weight of the gold and the silver:

$$W_a = W_{a,Au} + W_{a,Ag} = (21.3 \text{ N}) + (2.26 \text{ N}) = 23.6 \text{ N}$$

REFLECT   Archimedes' suggested method would require a very precise scale to be put into practice.

95. **ORGANIZE AND PLAN**   The volume flow rate is the cross-sectional area times the flow speed, so we want to find the flow speed. Bernoulli's equation relates the flow speed to differences in elevation and in pressure for two points in the flow. We will choose these two points to be where the thin tube connects to the pipe, as shown in the figure below. The elevation difference between these two points is known. The pressure difference can be calculated be comparing the gauge pressures of the columns of fluid above each point.

Known: $d_A = 1.9$ cm; $d_B = 0.64$ cm; $\rho_{oil}/\rho_{water} = 0.82$; $h_2 = 1.4$ cm.

SOLVE   Bernoulli's equation for points A and B is:

$$P_A + \tfrac{1}{2}\rho_{water}v_A^2 + \rho_{water}gy_A = P_B + \tfrac{1}{2}\rho_{water}v_B^2 + \rho_{water}gy_B$$

where the difference in elevation between the two points is

$$h_4 = y_A - y_B = \tfrac{1}{2}(d_A - d_B) = \tfrac{1}{2}\big((1.9\text{ cm}) - (0.64\text{ cm})\big) = 0.63\text{ cm}$$

as shown in the figure below. The gauge pressure at the top of the thin tube can be used to relate the pressures at points A and B. The gauge pressure at the top of the thin tube equals:

$$P_A + \rho_{water}g(h_3 + h_2) + \rho_{oil}gh_1 = P_B + \rho_{water}g(h_4 + h_3) + \rho_{oil}g(h_2 + h_1)$$

This expression can be simplified by removing identical terms from either side of the equal sign:

$$P_A + \rho_{water}gh_2 = P_B + \rho_{water}gh_4 + \rho_{oil}gh_2$$

Rewrite this expression to obtain the difference between the pressures at the points A and B:

$$P_A - P_B = \rho_{water}gh_4 - (\rho_{water} - \rho_{oil})gh_2$$

Substitute this expression in Bernoulli's equation above:

$$P_A + \tfrac{1}{2}\rho_{water}v_A^2 + \rho_{water}gy_A = P_B + \tfrac{1}{2}\rho_{water}v_B^2 + \rho_{water}gy_B$$
$$P_A - P_B + \tfrac{1}{2}\rho_{water}v_A^2 + \rho_{water}gh_4 = \tfrac{1}{2}\rho_{water}v_B^2$$
$$2\rho_{water}gh_4 - (\rho_{water} - \rho_{oil})gh_2 + \tfrac{1}{2}\rho_{water}v_A^2 = \tfrac{1}{2}\rho_{water}v_B^2$$

The only unknowns in this expression are the flow speeds at points A and B. These are related because the volume flow rate is the same at both points:

$$Q = \tfrac{\pi}{4}d_A^2 v_A = \tfrac{\pi}{4}d_B^2 v_B$$
$$v_A = \frac{d_B^2}{d_A^2}v_B$$

Substitute for v_A:

$$2\rho_{water}gh_4 - (\rho_{water} - \rho_{oil})gh_2 + \tfrac{1}{2}\rho_{water}\left(v_B\frac{d_B^2}{d_A^2}\right)^2 = \tfrac{1}{2}\rho_{water}v_B^2$$

We can solve this expression for the flow speed at point B:

$$2\rho_{water}gh_4 - (\rho_{water} - \rho_{oil})gh_2 = \tfrac{1}{2}\rho_{water}v_B^2\left(1 - \frac{d_B^4}{d_A^4}\right)$$
$$v_B = \sqrt{2g\left(2h_4 - \left(1 - \frac{\rho_{oil}}{\rho_{water}}\right)h_2\right)\left(1 - \frac{d_B^4}{d_A^4}\right)^{-1}}$$

Insert our known values to calculate the flow speed:

$$v_B = \sqrt{2(9.80\text{ m/s}^2)(2(0.63\text{ cm}) - (1 - 0.82)(1.4\text{ cm}))\left(1 - \frac{(0.64\text{ cm})^4}{(1.9\text{ cm})^4}\right)^{-1}} = 0.45\text{ m/s}$$

Finally, from the flow speed we can calculate the volume flow rate:

$$Q = \tfrac{\pi}{4}d_B^2 v_B = \tfrac{\pi}{4}(0.64\text{ cm})^2(0.45\text{ m/s}) = 1.4\times10^{-5}\text{ m/s}$$

REFLECT   The higher flow speed at the narrow part of the pipe is sucking the oil in the thin tube toward point B. When this suction has moved the oil 0.7 cm (so that the difference in between the oil levels on the two sides of the tube is 1.4 cm) the net gravitational force on the two fluid columns is in equilibrium with the net suction force.

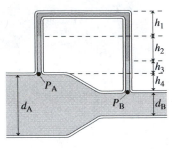

11

WAVES AND SOUND

CONCEPTUAL QUESTIONS

1. **SOLVE** If the frequency doubles in a medium where the speed of the wave is independant of the frequency then the relationship between speed frequency and wavelength determines the change in wavelength.

$$v_w = \lambda_o f_o$$

where v_w is the velocity of waves in water.
If the frequency $f_1 = 2f_o$ is doubled the wavelength λ_1 is given by:

$$\lambda_1 = \frac{v_w}{f_1} = \frac{v_w}{2f_o} = \frac{1}{2}\lambda_o$$

The wavelength is halved if the frequency is doubled.
REFLECT The relationship between frequency, velocity, and wavelength places each of these quantities on the same level of mathematical importance. Physically, however, in most cases the velocity is fixed in homogeneous media so the equation reflects an inverse relationship between frequency and wavelength.

$$\lambda \propto \frac{1}{f}$$

While this relationship is generic for all wave phenomena, it cannot always be used in the way it is used above. Consider the example of light of different frequencies traveling in glass. The speed of the light through the glass is dependent upon frequency, which is the reason we have rainbows. For each color (frequency) the fundamental relationship above still applies but because the velocity is dependant upon frequency, doubling the frequency will not correspond to light with half the wavelength.

3. **SOLVE** The frequency and wavelength do not depend upon the amplitude of the wave, therefore the frequency and wavelength will be unaffected by an increase in amplitude.
REFLECT The frequency is simply dependent upon how quickly the source in the medium is oscillating, and the wavelength is dependent upon the frequency and the speed that the disturbance travels through the medium. The amplitude may determine how far the wave travels before being dampened out by dispersive mechanisms, but it does not affect the speed of travel of disturbances in the medium.

5. **SOLVE** The "energy" in a wave is not described by one point in space. The energy of a wave is distributed over a volume containing the wave. If there is a point in space that destructive interference is observed there will be points of constructive interference observed (i.e., where the local energy density is greater than each of the waves individually). These points of constructive interference are where the energy "goes." The energy of the wave is a sum over all these points in space. Therefore, when considering all constructive and destructive interference, the total energy of the wave is the sum of the energies radiating from the two sources of the wave.
REFLECT The principle of conservation of energy is one of the most important tools in our understanding of nature. If you ever come up to a puzzle that appears to contradict this principle you can rest assured that if you look hard enough you will find the missing energy. If you cannot find it you could be on your way to a Nobel prize.

7. **SOLVE** The velocity of a wave pulse on a rope increases as the tension in the rope increases. The tension in the rope is maximum where it is fastened to the ceiling because at that point the rope must support the weight of the entire rope below it. Therefore, assuming the rope is hanging under its own weight alone, the wave will increase in

speed as it travels up the rope because the tension in the rope increases as the position on the rope approaches the ceiling. As the reflected wave travels back down the rope, the pulse will slow down because the tension on the rope decreases as the rope has to support less of a percentage of the total weight of the rope.

REFLECT   Pulse slowing down the rope is an easy effect to observe. Try it! If the rope supports a weight that is much greater than the weight of the rope itself, this effect would diminish because the tension in the rope would be due almost entirely to the weight hanging from the rope causing an effectively uniform wave velocity along the entire rope.

9.   **SOLVE**   An octave difference between string 1 and string 2 (S_1 and S_2) corresponds to a factor of 2 difference between the fundamental frequencies of the two strings $f_{o2} = 2f_{o1}$ (we shall assume that string two has a fundamental frequency an octave above S_1). The fundamental frequency is proportional to the velocity of the wave on the string. Since the velocity with fixed tension is inversely proportional to the square root of the mass per unit length (μ), the frequency is inversely proportional to the mass per unit length.

$$f_o \propto \frac{1}{\sqrt{\mu}}$$

Assuming the volume density of the material of the two strings is the same, the mass per unit length is proportional to the volume of the string which is proportional to D^2 where D is the diameter of the string (the strings have the same length).

$$\mu \propto D^2$$

Combining the above proportionalities yields:

$$f_o \propto \frac{1}{D}$$

If $f_{o1} = 2f_{o2}$ then $\frac{1}{D_1} = \frac{2}{D_2}$ which yields:

$$D_1 = \frac{D_2}{2}$$

Consequently, a string whose fundamental frequency is twice that of another possesses, half the diameter.

REFLECT   The high E and low E string of a guitar are under approximately the same tension and have approximately the same volume mass density (larger steel strings are wound while the smaller ones are not). The fundamental frequency of these two strings differs by two octaves (so differs by a factor of 4). The thicknesses (or gauge) of the strings listed in Wikipedia also differs by a factor of 4 (approximately).

11.   **SOLVE**   Assuming a point source of the sound, the sound intensity will decrease as a function of $1/r^2$ where r is the distance of the observer from the source. Doubling the distance will decrease the intensity by a factor of 4. The sound intensity level is measured in units of decibels on a logarithmic scale given by:

$$SIL = 10 \ \log(I/I_o)$$

Let the SIL at one distance be $SIL_1 = 10 \ \log(I_1/I_o)$. Doubling the distance produces
$$SIL_2 = 10 \ \log(I_1/4I_o) = 10 \ \log(I_1/I_o) - 10 \ \log 4 = SIL_1 - 6 \ \text{dB}$$

Therefore, every time the distance is doubled the sound intensity level decreases by 6 decibels.

REFLECT   The SIL is constructed to produce a value of 0 dB when a sound cannot be heard anymore by the typical human auditory system. Normal talking dB levels at 1 m distant range from 40–60 dB. By our calculation, at 2 m normal talking would have a range of 34–54 dB and so on until at about 100 meters the above range is 0–20 dB. Which means that with a soft talker in a very quiet environment you could just barely tell that they were talking if you were 100 meters away, and the loud talker would sound like a whisper.

13.   **SOLVE**   Sliding a trombone changes the length of the air column supporting the standing wave mode which ultimately produces the sound emanating from the instrument. The longer the air column, the lower the pitch of the fundamental mode. The same effect is seen in a slide whistle.

REFLECT   Directly changing the length of the tube is the principle behind all brass wind instruments. However, while the trombone allows for a continuous change in the lengths (producing a distinctive glissando capability), other instruments—like the trumpet—produce discrete changes in length by rerouting through paths of different length.

MULTIPLE-CHOICE PROBLEMS

15.   ORGANIZE AND PLAN   Given values: frequency f and wavelength λ.
We are asked to find speed v. The fundamental relationship connecting these three values is:
$$v = \lambda f$$

SOLVE   Plugging in values yields:
$$v = 0.40 \text{ Hz} \times 2.0 \text{ m} = 0.80 \text{ m/s}$$

Choice (d)

REFLECT   A straightforward application of the fundamental relationship between v, f, and λ.

17.   ORGANIZE AND PLAN   The wavelength is independent from the amplitude of the wave so changes in amplitude will produce no change in wavelength.

SOLVE   Doubling the amplitude does not affect the original wavelength of $\lambda = 2.67$ m
Choice (b)

REFLECT   If amplitude had an affect upon wavelength musical instruments would be even more challenging to play than they are. Pianos would change pitch with loudness so playing a specific pitch would require both a dynamic level and key. Music would change in pitch on your stereo depending on volume. Experience tells us this simply is not the case.

19.   ORGANIZE AND PLAN   The first harmonic is the fundamental frequency which has 2 nodes and 1 anti-nodes. From Conceptual Question 6 we deduced that for the nth harmonic there are $n+1$ nodes and n anti-nodes.

SOLVE   For the third harmonic there are $3+1$ nodes and 3 anti-nodes.
Choice (b)

REFLECT   A straightforward application of previous result and a little vocabulary.

21.   ORGANIZE AND PLAN   The sound intensity level is given by:
$$SIL = 10 \, \log(I/I_o)$$

Where the intensity $I_o = 10^{-12}$ W/m^2 is chosen to be the threshold intensity that can be heard by typical human ears.

If we are given the intensity of a sound as we are in this problem, the determination of the SIL is a one-step calculation.

SOLVE   Plugging in values:
$$SIL = 10 \times \log(2.0 \times 10^{-7}/10^{-12}) = 10 \times (\log 2.0 + 5) = 53 \text{ dB}$$

Choice (b)

REFLECT   As a student, you may ask why not just use the intensity. One reason is the operational intensity range for human hearing spans approximately 12 orders of magnitude from 0.000000000001 W/m^2 to 1 W/m^2. It is common to use a logarithmic scale when dealing with such widely ranging values.

23.   ORGANIZE AND PLAN   We are given an approach velocity of the train and the source frequency of a sound and asked to find the frequency perceived by a stationary observer. We call upon Equation 11.7:
$$f' = \frac{f}{1 - v_s/v}$$

where f is the frequency as heard from the train, v_s is the velocity of the train approaching the stationary observer, v is the speed of the wave (in this case the speed of sound 343 m/s).

Solve   Plugging in values:

$$f' = \frac{1.13 \ \text{kHz}}{1-(20 \ \text{m/s})/(343 \ \text{m/s})} = 1.2 \ \text{kHz}$$

Choice (b)

Reflect   This is a significant difference in pitch and corresponds to roughly the pitch difference between any two adjacent keys on a piano.

25.   **Organize and Plan**   From Figure 11.15b the fundamental and first overtone (third harmonic) is depicted. The first overtone in the half-opened pipe is 3 times the frequency of the first harmonic since the first harmonic allows for 1/4 of a wavelength in the tube while the first overtone allows for 3/4 of a wavelength.

We are given the fundamental frequency f_o. To find the frequency of the first overtone we simply multiply by three.

Solve   The frequency of the first overtone is $3 \times f_o = 3 \times 220 \ \text{Hz} = 660 \ \text{Hz}$

Choice (d)

Reflect   One end closed pipe produces only odd harmonics because the standing wave modes require a node at one end and an anti-node at the other. The one end closed pipe corresponds to instruments such as trumpets, clarinets, oboes. If both ends are open (in a flute, for example) then all harmonics are produced. This distinct difference in harmonic content is responsible for the distinctive sounds of the two types of instruments.

PROBLEMS

27.   **Organize and Plan**   We are given values for wavelength and frequency and asked to determine wave speed. We will employ the fundamental relationship: $v = \lambda f$

Solve   Plugging in values: Part (a): $v = 1.55 \ \text{m} \times 0.365 \ \text{Hz} = 0.566 \ \text{m/s}$

Part (b): $v = 1.55 \ \text{m} \times 0.730 \ \text{Hz} = 1.13 \ \text{m/s}$

Reflect   Proportionality between frequency and velocity for fixed wavelength yields a doubling of the velocity with a doubling of the frequency.

29.   **Organize and Plan**   We assume the waves originated from the same spot at $t = 0$. The time of travel between source and observer is $t = d/v$ where d is the distance between source and observer and v is the velocity of the wave.

The time of travel for the p waves is $t_p = d/v_p$ while the time of travel for the s waves is $t_s = d/v_s$. Since the p waves are faster than the s waves $t_p < t_s$.

We are given the difference in arrival times $\Delta t = 24 \ s$. We can derive the relationship between d and Δt as follows:

$$\Delta t = t_s - t_p = d \times (\frac{1}{v_s} - \frac{1}{v_p})$$

Solving for d yields:

$$\Delta t \frac{v_p v_s}{v_p - v_s} = d$$

Solve   Plugging in values:

$$d = 24 \ \text{s} \frac{6 \times 4 \frac{\text{km}^2}{\text{s}^2}}{6 \frac{\text{km}}{\text{s}} - 4 \frac{\text{km}}{\text{s}}} = 288 \ \text{km}$$

Reflect   P waves (primary waves) and s waves (secondary waves) are used to deduce many things including location of earthquakes and the structure of the interior of the earth. Fundamental in making the connections between the phenomena and the deduced information is an understanding of waves.

31. **ORGANIZE AND PLAN** Establishing the requested relationships require study of the arguments and understanding the periodicity requirements of these arguments.

SOLVE The wave form $y(x, y) = A\cos(kx - \omega t)$ describes the displacement as a function of x for all time. At $t = 0$ the wave completes one full cycle when the argument equals 2π (shown in figure below). We shall use this fact to define the value of k as follows:

$$k\lambda = 2\pi$$

Solving for k yields: $k = \dfrac{2\pi}{\lambda}$

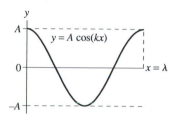

If we fix the position $x = 0$ and watch the point oscillate in time. One full oscillation occurs over one period $t = T$. As above, one full oscillation corresponds to the argument equaling 2π. We use this fact to define the value of ω as follows:

$$\omega T = 2\pi$$

Solving for ω yields: $\omega = \dfrac{2\pi}{T} = 2\pi f$

We shall determine the wave speed by looking at how the position of the peak changes with time. The location of one peak is obtained by setting the argument of the sinusoid equal to zero and finding x as a function of time.

$$kx - \omega t = 0$$

Solving for x yields:

$$x = \frac{\omega}{k}t$$

The slope of the position as a function of time is the velocity. Consequently the velocity is:

$$v = \frac{\omega}{k}$$

REFLECT The wave number and the angular frequency are the most natural numbers to describe waves, and in more advanced physics these numbers dominate the theory. It is useful to be able to derive these relationships. For quick recall these relationships can also be derived using unit analysis.

33. **ORGANIZE AND PLAN** The direction of the velocity of the wave is determined by the relative signs of the spatial and temporal arguments in the expression for the waveform.

Recall from Problem 36 that the velocity was determined by setting the argument to zero:

$$kx \pm \omega t = 0$$

Isolating for x yields:

$$x = \mp \frac{\omega}{k}t$$

SOLVE If the slope of the position as a function of time is negative then the wave travels in the negative x direction. The sign in the argument that corresponds to a negative velocity is positive. Consequently, the sinusoidal form is:

$$y(x,t) = A\cos(kx + \omega t) = A\cos\left(2\pi\left(\frac{x}{\lambda} + \frac{t}{T}\right)\right)$$

REFLECT Knowing the functional form for a wave traveling in both directions is required to describe waves arising out of any disturbance. The preceding three problems are important prerequisites for more advanced mathematical treatments of waves.

35   **ORGANIZE AND PLAN**   Our goal is to determine the units of the quantity representing the velocity of a wave on a string given by $\sqrt{T/\mu}$, where T is the tension in a string (units of force) and μ is the linear mass density (units of mass per unit length). In mks, the units of force are $[F]=\dfrac{kg \times m}{s^2}$ and the linear mass density is $[\mu]=\dfrac{kg}{m}$.

SOLVE   Inserting the constituent units yields

$$[v]=\sqrt{\dfrac{\cancel{kg}\,m^2}{\cancel{kg}\,s^2}}=\dfrac{m}{s}$$

REFLECT   Unit analysis can be a very powerful tool. Students should get in the habit of always checking units to check solutions.

37.   **ORGANIZE AND PLAN**   Recall from Problem 40 the condition for total destructive interference (the dead spot) occurs when

$$d_1-d_2=\dfrac{(2n+1)\lambda}{2} \qquad n=0,\ 1,\ 2,...$$

where d_1 and d_2 are the distances between the observer and the two speakers.
At the midpoint the difference is $d_1-d_2=0$. The location of the first dead spot is when $d_1-d_2=\dfrac{\lambda}{2}$.

The amount that you have to move away from the midpoint $\Delta x=\dfrac{d_1-d_2}{2}$. In terms of the given λ

$$\Delta x=\dfrac{\lambda}{4}$$

SOLVE   Plugging in values:
You must move

$$\Delta x=\dfrac{86.0\ cm}{4}=21.5\ cm$$

to reach the first dead spot.

REFLECT   Another hot spot occurs at 43 cm, dead spot at 64.5, etc. For a tone that is an octave above the given wavelength, the first dead spot will occur at 10.8 cm, hot spot at 21.5 cm. Different frequencies will have different hot spots and dead spots.

39.   **ORGANIZE AND PLAN**   We are given the length and the fundamental frequency of a violin string and asked to find the velocity of the wave on the string. We can use the expression for the fundamental frequency, $f_1=\dfrac{v}{2L}$ and isolate for v to find:

$$v=2Lf_1$$

Given the velocity determined above and the tension in the string we can use the relationship $v=\sqrt{T/\mu}$ and isolate for μ to find:

$$\mu=T/v^2=\dfrac{T}{4L^2f_1^2}$$

SOLVE   Plugging in values:
Part (a): The velocity is $v=2\times0.60\ m\times196\ Hz=235\ m/s$
Part b: The linear mass density is $\mu=49\ N/(235\ m/s)^2=8.9\times10^{-4}\ kg/m$
REFLECT   Tuning, or changing the fundamental frequency of the a string, consists of changing the tension of the string. The design of the violin has roughly similar tensions on each of the four strings so the different open string fundamental frequencies must come primarily from differing linear mass densities. See the next problem for a quantitative description of this.

41.   **ORGANIZE AND PLAN**   If the fundamental frequency is given by f_1, the first two overtones of a vibrating string are given by $2f_1$ and $3f_1$.

SOLVE   Plugging in values:

G String: Second harmonic is: $f_2 = 2 \times 196$ Hz $= 392$ Hz. Third harmonic is: $f_3 = 3f_1 = 588$ Hz

D String: Second harmonic is: $f_2 = 2 \times 294$ Hz $= 588$ Hz. Third harmonic is: $f_3 = 3f_1 = 882$ Hz

A String: Second harmonic is: $f_2 = 2 \times 440$ Hz $= 880$ Hz. Third harmonic is: $f_3 = 3f_1 = 1320$ Hz

E String: Second harmonic is: $f_2 = 2 \times 659$ Hz $= 1318$ Hz. Third harmonic is: $f_3 = 3f_1 = 1977$ Hz

REFLECT   The harmonic content of musical instruments produces the tone of a note beyond just the note on the music (which corresponds only to the fundamental frequency). Skilled violinist can shape the harmonic content by positioning the bow or using light pressure on the strings to dampen certain harmonics.

43.   **ORGANIZE AND PLAN**   The velocity as a function of tension and linear mass density is $v = \sqrt{\frac{T}{\mu}}$. The tension is the total force applied to the wire. The total force per unit area is the pressure $P = T/A$ where A is the cross-sectional area of the cable. The equation of the velocity in terms of the pressure is then:

$$v = \sqrt{\frac{P \times A}{\mu}} = \sqrt{\frac{P}{\mu/A}} = \sqrt{\frac{P}{\rho}}$$

where ρ is the volume mass density.

The function of the velocity is now in the form given by the problem.

Note on units: The MPa is 10^6 N/m².

SOLVE   The velocity of the wave on the copper wire right at the threshold pressure is

$$v = \sqrt{\frac{331 \times 10^6 \text{ N/m}^2}{8890 \text{ kg/m}^3}} = 193 \text{ m/s}$$

The answer is independent of the diameter because as the diameter increases, more force would have to be applied to break the wire. This would increase the tension and therefore increase the velocity. However, as the diameter is increased, the linear mass density increases by the same factor as the tension. This would decrease the velocity. These two effects cancel out.

REFLECT   This velocity is the maximum velocity possible on a copper wire. Previous problems have shown wave velocities on wires greater than 193 m/s. We can then conclude that guitar and violin wires cannot be made of copper.

45.   **ORGANIZE AND PLAN**   As is discussed in the text the beat frequency is half the difference in the individual wave's frequencies (proved in Problem 102). Therefore, the beat frequency will be $f_b = \frac{f_1 - f_2}{2}$

SOLVE   Plugging in values:

$$f_b = \frac{440 \text{ Hz} - 439.6 \text{ Hz}}{2} = 0.2 \text{ Hz}$$

REFLECT   This frequency will be heard as an oscillation with about 5 periods per second. As the pitches get closer, the period increases. Musicians get around this high tolerance required for a unnoticeable period by employing vibrato, which is a manual modulation of the pitch frequency. The interested student can investigate what sort of resolution is required in adjusting the tension of the string to actually be able to tune the instruments (the A string of the violin) to even this precision.

47.   **ORGANIZE AND PLAN**   The distance traveled over the time it took the signal to reflect off the bottom and come back is twice the distance the submarine is from the bottom. Recalling that rate times time equals distance we obtain the required distance as follows:

$$2h = vt$$

The depth of the ocean D at that point is simply the depth of the submarine d plus the distance between the sub and the sea floor.

$$D = d + h = d + \frac{vt}{2}$$

Note: The velocity of sound in water is $v = 1480$ m/s

SOLVE   Plugging in values:
The depth of the ocean is

$$D = 65 \text{ m} + \frac{1480 \text{ m/s} \times 0.86 \text{ s}}{2} = 700 \text{ m}$$

REFLECT   The Abyssal Plain (the deepest flat part of the ocean) is around 4000 m so this part of the ocean is one of the shallower parts. For example, about 30 miles west of the San Francisco coast has an ocean depth of about 700 m.

49.   **ORGANIZE AND PLAN**   Again, we recall $vt = d$. At $20°C$ the speed of sound is 343 m/s while at $0°C$ the speed is 331 m/s. Therefore, it will take longer for the sound to travel a fixed distance in colder air.

SOLVE   The distance traveled in 35 s is

$$343 \text{ m/s} \times 35 \text{ s} = 12 \times 10^3 \text{ m} = 12 \text{ km}$$

It will take $t = \dfrac{12 \times 10^3 \text{ m}}{331 \text{ m/s}} = 36$ s

REFLECT   As expected, it takes longer for sound to travel a fixed distance d when it is colder but only a second difference over 12 km.

51.   **ORGANIZE AND PLAN**   Total power P_t will be obtained from the impinging intensity I_1 since intensity is power per unit area.

$$P_t = I_1 \times A$$

Determination of I_1 is achieved by inverting the SIL for intensity:

$$I_1 = I_o 10^{\frac{SIL}{10}}$$

Preliminary calculations:

$$I_1 = 10^{-12} \text{ W/m}^2 10^{8.5} = 10^{-3.5} \text{ W/m}^2 = 0.000316 \text{ W/m}^2$$

SOLVE   The power incident on the eardrum for an 85 dB sound is
$$P_t = 0.000316 \text{ W/m}^2 \times \pi \times (0.005 \text{ m})^2 = 25 \times 10^{-9} \text{ W} = 25 \text{ nW}$$

REFLECT   There is a very small amount of power incident upon the eardrum. It is about the same power that would be required to lift a flea in the Earth's gravitational field at a rate of about 5/10 of an inch per minute.

53.   **ORGANIZE AND PLAN**   As we derived in Problem 57, the SIL can be given as

$$SIL = 10 \, \log\frac{I_1}{\gamma I_o} = SIL_r - 10 \, \log\gamma$$

when there is a known reference SIL_r. The decrease in SIL levels when the intensity diminishes by a factor of $1/\gamma$ is $10 \, \log\gamma$.
We are given a drop of x dB and asked to find the corresponding value of γ. This can be found as follows: The desired drop is

$$x \text{dB} = 10 \, \log\gamma$$

Isolating for γ

$$\gamma = 10^{x/10}$$

SOLVE   If $x = 3$ then $\gamma \approx 2$. In words, if the intensity decreases by a factor of 1/2 then the *SIL* diminishes by 3 dB

REFLECT   These types of rules of thumb are very handy when dealing with measurement scales that are not immediately intuitive.

55.   **ORGANIZE AND PLAN**   The power output from the band is P_o. The intensity level a distance d away is determined by remembering that the total power is spread over the surface of a sphere with the radius given by d. Intensity is power per unit area or

$$I = P/A = \frac{P_o}{4\pi d^2}$$

If we want to decrease the *SIL* by 15 dB we can recall the derivation in Problem 60:
We are given a drop of x dB and asked to find the corresponding value of γ (where $1/\gamma$ is the fraction by which the intensity is diminished). This can be found as follows: The desired drop is
$$x \, \mathrm{dB} = 10 \, \log \gamma$$

Isolating for γ

$$\gamma = 10^{x/10}$$

SOLVE   Part (a): The intensity level at $d = 25$ m is

$$I = \frac{6.5 \text{ W}}{4\pi(25 \text{ m})^2} = 0.00083 \text{ W/m}^3$$

which corresponds to the sound intensity level

$$SIL = \log \frac{I}{I_0} = \log \frac{8.3 \times 10^{-4} \text{ W/m}^2}{10^{-12} \text{ W/m}^2} = 89.2 \text{ dB}$$

Part (b): The fractional change in intensity required to diminish the *SIL* level by 15 dB is

$$1/\gamma = 10^{-1.5} = 0.0316$$

REFLECT   A difference of 15 dB can also be seen as a decrease of 3 dB five times. In Problem 60 we discovered that the intensity decreases by a factor of roughly 1/2 for every 3 db of diminished *SIL*. So, a decrease of 15 dB corresponds to a decrease by a factor of $\frac{1}{2^5}$ which is approximately equal to the value obtained above.

57.   **ORGANIZE AND PLAN**   Because the intensity is inversely proportional to the distance squared a factor of p increase in the distance results in a factor of $1/p^2$ change in the intensity (derived in problem 57 where $p^2 = \gamma$). With a factor change of $1/p^2$ in the intensity the original sound intensity level before the change in distance SIL_o is changed as follows:

$$SIL = SIL_o - 20 \, \log p \, \mathrm{dB}$$

The value $\Delta SIL = -20 \, \log p \, \mathrm{dB}$ is the intensity level change.

SOLVE   Plugging in values:
Part (a): If $p = 2$ $\Delta SIL = -20 \, \log 2 \, \mathrm{dB} = -6 \text{ dB}$
Part (a): If $p = 10$ $\Delta SIL = -20 \, \log 10 \, \mathrm{dB} = -20 \text{ dB}$
Part (a): If $p = 100$ $\Delta SIL = -20 \, \log 100 \, \mathrm{dB} = -40 \text{ dB}$
REFLECT   Because of the logarithmic scale notice that changes in orders of magnitude of the intensity produce linear changes in the *SIL*. In the previous problem at the rock concert if we moved 100 times farther away from the original distance of 15 m (which is 1.5 km away), the very loud 105 dB sound diminishes to 65 dB, one could still hear the sound as if watching the television.

59. **ORGANIZE AND PLAN** The intensity of sound is the power per unit area. If you are a distance d away from a speaker, the total power P from the speaker is spread over the surface of a sphere with the radius d, so the intensity at d is

$$I(d) = \frac{P}{4\pi d^2}$$

The threshold of hearing of the human ear depends upon the frequency of the sound (this is obvious and implicit in the fact that a frequency range of human hearing exists). In Figure 11.14 we see the response curve for the human ear. We read of the curve that the threshold of hearing at 1000-Hz is $I_{o,1000\ Hz} = 10^{-12}$ W/m^2 and interpolate the threshold of hearing at 100-Hz to be $I_{o,100\ Hz} = 10^{-8.8}$ W/m^2.

Given the intensity and power we can isolate for the required distance:

$$d = \sqrt{\frac{P}{4\pi I}}$$

SOLVE Plugging in values:

Distance required to reach threshold of hearing for 1000 Hz frequencies:

$$d = \sqrt{\frac{0.1\ W}{4\pi \times 10^{-12}\ W/m^2}} = 90,000\ m = 90\ km$$

Distance required to reach threshold of hearing for 100 Hz frequencies:

$$d = \sqrt{\frac{0.1\ W}{4\pi \times 10^{-8.8}\ W/m^2}} = 2,000\ m = 2\ km$$

REFLECT This is an interesting result, because of the response curve of the human ear we will hear frequencies around 1000-Hz for much greater distances than other frequencies. It turns out that the human vocal range lies between about 300-Hz to 3000-Hz. The apparent coincidence between these two facts is not surprising.

61. **ORGANIZE AND PLAN** The difference in dB between person A and person B is $9.4\ dB - 2.4\ dB = 7\ dB$. As in Problem 67 we find the value of

$$\frac{I_B}{I_A} = \gamma = 10^{x/10}$$

where x is the change in dB level.

SOLVE The ratio of sound intensities is $\gamma = 10^{0.7} = 5$. In other words, the threshold intensity for person B is five times that of person A.

REFLECT Converting between changes in dB levels and changes in intensity is good practice for other log relationships such as pH values in chemistry. There is almost always another reason to do your physics homework than simply answering the question.

63. **ORGANIZE AND PLAN** The fundamental frequency of a flute is $f_1 = v/2L$, where v is the velocity of sound in air and L is the length of the pipe. As shown in the text, playing a low B of frequency 247 Hz at temperature 20°C requires a length of 0.694 m. If the length is fixed and the temperature suddenly becomes chilly the speed of sound will decrease which will decrease the fundamental frequency and thereby the pitch of the instrument. The speed of sound at 0°C is $v_0 = 331$ m/s.

SOLVE The fundamental frequency of the flute in the B position tuned for 20°C taken to 0°C temperatures is:

$$f_1 = \frac{331\ m/s}{2(0.694\ m)} = 238\ Hz$$

REFLECT As expected, the instrument becomes flat.

65. **ORGANIZE AND PLAN** If the pipe has one end open and the other closed the fundamental frequency is $f_1 = v/4L$. In the open-closed pipe only the odd harmonics are present so the first 3 overtones are $f_3 = 3f_1$, $f_5 = 5f_1$, and $f_7 = 7f_1$.

SOLVE The fundamental frequency is $f_1 = 343 \text{ m/s}/(4 \times (4.3 \text{ m})) = 20 \text{ Hz}$ with overtones $f_3 = 60$ Hz, $f_5 = 100$ Hz, and $f_7 = 140$ Hz

REFLECT The flute and the clarinet are approximately the same length but the flute has a higher pitch and distinctly different tonal flavor. The root of these fundamental differences is simply that fact that a flute is open at both ends and the clarinet is closed at one end and open at the other. (The reed in the clarinet acts as a driver to the oscillating cavity and is technically open sometimes but the pressure at the mouth piece is not the ambient pressure. It is effectively a closed end).

67. **ORGANIZE AND PLAN** The wavelength required for a particular frequency is given by the fundamental relationship:

$$\lambda = v/f$$

For open-closed pipes the wavelength of the fundamental is 4 times the length of the pipe. In other words, the length of the pipe $L = \frac{\lambda}{4}$. Combining the relationships:

$$L = \frac{v}{4f}$$

SOLVE Plugging in values for the different frequencies:

Part (a): L for 56 Hz: $L = \dfrac{343 \text{ m/s}}{4 \times 56 \text{ Hz}} = 1.53$ m

Part (b): L for 262 Hz: $L = \dfrac{343 \text{ m/s}}{4 \times 262 \text{ Hz}} = 0.33$ m

Part (c): L for 523 Hz: $L = \dfrac{343 \text{ m/s}}{4 \times 523 \text{ Hz}} = 0.16$ m

Part (d): L for 1200 Hz: $L = \dfrac{343 \text{ m/s}}{4 \times 1200 \text{ Hz}} = 0.07$ m

REFLECT This is a very wide range of sizes. The details of how pipe organs are constructed and how they produce their sound is beyond the scope of this book but you know enough now to ask some of the right questions if you are interested in knowing more.

69. **ORGANIZE AND PLAN** The fundamental frequency of a closed-open pipe is 1/2 the fundamental frequency of an open pipe:

$$2f_{co} = f_o$$

The first overtone of an open pipe is twice the frequency of the fundamental, so the frequency of the first overtone in the now opened pipe is

$$f_{o2} = 2f_o = 4f_{co}$$

SOLVE Plugging in values:
The frequency of the first overtone of the now-opened pipe is $f_{o2} = 4 \times 512 \text{ Hz} = 2048$ Hz

REFLECT The first overtone of the now-opened pipe is two octaves above the fundamental of the closed-open pipe.

71. **ORGANIZE AND PLAN** In this problem the observer is moving relative to the air and the emitter is stationary. Following conceptual Example 11.13 the observed frequency f' is related to the velocity and emitted frequency as follows:

$$f' = f\left(1 + v_o/v\right)$$

We are given f, f' and we know the speed of sound in the medium $v = 343$ m/s. We are asked to deduce the velocity. Isolating the above equation for v_o yields:

$$v_o = v(f'/f - 1)$$

SOLVE Plugging in values:
The velocity of the runner is

$$v_o = 343 \text{ m/s}(\frac{359 \text{ Hz}}{352 \text{ Hz}} - 1) = 6.8 \text{ m/s}$$

REFLECT The result is a reasonable speed for a fast runner (100 m dash record of just under 10 seconds corresponds to about $10/unitm/s$).

73. **ORGANIZE AND PLAN** As in the previous problem, we use the relationship:

$$f' = \frac{f}{1 \mp v_s/v}$$

where f is the emitted frequency of sound, v is the speed of sound in the medium, $-v_s$ corresponds to the velocity of the source approaching the observer, and $+v_s$ corresponds to the velocity of the source receding away from the observer.
We are given v_s and f' in this problem. Isolating to find f if the source is approaching the observer:

$$f = f' \times (1 - v_s/v)$$

In the second part of the problem, we use the result from the first part and apply

$$f' = \frac{f}{1 + v_s/v}$$

SOLVE The emitted frequency of the squawk:
$$f = 257 \text{ Hz} \times (1 - (13 \text{ m/s})/(343 \text{ m/s})) = 247 \text{ Hz}$$

Observed frequency as goose is receding:

$$f' = \frac{247 \text{ Hz}}{1 + (13 \text{ m/s})/(343 \text{ m/s})} = 238 \text{ Hz}$$

REFLECT These differences are noticeable. The difference in frequency between approaching goose and receding goose corresponds approximately to adjacent keys on the piano.

75. **ORGANIZE AND PLAN** To determine the observed frequency from the stationary helicopter we must know the velocity of the source (the parachutist). Recall kinematic equations for free-fall. After time t the velocity is $v = gt$ for constant acceleration. With the deduced velocity we employ the Doppler effect relation for a receding source:

$$f' = \frac{f}{1 + v_s/v}$$

In this problem:

$$f' = \frac{f}{1 + gt/v}$$

SOLVE Plugging in values:
The observed frequency of the shout is

$$f' = \frac{425 \text{ Hz}}{1 + \dfrac{9.8 \text{ m/s}^2 \times 4 \text{ s}}{343 \text{ m/s}}} = 381 \text{ Hz}$$

REFLECT The frequency is diminished as expected and the effect is noticeable. The glissando of the falling man is embedded in our cultural experience. What you hear in movies (or mimic with your voice when you are pretending you are falling) is a mixture between the Doppler effect and free-fall under constant acceleration.

77. **ORGANIZE AND PLAN** The expression for the observed frequency f' when the source is in relative motion to the medium (where the observer is stationary with respect to the medium) is:

$$f' = \frac{f}{1 \mp v_s/v}$$

In this case we are given the combined value $v_s/v = 0.99$.

SOLVE Plugging in appropriate number and signs:
When the jet is flying toward us the observed frequency is

$$f' = \frac{1200 \text{ Hz}}{1 - 0.99} = 120,000 \text{ Hz}$$

When the jet is flying away from us the observed frequency is

$$f' = \frac{1200 \text{ Hz}}{1 + 0.99} = 603 \text{ Hz}$$

REFLECT The 1200 Hz sound is not audible to humans when the plane is approaching (human range of hearing is 20–20,000 Hz). After the jet passes the sound is approximately lowered a full octave from the emitting frequency. Of course, at and past the speed of sound, the sound of the jet approaches us only when the jet hits us in the same way that if you hear a supersonic bullet that has been successfully aimed at you, hearing it means you are already bleeding.

79. **ORGANIZE AND PLAN** From conceptual Exercise 11.13, the expression for the observed frequency f' when the source is stationary with respect to the medium and the observer is moving with respect to the source is:
$$f' = f(1 \pm v_o/v)$$

where v_o is the velocity of the observer and v is the velocity of the wave in the medium. The sign is positive for approaching and negative for receding.
To double the observed frequency the observer must be approaching the source and the following condition must be met:

$$(1 + v_o/v) = 2$$

SOLVE Solving for v_o yields: $v_o = v$ To double the observed frequency when moving with respect to a stationary source you must travel the speed of sound in the medium.
REFLECT The difference between moving source and moving observer in this and the preceding problems highlights the importance of knowing who is moving and who is not with respect to the medium.

81. **ORGANIZE AND PLAN** In this case we have both the source and the observer moving away from each other. We shall first determine the frequency that would be observed due to the moving source moving away from a stationary point in the medium:

$$f'' = \frac{f}{1 + v_s/v}$$

Now, from the perspective of the medium in this region there is no difference between a stationary source emitting a frequency of f' and the moving source emitting a frequency of f. Consequently, we may now treat the observer as if he is moving away from a stationary source emitting at frequency f'. The observed frequency in this case is:

$$f' = f''(1 - v_o/v)$$

SOLVE Subbing in the value of f'' in terms of the actual emitted frequency f yields:

$$f' = \frac{1 - v_o/v}{1 + v_s/v} f$$

REFLECT   In contrast to the previous problem, the unphysical results occur here when the observer is moving away at and beyond the speed of sound (negative frequency does not make sense). Of course, this is a reflection of the fact that the observer can actually outrun the sound if she is going away from it.

83.   **ORGANIZE AND PLAN**   Sound intensity is inversely proportional to the distance squared. Consequently, if you double your distance from a sound source the intensity decreases by a factor of 1/4.

As in Problem 89: Let the reference intensity be I_1. The sound intensity level of this sound is:

$$SIL_1 = 10 \ \log\frac{I_1}{I_o}$$

If we consider another sound with intensity $I_1/4$ the corresponding SIL_2 is:

$$SIL_1 = 10 \ \log\frac{I_1}{4I_o} = 10 \ \log\frac{I_1}{I_o} - 10 \ \log 4 = SIL_1 - 10 \ \log 4$$

SOLVE   Doubling the distance decreases the intensity by a factor of 1/4.

Since the value $10 \ \log 4 = 6.02$ dB, doubling the distance reduces the SIL by 6 dB

REFLECT   Going the other way we find that halving the distance doubles the intensity twice (quadruples) the intensity. Either way you look at it you get the same answer using the two rules of thumb found in Problems 89 and 90. Doubling twice increases SIL by $3+3$ dB. Quadrupling intensity increases the SIL by 6 dB.

85.   **ORGANIZE AND PLAN**   We have a measured value of intensity at a reference point:
$$I_1 = P_1/d^2$$

and the corresponding $SIL_1 = 80$ dB

If we change d from 10 m to 2600 m we increase the distance by a factor of 260. The new intensity level I_2 is then

$$I_2 = I_1/(260)^2$$

The corresponding SIL is then:

$$SIL_2 = SIL_1 - 20 \ \log 260$$

SOLVE   The SIL level at the removed location is
$$SIL_2 = 80 \ \text{dB} - 48 \ \text{dB} = 32 \ \text{dB}$$

This SIL is about 10 dB louder than a quiet whisper.

REFLECT   Considering that absorption effects were neglected and there is ambient noise outside even in the quietest of places, this noise level is quite good. In urban environments, 40 dB is the low limit of ambient sound, so 32 dB will likely be drowned out by the noise.

87.   **ORGANIZE AND PLAN**   The bat emits a frequency that is Doppler shifted due to its velocity directed toward the wall. The wall "sees" a frequency of

$$f' = \frac{f}{1 - v_s/v}$$

where v_s is the velocity of the bat and f is the emitted frequency of the sound from the bat.

Once the sound has left the bat, the bat is now an observer and the wall is the source. The wave "emitted" by the wall is Doppler shifted with respect to the bat. Since the bat is moving and the source is stationary the frequency that the bat receives is

$$f'' = f'(1 + v_o/v)$$

In terms of the primary frequency:

$$f'' = f\frac{1 + v_o/v}{1 - v_s/v}$$

SOLVE   Plugging in values:

Ultimately the bat receives the frequency

$$f'' = 52.0 \text{ kHz} \frac{1 + \dfrac{7.5 \text{ m/s}}{343 \text{ m/s}}}{1 - \dfrac{7.5 \text{ m/s}}{343 \text{ m/s}}} = 54.3 \text{ kHz}$$

REFLECT   That speed is a respectable clip (it would hurt if you hit a wall going that fast). It is interesting to contrast how we determine relative velocity with our eyes. We have to use some sort of geometrical calculation built into our brains. Bats map pitch into a dynamic map of what is going on around them.

89.   **ORGANIZE AND PLAN**   A closed-open pipe has a fundamental frequency of $f = \dfrac{v}{4L}$. The velocity of sound at body temperature ($37° \text{C}$) can be obtained by Equation 11.4:

$$v(T) = 331 \text{ m/s} + 0.60 \frac{\text{m}}{\text{sC}} \times 37C = 353 \text{ m/s}$$

Given the fundamental frequency and velocity the length of the pipe is:

$$L = \frac{v}{4f}$$

SOLVE   Plugging in values:

The predicted length of the pipe from the crude model is $L = \dfrac{353 \text{ m/s}}{4 \times 620 \text{ Hz}} = 0.142 \text{ m}$

REFLECT   An effective length of 14 cm does not seem unreasonable. The physics of speech is very interesting and this problem is but the tip of a significant iceberg.

91.   **ORGANIZE AND PLAN**   Given the *SIL* at a given point, the intensity of the sound I_1 can be directly deduced:
$$I_1 = I_o 10^{SIL/10}$$

Since the intensity is the total power per unit area ($I = P/A$) we can extract the total power by multiplying the intensity by the surface area of the sphere with a radius d where d is the distance from the source.
$$P = 4\pi d^2 I_1$$

Combining the relationships:

$$P = 4\pi d^2 I_o 10^{SIL/10}$$

SOLVE   The total power emitted from the speakers
$$P = 4\pi(8.0 \text{ m})^2 10^{-12} \text{ W/m}^2 10^{90/10} = 0.80 \text{ W}$$

REFLECT   What this demonstrates is that it does not take much power to hurt your eardrums. It is important to note that the power rating of a speaker is not the power that comes out in the form of sound. Compare this output (less than a watt) to the output of a lightbulb (60 watts).

93.   **ORGANIZE AND PLAN**   The moving heart wall acts as a moving observer of a stationary source as well as a moving source of sound with respect to a stationary observer.
The moving observer "observes" a frequency of
$$f' = f(1 \pm v_o/v)$$

where f is the frequency of the ultrasound machine emitter and v is the speed of sound in water.
The "observed" frequency is now the frequency emitted by a moving source. The reflected frequency is:

$$f'' = \frac{f'}{1 \mp v_s/v}$$

In this case, both v_o and v_s are the same, say v_w. Finding the reflected frequency in terms of the primary frequency of the ultrasound emitter f :

$$f'' = f\frac{1 \pm v_w/v}{1 \mp v_w/v}$$

Solving for v_w :

$$v_w = \pm v\frac{f'' - f}{f'' + f}$$

The frequency shift is small compared to the source frequency so $f'' + f \approx 2f$.

$$v_w = v\frac{f'' - f}{2f}$$

(n.b.: We are only concerned with the speed of the heart wall (not the velocity) so the sign is not needed.)

SOLVE   The speed of the heart wall is

$$v_w = 1480 \text{ m/s}\frac{100 \text{ Hz}}{2 \times 5.0 \times 10^6 \text{ Hz}} = 0.015 \text{ m/s}$$

REFLECT   If you model the oscillation of the heart wall as a harmonic oscillator, the maximum velocity is the amplitude of the oscillation times the angular frequency. The angular frequency is approximately 15 rad/s (typical heart rate is 140 bpm). If you assume the amplitude of the oscillation is on the order of 1 mm, the maximum velocity is 1.5 cm/s, exactly consistent with the obtained result.

95.   **ORGANIZE AND PLAN**   Take a wave formed by the superposition of two waves with frequencies f_1 and f_2. The wave at a given point is described by the following equation:

$$y(t) = A\cos 2\pi f_1 t + A\cos 2\pi f_2 t$$

Recall the trig identity

$$\cos\alpha + \cos\beta = 2\cos\left[\frac{1}{2}(\alpha - \beta)\right]\cos\left[\frac{1}{2}(\alpha - \beta)\right]$$

Replacing *alpha* with $2\pi f_1 t$ and β with $2\pi f_2 t$ in the trig identity we find:

$$\cos 2\pi f_1 t + \cos 2\pi f_2 t = 2\cos\left[\frac{1}{2}(2\pi f_1 t - 2\pi f_2 t)\right]\cos\left[\frac{1}{2}(2\pi f_1 t + 2\pi f_2 t)\right]$$

Cleaning up a little:

$$A\cos 2\pi f_1 t + A\cos 2\pi f_2 t = 2A\cos\left[\frac{f_1 - f_2}{2}2\pi t\right]\cos\left[\frac{f_1 + f_2}{2}2\pi t\right]$$

SOLVE   Finally:

$$y(t) = 2A\cos\left[2\pi f_b t\right]\cos\left[2\pi f_a t\right]$$

where $f_b = \frac{f_1 - f_2}{2}$ and $f_a = \frac{f_1 + f_2}{2}$. The beat frequency f_b is slower than the average frequency f_a. The slower frequency cosine function acts as an envelope of the higher frequency oscillation.

REFLECT   Beat phenomena is a very helpful little trick that allows the detection of small changes in frequency. For example, Problem 101. Radar detectors use a very high frequency wave that gets shifted by a very small fraction. However, by superimposing the reflected signal with the source signal beat phenomena allow for a direct measurement of the shift in the frequency by looking at the frequency of the beats of the superimposed wave.

TEMPERATURE, THERMAL EXPANSION, AND IDEAL GASES

12

CONCEPTUAL QUESTIONS

1. **ORGANIZE AND PLAN** We chose a certain temperature, e.g., 25°C, and express it in °C, °F, and K to establish the order. We use

$$T(^\circ C) = \frac{5}{9}\left(T(^\circ F) - 32^\circ\right) \Rightarrow T(^\circ F) = \frac{9}{5}T(^\circ C) + 32^\circ$$

and

$$T(K) = T(^\circ C) + 273.15$$

SOLVE $25^\circ C = 298.15 \text{ K} = 77^\circ F$

Therefore, the ordering of the values from largest to smallest is $298.15 \text{ K} > 77^\circ F > 25^\circ C$.

REFLECT The Fahrenheit scale was the primary temperature standard for climatic, industrial, and medical purposes in most English-speaking countries until the 1960s. In the late 1960s and 1970s, the Celsius (formerly *centigrade*) scale was adopted by most of these countries as part of the standardizing process called metrication. Only in the United States and a few other countries (such as Belize) does the Fahrenheit system continue to be used, and only for nonscientific use. Most other countries have adopted Celsius as the primary scale in all use, although Fahrenheit continues to be the scale of preference for a minority of people in the UK, particularly when referring to summer temperatures. A lot of older Britons are conversant with both Celsius and Fahrenheit.

3. **SOLVE** Thermal energy is the sum of kinetic and potential energy associated with atoms and molecules.
   REFLECT The kinetic energy of an object is the extra energy which it possesses due to its motion. It is defined as *the work needed to accelerate a body of a given mass from rest to its current velocity*. Having gained this energy during its acceleration, the body maintains this kinetic energy unless its speed changes. Negative work of the same magnitude would be required to return the body to a state of rest from that velocity. Potential energy can be thought of as energy stored within a physical system. It is called *potential* energy because it has the potential to be converted into other forms of energy, such as kinetic energy, and to do work in the process. The standard (SI) unit of measure for potential energy is the joule, the same as for work or energy in general.

5. **ORGANIZE AND PLAN** We have to consider thermal expansion in three dimensions.
   SOLVE The whole will get larger since the material expands in all directions.
   REFLECT The thermal expansion coefficient for solids is usually significantly smaller than for liquids.

7. **ORGANIZE AND PLAN** We can use the thermal expansion coefficient data in Table 12.1.
   SOLVE Combining different materials in a construction can cause problems since thermal expansion coefficients can vary significantly between different materials.
   REFLECT When the temperature of a substance changes, the energy that is stored in the intermolecular bonds between atoms changes. When the stored energy increases, so does the length of the molecular bonds. As a result, solids typically expand in response to heating and contract on cooling; this dimensional response to temperature change is expressed by its coefficient of thermal expansion (CTE).

9.   **SOLVE**   Gases cease to follow ideal behavior when they are cooled toward their boiling point, since the intermolecular interactions between the gas molecules become more and more significant.

REFLECT   An ideal gas is a theoretical gas composed of a set of randomly moving point particles that interact only through inelastic collisions. The ideal gas concept is useful because it obeys the ideal gas law, a simplified equation of state, and is amenable to analysis under statistical mechanics. At normal ambient conditions such as standard temperature and pressure, most real gases behave qualitatively like an ideal gas. Generally, deviation from an ideal gas tends to decrease with higher temperature and lower density, as the work performed by intermolecular forces becomes less significant compared with the particles' kinetic energy, and the size of the molecules becomes less significant compared to the empty space between them. The ideal gas model tends to fail at lower temperatures or higher pressures, when the molecules come close enough that they start interacting with each other, and not just with their surroundings. This is usually associated with a phase transition. For example, clouds form when the gas of water molecules in the sky drops below the dew point, which causes the water molecules to "stick together" into little droplets. By contrast, at high temperatures and low pressures, the vast majority of familiar substances can be vaporized and will behave approximately as ideal gases.

11.   **ORGANIZE AND PLAN**   We use the ideal gas law to predict what happens to the volume of the bubble, assuming that the temperature does not change. With less depth, the pressure on the bubble decreases.

SOLVE   From $P = nRT\dfrac{1}{V}$ we can see that the volume of the bubble increases with decreasing pressure.

REFLECT   Hydrostatic pressure can be visualized as follows. Considering a small cube of liquid at rest below a free surface, pressure caused by the height of the liquid above must be balanced by a resisting pressure in this small cube. For an infinitely small cube the stress is the same in all directions, and liquid weight or equivalent pressure can be expressed as

$$p = \rho g h + p_a$$

where, p is the hydrostatic pressure (Pa), ρ is the liquid density (kg/m^3), g is gravitational acceleration (m/s^2), h is the height of liquid above (m), and p_a is the atmospheric pressure (Pa).

13.   **ORGANIZE AND PLAN**   We have to consider the random movement of the gas molecules.

SOLVE   The gas molecules undergo collisions with each other and interact with each other, which blocks the direct path for them, and results in travel times across the room in the order of seconds.

REFLECT   The Collision theory, proposed by Max Trautz and William Lewis in 1916 and 1918, qualitatively explains how chemical reactions occur and why reaction rates differ for different reactions. This theory is based on the idea that reactant particles must collide for a reaction to occur, but only a certain fraction of the total collisions have the energy to connect effectively and cause the reactants to transform into products. This is because only a portion of the molecules have enough energy and the right orientation (or "angle") at the moment of impact to break any existing bonds and form new ones. The minimal amount of energy needed for this to occur is known as activation energy. Particles from different elements react with each other by releasing activation energy as they hit each other. If the elements react with each other, the collision is called successful, but if the concentration of at least one of the elements is too low, there will be fewer particles for the other elements to react with and the reaction will happen much more slowly. As temperature increases, the average kinetic energy and speed of the molecules increases but this only slightly increases the number of collisions. The rate of the reaction increases with temperature increase because a higher fraction of the collisions overcome the activation energy.

MULTIPLE-CHOICE PROBLEMS

15.   **ORGANIZE AND PLAN**   We use the equation relating °C and K:

$$T(\text{K}) = T(°\text{C}) + 273.15 \;\Rightarrow\; T(°\text{C}) = T(\text{K}) - 273.15$$

SOLVE   We obtain

$$T(°C) = 77 \text{ K} - 273.15 = -196.15°C$$

The answer is (d).

REFLECT   Liquid nitrogen is a liquefied atmospheric gas produced industrially in large quantities by fractional distillation of liquid air. It is pure nitrogen in a liquid state at a very low temperature. Liquid nitrogen is a colorless, clear liquid with density at its boiling point of 0.807 g/mL and a dielectric constant of 1.4. Liquid nitrogen is often referred to by the abbreviation LN_2 and has the UN number 1977. At atmospheric pressure, liquid nitrogen boils at 77°K (−196°C; −321°F) and is a cryogenic fluid which can cause rapid freezing on contact with living tissue, which may lead to frostbite. When appropriately insulated from ambient heat, liquid nitrogen can be stored and transported, for example in vacuum flasks. Here, the very low temperature is held constant at 77°K by slow boiling of the liquid, resulting in the evolution of nitrogen gas. Depending on the size and design, the holding time of vacuum flasks ranges from a few hours to a few weeks.

17.   **ORGANIZE AND PLAN**   We use the equation for the linear thermal expansion and realize that the change in length, ΔL, over the initial length, L, is 1% = 0.01. We use the linear expansion coefficient, α, for steel from Table 12.1.

$$\frac{\Delta L}{L} = \alpha \, \Delta T = 0.01$$

SOLVE   We Solve for ΔT and obtain:

$$\Delta T = \frac{0.01}{\alpha} = \frac{0.01}{\left(1.2 \times 10^{-5} \text{ °C}^{-1}\right)} = 833.33°C$$

The answer is (b), about 850°C.

REFLECT   A Quartz rod heated by 850°C would only expand 0.34%.

19.   **SOLVE**   The density of water is greatest at 4°C.

REFLECT   The density of water is dependent on its temperature, but the relation is not linear and is not even monotonic. The solid form of most substances is denser than the liquid phase; thus, a block of the solid will sink in the liquid. But, by contrast, a block of common ice floats in liquid water because ice is *less* dense than liquid water. When cooled from room temperature liquid water becomes increasingly denser, just like other substances. But at approximately 4°C, water reaches its maximum density. As it is cooled further under ambient conditions, it expands to become *less* dense. Upon freezing, the density of ice decreases by about 9%. The reason of this is the "cooling" of intermolecular vibrations, allowing the molecules to form steady hydrogen bonds with their neighbors and thereby gradually locking into positions reminiscent of the hexagonal packing achieved upon freezing to ice Ih. While the hydrogen bonds are shorter in the crystal than in the liquid, this locking effect reduces the average coordination number of molecules as the liquid approaches nucleation.

21.   **ORGANIZE AND PLAN**   As in Problem 20, we use the ideal gas law for two different conditions and solve appropriately.

$$P_1V = nRT_1 \Rightarrow P_1 = \frac{nRT}{V_1}$$

$$P_2V = nRT_2 \Rightarrow P_2 = \frac{nRT}{V_2}$$

SOLVE

$$P_2 = \frac{nRT}{V_2} \Rightarrow P_2 = P_1\frac{V_1}{V_2} = (1 \text{ atm}) \times \left(\frac{120 \text{ m}^3}{110 \text{ m}^3}\right) = 1.091 \text{ atm}$$

The answer is (c).

REFLECT   The pressure and volume are inversely proportional to each other.

23.   **ORGANIZE AND PLAN**   We use the equation for the rms speed of an ideal gas molecule in an approximation.

$$v_{rms} = \sqrt{\frac{3\,k_B T}{m}}$$

SOLVE

$$v_{rms} = \sqrt{\frac{3k_B T}{m}} = \sqrt{\frac{3 \times \left(1.38 \times 10^{-23}\ \text{J K}^{-1}\right) \times \left(273\ \text{K}\right)}{\left(28\ \text{g mol}^{-1}\right) \times \left(\dfrac{1}{6.022 \times 10^{23}\ \text{mol}^{-1}}\right) \times \left(\dfrac{10^{-3}\ \text{kg}}{\text{g}}\right)}} = 493.03\ \text{ms}^{-1}$$

The answer(b) is closest.

REFLECT   Do not confuse the rms speed with the most probable speed!

PROBLEMS

25.   **ORGANIZE AND PLAN**   We convert °F to °C using:

$$T(^\circ C) = \frac{5}{9}\left(T(^\circ F) - 32^\circ\right)$$

SOLVE

$$T(^\circ C) = \frac{5}{9}\left(5^\circ F - 32^\circ\right) = -15^\circ C$$

REFLECT   32°F corresponds to 0°C, the temperature at which water freezes.

27.   **ORGANIZE AND PLAN**   We use the equations to convert from °C to °F and K.

$$T(^\circ F) = \frac{9}{5}T(^\circ C) + 32^\circ \quad \text{and} \quad T(K) = T(^\circ C) + 273.15$$

SOLVE
(a)

$$T_{boil}(^\circ F) = \frac{9}{5}T(^\circ C) + 32^\circ = \frac{9}{5}\left(-196^\circ C\right) + 32^\circ = -320.8^\circ F$$

$$T_{boil}(K) = T(^\circ C) + 273.15 = -196^\circ C + 273.15 = 77.15\ K$$

(b)

$$T_{melt}(^\circ F) = \frac{9}{5}T(^\circ C) + 32^\circ = \frac{9}{5}\left(327^\circ C\right) + 32^\circ = 620.6^\circ F$$

$$T_{melt}(K) = T(^\circ C) + 273.15 = 327^\circ C + 273.15 = 600.15\ K$$

REFLECT   The element with the lowest boiling point is helium. Both the boiling points of rhenium and tungsten exceed 5000 K at standard pressure. Due to the experimental difficulty of precisely measuring extreme temperatures without bias, there is some discrepancy in the literature as to whether tungsten or rhenium has the higher boiling point.

29.   **ORGANIZE AND PLAN**   We use the equation relating °C and °F for two different temperatures in °C and subtract them from each other:

$$T_2(°F) - T_1(°F) = \frac{9}{5}T(°C_2) + 32° - \frac{9}{5}T(°C_1) + 32° = \Delta T(°F) = \frac{9}{5}\Delta T(°C)$$

SOLVE

$$\Delta T(°F) = \frac{9}{5}\Delta T(°C) = \frac{9}{5}(3°C) = 5.4°F$$

REFLECT   Global warming is the increase in the average temperature of the Earth's near-surface air and oceans since the mid-20th century and its projected continuation. Global surface temperature increased 0.74 ± 0.18 C (1.33 ± 0.32 F) during the last century. The Intergovernmental Panel on Climate Change (IPCC) concludes that increasing greenhouse gas concentrations resulting from human activity such as fossil fuel burning and deforestation are responsible for most of the observed temperature increase since the middle of the 20th century. The IPCC also concludes that natural phenomena such as solar variation and volcanoes produced most of the warming from preindustrial times to 1950 and had a small cooling effect afterward. These basic conclusions have been endorsed by more than 40 scientific societies and academies of science, including all of the national academies of science of the major industrialized countries. Climate model projections summarized in the latest IPCC report indicate that global surface temperature will probably rise a further 1.1 to 6.4°C (2.0 to 11.5°F) during the 21st century. The uncertainty in this estimate arises from the use of models with differing sensitivity to greenhouse gas concentrations and the use of differing estimates of future greenhouse gas emissions. Some other uncertainties include how warming and related changes will vary from region to region around the globe. Most studies focus on the period up to 2100. However, warming is expected to continue beyond 2100 even if emissions stop, because of the large heat capacity of the oceans and the long lifetime of carbon dioxide in the atmosphere. Increasing global temperature will cause sea levels to rise and will change the amount and pattern of precipitation, probably including expansion of subtropical deserts. The continuing retreat of glaciers, permafrost, and sea ice is expected, with the Arctic region being particularly affected. Other likely effects include shrinkage of the Amazon rainforest and Boreal forests, increases in the intensity of extreme weather events, species extinctions, and changes in agricultural yields. Political and public debate continues regarding what actions (if any) to take in response to global warming. The available options are mitigation to reduce further emissions; adaptation to reduce the damage caused by warming; and, more speculatively, geoengineering to reverse global warming. Most national governments have signed and ratified the Kyoto Protocol aimed at reducing greenhouse gas emissions.

31.   **ORGANIZE AND PLAN**   For part (a), we use the equation from Problem 26,

$$T_2(°F) - T_1(°F) = \frac{9}{5}T(°C_2) + 32° - \left(\frac{9}{5}T(°C_1) + 32°\right) = \Delta T(°F) = \frac{9}{5}\Delta T(°C)$$

For part (b), we convert 15°C to °F and then subtract the natural green house temperatures in °C and °F form the values.

SOLVE

(a)

$$\Delta T(°F) = \frac{9}{5}\Delta T(°C) = \frac{9}{5}(33°C) = 59.4°F$$

(b) Converting °C to °F:

$$T(°F) = \frac{9}{5}T(°C) + 32° = \frac{9}{5}(15°C) + 32° = 59°F$$

The natural green house effect expressed in °F is 59°F.

$$T_{\text{Earth}}\left(°F\right)=59°F-59.4°F=-0.4°F$$
$$T_{\text{Earth}}\left(°C\right)=15°C-33°C=-18°C$$

For an average temperature of 15°C without the natural green house effect, Earth's temperature would be $-0.4°F$ or $-18°C$.

REFLECT   Without the natural green house effect, life as we know it would be sustainable.

33.   **ORGANIZE AND PLAN**   We use the equation to convert from °C to °F:

$$T\left(°F\right)=\frac{9}{5}T\left(°C\right)+32$$

SOLVE

$$T\left(°F\right)=\frac{9}{5}\left(38.2°C\right)+32=100.76°F$$

REFLECT   Fever (also known as pyrexia, from the Greek *pyretos* meaning fire, or a febrile response, from the Latin word *febris*, meaning fever, and archaically known as ague) is a frequent medical sign that describes an increase in internal body temperature to levels above normal. Fever is most accurately characterized as a temporary elevation in the body's thermoregulatory set-point, usually by about 1–2°C (1.8–3.6°F). Fever is caused by an elevation in the thermoregulatory set-point, causing typical body temperature (generally and problematically considered to be 37°C or 98.6°F; see below for specifics) to rise, and effector mechanisms are enacted as a result. A feverish individual has a general feeling of cold despite an increased body temperature, and increases in heart rate, muscle tone, and shivering, all of which are caused by the body's attempts to counteract the newly perceived hypothermia and reach the new thermoregulatory set-point. Fever differs from hyperthermia in that hyperthermia is an increase in body temperature over the body's thermoregulatory set-point, due to excessive heat production or insufficient thermoregulation, or both. A fever is considered one of the body's immune mechanisms to attempt a neutralization of a perceived threat inside the body, be it bacterial or viral. Carl Wunderlich discovered that fever is not a disease, but the body's response to a disease.

35.   **ORGANIZE AND PLAN**   For part (a), we use the equation to convert from °F to °C and set both sides of the equation equal by replacing the temperatures with a variable, i.e., x, and Solve for x. To answer (b), we use the equation to convert °C to K.

SOLVE

(a)

$$T\left(°C\right)=\frac{5}{9}\left(T\left(°F\right)-32\right)$$
$$x=\frac{5}{9}\left(x-32\right)=\frac{5}{9}x-\frac{5}{9}32$$
$$x-\frac{5}{9}x=-\frac{5}{9}32$$
$$x=-40$$

(b)

$$T\left(K\right)=T\left(°C\right)+273.15=\left(-40°C\right)+273.15=233.15\ K$$

REFLECT   Fahrenheit usually refers to a temperature scale proposed in 1724 by, and named after, the physicist Daniel Gabriel Fahrenheit (1686–1736). Today, the scale has been replaced by the Celsius scale in most countries; it is still in use for nonscientific purposes in the United States and a few other nations, such as Belize.

37.   **ORGANIZE AND PLAN**   We use the equation for linear expansion, $\dfrac{\Delta L}{L} = \alpha \Delta T$, **solved** for ΔL, and with the expansion coefficient of steel, α, from Table 12.1, to calculate the change in length due to the different temperature. The temperature difference ΔT can be positive or negative. Then we add the change in length to the original length.

SOLVE

(a)

$$\Delta L = \alpha \ \Delta T L = \left(1.2 \times 10^{-5} \, {}^\circ\text{C}^{-1}\right) \times \left(12 \, {}^\circ\text{C}\right) \times \left(50 \text{ m}\right) = 0.0072 \text{ m}$$

$$L\left(32 \, {}^\circ\text{C}\right) = 0.0072 \text{ m} + 50 \text{ m} = 50.0072 \text{ m}$$

(b)

$$\Delta L = \alpha \Delta T L = \left(1.2 \times 10^{-5} \, {}^\circ\text{C}^{-1}\right) \times \left(-30 \, {}^\circ\text{C}\right) \times \left(50 \text{ m}\right) = -0.018 \text{ m}$$

$$L\left(-10 \, {}^\circ\text{C}\right) = -0.018 \text{ m} + 50 \text{ m} = 49.982 \text{ m}$$

REFLECT   Using the tape at temperatures significantly different from 20°C can result in relevant measuring errors.

39.   **ORGANIZE AND PLAN**   We use the equation for linear expansion, $\dfrac{\Delta L}{L} = \alpha \Delta T$, and apply it in the bone's long and short dimensions with α_{long} and α_{short}, and the corresponding expansion coefficients. We need to convert the fever temperature to °C and subtract the normal body temperature of 37.0°C from it, to get the temperature difference.

SOLVE

$$T\left({}^\circ\text{C}\right) = \frac{5}{9}\left(T\left({}^\circ\text{F}\right) - 32\right) = \frac{5}{9}\left(104.5 \, {}^\circ\text{F} - 32\right) = 40.3 \, {}^\circ\text{C}$$

$$\Delta L_{\text{long}} = \alpha_{\text{long}} \ \Delta T_f \ L_{\text{long}} = \left(8.9 \times 10^{-5} \, {}^\circ\text{C}^{-1}\right) \times \left(40.3 \, {}^\circ\text{C} - 37.0 \, {}^\circ\text{C}\right) \times \left(43.2 \times 10^{-2} \text{ m}\right) = 1.3 \times 10^{-4} \text{ m}$$

$$\Delta L_{\text{short}} = \alpha_{\text{short}} \ \Delta T_f \ L_{\text{short}} = \left(5.4 \times 10^{-5} \, {}^\circ\text{C}^{-1}\right) \times \left(40.3 \, {}^\circ\text{C} - 37.0 \, {}^\circ\text{C}\right) \times \left(2.75 \times 10^{-2} \text{ m}\right) = 4.9 \times 10^{-6} \text{ m}$$

Reflect   Both expansion coefficients for bone are significantly higher than the coefficient for steel! When a person has a fever, the bone expands in its long dimension more than a centimeter!

41.   **ORGANIZE AND PLAN**   We use the equation for volume thermal expansion, $\dfrac{\Delta V}{V} = \beta \ \Delta T$, and **solve** for the change in volume. For use the expansion coefficient of water at 20°C. Furthermore, we assume a spherical shape of the cell, and calculate the cells volume as $V = \dfrac{4}{3}\pi r^3$. After we calculated the change in volume we solve this equation for the radius, r, to get the change in radius.

SOLVE

$$V_{\text{sphere}} = \frac{4}{3}\pi \ r_{\text{sphere}}^3 = \frac{4}{3}\pi \ \left(5.0 \times 10^{-6} \text{ m}\right)^3 = 5.24 \times 10^{-16} \text{ m}^3$$

$$\Delta V = \beta \ \Delta T_f \ V = \left(2.1 \times 10^{-4} \, {}^\circ\text{C}^{-1}\right) \times \left(32 \, {}^\circ\text{C} - 37.0 \, {}^\circ\text{C}\right) \times \left(5.24 \times 10^{-16} \text{ m}^3\right) = -5.50 \times 10^{-19} \text{ m}^3$$

$$\Delta r = -\sqrt[3]{\frac{3}{4}\frac{\left(5.50 \times 10^{-19} \text{ m}^3\right)}{\pi}} = -5.08 \times 10^{-7} \text{ m} = -0.508 \ \mu\text{m}$$

REFLECT   The cell's diameter decreases by about 10%!

43.   **ORGANIZE AND PLAN**   We use the equation for linear thermal expansion to calculate the change in length of the pendulum. Then we can use the equation that relates the length of the pendulum's period (the time it takes the pendulum to move from its highest point on one side to its highest point on the other side) to its length to predict if it will go slower or faster with the calculated change in pendulum length.

SOLVE

$$\Delta L = \alpha \ \Delta T L = \left(2.4 \times 10^{-5} \, {}^\circ\text{C}^{-1}\right) \times \left(5 \, {}^\circ\text{C}\right) \times \left(1 \text{ m}\right) = 1.2 \times 10^{-4} \text{ m}$$

Since the period of the pendulum is given by:

$$T = 2\pi\sqrt{\frac{L}{g}}$$

where L is the length of the pendulum and g is Earth's gravitational constant, the pendulum will swing slower (period T gets longer) with an increase in the pendulum length.

The period of the pendulum changes from 2.006067 s to 2.006187 s.

The number of times, N, the pendulum swings through one period per day before and after the temperature change are:

$$N_{before} = \frac{(24 \text{ h d}^{-1} \times 60 \text{ min h}^{-1} \times 60 \text{ s min}^{-1})}{(2.006067 \text{ s})} = 43069.3 \text{ d}^{-1}$$

$$N_{after} = \frac{(24 \text{ h d}^{-1} \times 60 \text{ min h}^{-1} \times 60 \text{ s min}^{-1})}{(2.006187 \text{ s})} = 43066.8 \text{ d}^{-1}$$

That means the pendulum swings fewer times through one period after the temperature change and is "behind" in time:

$$\Delta N = N_{after} - N_{before} = 43066.8 \text{ d}^{-1} - 43069.3 \text{ d}^{-1} = -2.58 \text{ d}^{-1}$$

Finally, the number of seconds per day it is behind can be calculated as:

$$\Delta t = \Delta N \times N_{after} = (-2.58 \text{ d}^{-1}) \times (2.006187 \text{ s}) = -5.17 \text{ s d}^{-1}$$

REFLECT   The change in the period is about 60 ppm!

45.   **ORGANIZE AND PLAN**   We have to consider thermal expansion in all dimensions.
SOLVE   Every linear dimension increases by the same percentage with a change in temperature outwards, including holes. The hole will get larger.
REFLECT   This assumes that the expanding material is uniform.

47.   **ORGANIZE AND PLAN**   We use the equation for a two dimensional expansion, $\frac{\Delta A}{A} = 2\alpha \Delta T$, and assume room temperature as 25°C. Since the temperature is increased, the area of the hole will increase.
SOLVE   The difference in surface is

$$\frac{\Delta A}{A} = 2\alpha \Delta T$$
$$\Delta A = 2\alpha \Delta TA$$
$$\Delta A = 2\alpha \Delta T A = (0.250 \text{ m}^2) \times 2 \times (1.7 \times 10 - 5°\text{C}^{-1})(400°\text{C} - 25°\text{C}) = 3.19 \times 10^{-3} \text{ m}^2$$

The new hole size is then:

$$A(\text{hole}) = 2\alpha \Delta TA = 0.250 \text{ m}^2 + 3.19 \times 10^{-3} \text{ m}^2 = 0.253 \text{ m}^2$$

REFLECT   The hole increased by a little over 1% in area.

49.   **ORGANIZE AND PLAN**   First, we determine the molecular weights of the atoms and molecules under questions. Then we use $m = Mn$, where M is the molecular weight and n is the number of moles.
SOLVE
(a) $m(\text{Ar}) = M(\text{Ar}) \, n = (40 \text{ g mol}^{-1}) \times (1 \text{ mol}) = 40 \text{ g}$
(b) $m(\text{CO}_2) = M(\text{CO}_2) \, n = (44 \text{ g mol}^{-1}) \times (0.25 \text{ mol}) = 11 \text{ g}$
(c) $m(\text{Ne}) = M(\text{Ne}) \, n = (20 \text{ g mol}^{-1}) \times (2.6 \text{ mol}) = 52 \text{ g}$
(d) $m(\text{UF}_6) = M(\text{UF}_6) \, n = (352 \text{ g mol}^{-1}) \times (1.5 \text{ mol}) = 528 \text{ g}$

REFLECT   The molecular mass (abbreviated M) of a substance, frequently referred by the older term molecular weight and abbreviated as MW, is the mass of one molecule of that substance, relative to the unified atomic mass unit u (equal to 1/12 the mass of one isotope of carbon-12). This is distinct from the relative molecular mass of a molecule, which is the ratio of the mass of that molecule to 1/12 of the mass of carbon 12 and is a dimensionless number. Relative molecular mass is abbreviated to M_r. Molecular mass differs from more common measurements of the mass of chemicals, such as molar mass, by taking into account the isotopic composition of a molecule rather than the average isotopic distribution of many molecules. As a result, molecular mass is a more precise number than molar mass; however it is more accurate to use molar mass on bulk samples. This means that molar mass is appropriate most of the time except when dealing with single molecules.

51.   **ORGANIZE AND PLAN**   We use the ideal gas law, solved for the number of moles, n, and multiply by Avogadro's number, N_A, to get the actual number of molecules, N. We need to calculate the volume and express all numbers in SI units.

SOLVE $PV = nRT$

$$n = \frac{PV}{RT}$$

$$N = nN_A = \frac{PV}{RT}N_A = \frac{(101325\ \text{Pa})\times(8.0\ \text{m}\times7.0\ \text{m}\times2.8\ \text{m})}{(8.314\ \text{J K}^{-1}\ \text{mol}^{-1})\times(293.15\ \text{K})}\times(6.022\times10^{23}\ \text{mol}^{-1}) = 3.93\times10^{27}$$

REFLECT   The Avogadro constant (symbols: L, N_A) is the number of "elementary entities" (usually atoms or molecules) in one mole, that is (from the definition of the mole), the number of atoms in exactly 12 grams of carbon-12. It was originally called Avogadro's number.

53.   **ORGANIZE AND PLAN**   As in Problem 47, we use Boyle's law, which states that the product of p and V is constant for a constant temperature and amount of gas.

SOLVE   Since the pressure increases, the volume, and therefore the radius, has to decrease.

$$P_{before}\ r_{before}^3 = P_{after}\ r_{after}^3$$

$$r_{after} = r_{before}\ \sqrt[3]{\frac{P_{before}}{P_{after}}} = (0.10\ \text{m})\times\sqrt[3]{\frac{(1.05\ \text{atm})}{(1.75\ \text{atm})}} = 0.08\ \text{m}$$

REFLECT   The balloon decreased by about 20%.

55.   **ORGANIZE AND PLAN**   We combine the ideal gas law, $PV = nRT$, and the definitions of molecular weight, $M = \dfrac{m}{n}$, and density $\rho = \dfrac{m}{V}$, to derive an equation that allows us to calculate the density of the noble gases with the information given. We then use the result from Problem 52 to determine which of the gases is lighter than air.

SOLVE

$$\frac{PV}{RT} = n$$

$$\frac{pV}{RT} = \frac{m}{M}$$

Therefore:

$$\rho = \frac{m}{V} = \frac{PM}{RT}$$

$$\rho(\text{He}) = \frac{PM}{RT} = \frac{(101325\ \text{Pa})\times(4.0\times10^{-3}\ \text{kg mol}^{-1})}{(8.314\ \text{J K}^{-1}\ \text{mol}^{-1})\times(298.15\ \text{K})} = 0.1\ \text{kg m}^3$$

$$\rho(\text{Ne}) = \frac{PM}{RT} = \frac{(101325 \text{ Pa}) \times (20.2 \times 10^{-3} \text{ kg mol}^{-1})}{(8.314 \text{ J K}^{-1} \text{ mol}^{-1}) \times (298.15 \text{ K})} = 0.8 \text{ kg m}^3$$

$$\rho(\text{Ar}) = \frac{PM}{RT} = \frac{(101325 \text{ Pa}) \times (40.0 \times 10^{-3} \text{ kg mol}^{-1})}{(8.314 \text{ J K}^{-1} \text{ mol}^{-1}) \times (298.15 \text{ K})} = 1.6 \text{ kg m}^3$$

$$\rho(\text{Kr}) = \frac{PM}{RT} = \frac{(101325 \text{ Pa}) \times (83.8 \times 10^{-3} \text{ kg mol}^{-1})}{(8.314 \text{ J kg mol}^{-1}) \times (298.15 \text{ K})} = 3.43 \text{ kg m}^3$$

$$\rho(\text{Xe}) = \frac{PM}{RT} = \frac{(101325 \text{ Pa}) \times (131.3 \times 10^{-3} \text{ kg mol}^{-1})}{(8.314 \text{ J K}^{-1} \text{ mol}^{-1}) \times (298.15 \text{ K})} = 5.4 \text{ kg m}^3$$

$$\rho(\text{Rn}) = \frac{PM}{RT} = \frac{(101325 \text{ Pa}) \times (222.0 \times 10^{-3} \text{ kg mol}^{-1})}{(8.314 \text{ J K}^{-1} \text{ mol}^{-1}) \times (298.15 \text{ K})} = 9.1 \text{ kg m}^3$$

From Problem 49 we know that $6.72 \text{ g} (5.07 \text{ g} + 1.56 \text{ g} + 0.09 \text{ g})$ of air are in 5.5 L. We convert that to density:

$$\rho(\text{Air}) = \frac{m}{V} = \frac{(6.72 \times 10^{-3} \text{ kg})}{(0.0055 \text{ m}^3)} = 1.22 \text{ kg m}^3$$

Helium and Neon are lighter than air.

REFLECT   The density of air, ρ (Greek: rho) (air density), is the mass per unit volume of Earth's atmosphere, and is a useful value in aeronautics and other sciences. Air density decreases with increasing altitude, as does air pressure. It also changes with variances in temperature or humidity. At sea level and 20°C, air has a density of approximately 1.2 kg/m^3.

57.   **ORGANIZE AND PLAN**   (a) We use the ideal gas law at different temperatures and pressures, assuming constant volume and number of moles. For part (b), we need to incorporate the volume increase of 3%.
SOLVE
(a)

$$n_{\text{before}} = n_{\text{after}}$$

$$\frac{P_{\text{before}} V}{RT_{\text{before}}} = \frac{P_{\text{after}} V}{RT_{\text{after}}}$$

$$\frac{P_{\text{before}}}{T_{\text{before}}} = \frac{P_{\text{after}}}{T_{\text{after}}}$$

$$P_{\text{after}} = \frac{P_{\text{before}}}{T_{\text{before}}} T_{\text{after}} = \frac{(220 \text{ kPa})}{(268.15°C)} \times (305.15°C) = 250.4 \text{ kPa}$$

(b)

$$n_{\text{before}} = n_{\text{after}}$$

$$\frac{P_{\text{before}} V_{\text{before}}}{RT_{\text{before}}} = \frac{P_{\text{after}} (1.03 \, V_{\text{before}})}{RT_{\text{after}}}$$

$$\frac{P_{\text{before}}}{T_{\text{before}}} = \frac{1.03 \, P_{\text{after}}}{T_{\text{after}}}$$

$$P_{\text{after}} = \frac{P_{\text{before}}}{1.03 \, T_{\text{before}}} T_{\text{after}} = \frac{(220 \text{ kPa})}{1.03 \, (268.15°C)} \times (305.15°C) = 243.0 \text{ kPa}$$

REFLECT   When a volume increase is considered, the increase of pressure due to the increase in temperature is notably lower.

59.   ORGANIZE AND PLAN   (a) We substitute the definition of molecular mass into the ideal gas law and solve for the mass. In the calculation we use a mean molecular weight of air, $\overline{M}$, composed by oxygen and nitrogen. We only use one temperature, 15°C. In part (b) we calculate the mass of air as described in (a) for the two temperatures given, and then form the difference.

SOLVE   (a)

$$PV = nRT \text{ gives } n = \frac{PV}{RT}$$

$$M = \frac{m}{n} \text{ gives } m = Mn$$

Therefore:

$$m\left(\text{Air, } 15°\text{C}\right) = \frac{\overline{M}PV}{RT}$$

$$= \frac{\left(0.78 \times 28 \times 10^{-3} \text{ kg mol}^{-1} + 0.21 \times 32 \times 10^{-3} \text{ kg mol}^{-1}\right) \times \left(350 \times 10^{3} \text{ Pa}\right) \times \left(3.1 \times 10^{-4} \text{ m}^{3}\right)}{\left(8.314 \text{ J K}^{-1} \text{ mol}^{-1}\right) \times \left(288.15 \text{ K}\right)} = 1.29 \times 10^{-3} \text{ kg}$$

(b)

$$m\left(\text{Air, } 15°\text{C}\right) = \frac{\overline{M}PV}{RT}$$

$$= \frac{\left(0.78 \times 28 \times 10^{-3} \text{ kg mol}^{-1} + 0.21 \times 32 \times 10^{-3} \text{ kg mol}^{-1}\right) \times \left(200 \times 10^{3} \text{ Pa}\right) \times \left(3.1 \times 10^{-4} \text{ m}^{3}\right)}{\left(8.314 \text{ J K}^{-1} \text{ mol}^{-1}\right) \times \left(288.15 \text{ K}\right)} = 9.24 \times 10^{-4} \text{ kg}$$

$$m\left(\text{Air, } 22°\text{C}\right) = \frac{\overline{M}PV}{RT}$$

$$= \frac{\left(0.78 \times 28 \times 10^{-3} \text{ kg mol}^{-1} + 0.21 \times 32 \times 10^{-3} \text{ kg mol}^{-1}\right) \times \left(600 \times 10^{3} \text{ Pa}\right) \times \left(3.1 \times 10^{-4} \text{ m}^{3}\right)}{\left(8.314 \text{ J K}^{-1} \text{ mol}^{-1}\right) \times \left(295.15 \text{ K}\right)} = 2.16 \times 10^{-3} \text{ kg}$$

$$\Delta m = m\left(\text{Air, } 22°\text{C}\right) - m\left(\text{Air, } 15°\text{C}\right) = \left(2.16 \times 10^{-3} \text{ kg}\right) - \left(9.24 \times 10^{-4} \text{ kg}\right) = 1.24 \times 10^{-3} \text{ kg}$$

REFLECT   When a temperature change is included in the calculation, the mass of air that needs to be added is smaller, since a temperature increase adds to a pressure increase.

61.   ORGANIZE AND PLAN   We use the definition of the hydrostatic pressure, the pressure at a certain depth, inserted into the ideal gas law, and we can calculate the amount air exhaled by the diver. Then, we use this result to calculate the volume of the bubble at the surface using the ideal gas law. We assume a water temperature of 20°C.

SOLVE   With the definition of the hydrostatic pressure, $P = \rho g h$, where g, r, and h are the density of water, the gravitational constant, and the depth, we can calculate number of moles of air exhaled at 12.5 m:

$$n = \frac{PV}{RT} = \frac{\left(\rho g h + P_0\right) V}{RT}$$

$$n = \frac{\left[\left(1030 \text{ kg m}^{3}\right) \times \left(9.81 \text{ m/s}^{2}\right) \times \left(12.5 \text{ m}\right) + \left(101325 \text{ Pa}\right)\right] \times \left(0.000\,025 \text{ m}^{3}\right)}{\left(8.314 \text{ J mol}^{-1} \text{ K}^{-1}\right) \times \left(293.15 \text{ K}\right)} = 2.33 \times 10^{-3} \text{ mol}$$

Using the number of moles, the volume at the surface is then:

$$V = \frac{nRT}{P} = \frac{\left(2.33 \times 10^{-3} \text{ mol}\right) \times \left(8.314 \text{ J K}^{-1} \text{ mol}^{-1}\right) \times \left(293.15 \text{ K}\right)}{\left(101325 \text{ Pa}\right)} = 5.61 \times 10^{-5} \text{ m}^{3} = 56.1 \text{ cm}^{3}$$

REFLECT   Scuba diving (scuba originally being an acronym for Self-Contained Underwater Breathing Apparatus, although now widely considered a word in its own right) is a form of underwater diving in which a diver uses a scuba set to breathe underwater for recreation, commercial, or industrial reasons. Unlike early diving, which relied exclusively on air pumped from the surface, scuba divers carry their own source of breathing gas (usually compressed air), allowing them greater freedom than with an air line. Both surface supplied and scuba diving allow divers to stay underwater significantly longer than with breath-holding techniques as used in snorkeling and free-diving.

According to the purpose of the dive, a diver usually moves underwater by swim fins attached to his feet, but external propulsion can come from an underwater vehicle, or a sled pulled from the surface.

63.   **ORGANIZE AND PLAN**   We use the ideal gas law combined with the definition of molecular mass. To obtain the same buoyancy with He as with H_2, the blimp has to have the same volume, since the mass of the replaced air volume is providing the force to keep the blimp in the air.

SOLVE

(a)

$$n = \frac{PV}{RT} = \frac{m}{M}$$

The mass of H_2 in the Hindenburg is:

$$m(H_2) = \frac{MPV}{RT} = \frac{(2.02 \times 10^{-3} \text{ kg mol}^{-1}) \times (101325 \text{ Pa}) \times (2.12 \times 10^5 \text{ mol}^3)}{(8.314 \text{ J mol}^{-1} \text{ K}^{-1}) \times (293.15 \text{ K})} = 17803.4 \text{ kg}$$

The mass of He with the same buoyancy:

$$m(He) = \frac{MPV}{RT} = \frac{(4.00 \times 10^{-3} \text{ kg mol}^{-1}) \times (101325 \text{ Pa}) \times (2.12 \times 10^5 \text{ mol}^3)}{(8.314 \text{ J mol}^{-1} \text{ K}^{-1}) \times (293.15 \text{ K})} = 35254.3 \text{ kg}$$

The mass of the H_2 in the Hindenburg is less.

(b) The mass of H_2 under the different conditions is:

$$m(H_2) = \frac{MPV}{RT} = \frac{(2.02 \times 10^{-3} \text{ kg mol}^{-1}) \times (1.05 \times 10^5 \text{ Pa}) \times (2.12 \times 10^5 \text{ mol}^3)}{(8.314 \text{ J mol}^{-1} \text{ K}^{-1}) \times (283.15 \text{ K})} = 19098.4 \text{ kg} = 1.9 \times 10^4 \text{ kg}$$

REFLECT   LZ 129 *Hindenburg* (Deutsche Luftschiff Zeppelin #129; Registration: D-LZ 129) was a large German commercial passenger-carrying rigid airship, the lead ship of the *Hindenburg* class, the largest flying machines of any kind (by dimension) ever built. The airship flew from March 1936 until destroyed by fire 14 months later on May 6, 1937, at the end of the first North American transatlantic journey of its second season of service. Thirty-six people died in the accident, which occurred while landing at Lakehurst Naval Air Station in Manchester Township, New Jersey.

65.   **ORGANIZE AND PLAN**   (a) We use the equation for the rms speed as $v_{rms} = \sqrt{\dfrac{3 k_B T}{m}}$, where we find the mass of one H_2 molecule by dividing by Avogadro's number, N_A. For part (b), we repeat the calculation for the higher temperature.

SOLVE

(a)

$$v_{rms} = \sqrt{\frac{3 k_B T}{(M / N_A)}} = \sqrt{\frac{3 (1.38 \times 10^{-23} \text{ J K}^{-1}) \times (273.15 \text{ K})}{(2.02 \times 10^{-3} \text{ kg mol}^{-1} / 6.022 \times 10^{23} \text{ mol}^{-1})}} = 1836.1 \text{ m/s}^{-1}$$

(b)

$$v_{rms} = \sqrt{\frac{3 k_B T}{(M / N_A)}} = \sqrt{\frac{3 (1.38 \times 10^{-23} \text{ J K}^{-1}) \times (546.0 \text{ K})}{(2.02 \times 10^{-3} \text{ kg mol}^{-1} / 6.022 \times 10^{23} \text{ mol}^{-1})}} = 2595.9 \text{ m/s}^{-1}$$

REFLECT   The rms speed increases with the square root of the increase in temperature.

67. ORGANIZE AND PLAN We use the equation for the rms speed, $v_{rms} = \sqrt{\dfrac{3k_B T}{m}}$, to predict the temperature of double speed.

SOLVE The rms speed increases with the square root of the increase in temperature, therefore, the temperature has to increase fourfold, to 1172 K, in order for the speed to double.

REFLECT Root mean square speed is the measure of the speed of particles in a gas that is most convenient for problem solving within the kinetic theory of gases. It is given by the formula. This works well for both nearly ideal, atomic gases like helium and for molecular gases like diatomic oxygen. This is because despite the larger internal energy in many molecules (compared to that for an atom), $3RT/2$ is still the mean translational kinetic energy.

69. We use the equation for the rms speed of an ideal gas molecule.

$$v_{rms} = \sqrt{\frac{3k_B T}{m}}$$

The rms speed increases with the square root of the temperature. Therefore, the higher the temperature, the higher the rms speed.

SOLVE $v_{rms} \propto \sqrt{T}$

$$\frac{v_{rms}(1)}{v_{rms}(2)} = \sqrt{\frac{T_1}{T_2}} = \sqrt{\frac{(293.15 \text{ K})}{(353.15 \text{ K})}} = 0.91$$

REFLECT The temperature has to be quadrupled in order to double the rms speed.

71. ORGANIZE AND PLAN (a) We use the same equation as in Problem 67, $\bar{K} = \dfrac{3}{2}k_B T$, and treat H as an ideal gas. In part (b), we compute the percentage difference between the average thermal energy and the ionization energy.

SOLVE
(a)

$$\bar{K}(H) = \frac{3}{2}k_B T = \frac{3}{2}(1.38 \times 10^{-23} \text{ J K}^{-1}) \times (5800 \text{ K}) = 1.20 \times 10^{-19} \text{ J}$$

(b)

$$\text{percentage difference} = \frac{(1.20 \times 10^{-19} \text{ J})}{(2.18 \times 10^{-18} \text{ J})} = 0.055$$

REFLECT On the sun's surface the thermal energy amounts to 5.5% of hydrogen's ionization energy!

73. ORGANIZE AND PLAN (a) We use the equation for the most probable speed as $v^* = \sqrt{\dfrac{2k_B T}{m}}$, and for the rms speed as $v_{rms} = \sqrt{\dfrac{3k_B T}{(M/N_A)}}$. In both cases we find the mass of one He atom by dividing by Avogadro's number, N_A.

For part (b) we consider the distributions of speed in Figure 12.12.

SOLVE
(a)

$$v^*(He) = \sqrt{\frac{2k_B T}{(M/N_A)}} = \sqrt{\frac{2 \times (1.38 \times 10^{-23} \text{ J K}^{-1}) \times (293 \text{ K})}{(4 \times 10^{-3} \text{ kg mol}^{-1}/6.022 \times 10^{23} \text{ mol}^{-1})}} = 1103.4 \text{ m s}^{-1} = 1.1034 \text{ km s}^{-1}$$

$$v_{rms}(He) = \sqrt{\frac{3k_BT}{(M/N_A)}} = \sqrt{\frac{3\,(1.38\times10^{-23}\text{ J K}^{-1})\times(293\text{ K})}{(4\times10^{-3}\text{ kg mol}^{-1}/6.022\times10^{23}\text{ mol}^{-1})}} = 1351.4\text{ m/s}^{-1} = 1.3514\text{ km/s}^{-1}$$

(b) Based on the distribution of speeds, there are a number of molecules at a significant higher speed than the probable speed. Helium atoms are light and therefore have greater thermal speed so they could easily rise to the upper atmosphere of Earth. Over there, they continue to absorb energy from the Sun and eventually move fast enough to escape from Earth.

REFLECT   In physics, escape velocity is the speed where the kinetic energy of an object is equal to the magnitude of its gravitational potential energy, It is commonly described as the speed needed to "break free" from a gravitational field (without any additional impulse) and is theoretical, totally neglecting atmospheric friction. The term *escape velocity* can be considered a misnomer because it is actually a speed rather than a velocity, i.e., it specifies how fast the object must move but the direction of movement is irrelevant, unless "downward." In more technical terms, escape velocity is a scalar (and not a vector). Escape velocity gives a minimum delta-v budget for rockets when no benefit can be obtained from the speeds of other bodies for a particular mission; but it neglects losses such as air drag and gravity drag. However, in some cases it can be improved upon, for example, by use of gravitational slingshots.

GENERAL PROBLEMS

75.   **ORGANIZE AND PLAN**   As in Problem 54. we combine the ideal gas law, $PV = nRT$, and the definitions of molecular weight, $M = \dfrac{m}{n}$, and density $\rho = \dfrac{m}{V}$, to derive an equation that allows us to calculate the density of the noble gases with the information given. We can then determine the ratio of densities at different temperatures.

SOLVE   Ideal gas law $\dfrac{PV}{RT} = n$

$$\frac{PV}{RT} = \frac{m}{M}$$

Therefore:

$$\rho = \frac{m}{V} = \frac{PM}{RT}$$

With constant pressure we obtain:

$$\rho = C\frac{1}{T}$$

And for two different temperatures:

$$\frac{\rho_1}{\rho_2} = \frac{T_2}{T_1} = \frac{263.15\text{ K}}{305.15\text{ K}} = 0.86$$

The density at 32°C is 14% lower than the density at −10°C.

REFLECT   The addition of water vapor to air (making the air humid) reduces the density of the air, which may at first appear contrary to logic. This occurs because the molecular mass of water (18) is less than the molecular mass of air (around 29). For any gas, at a given temperature and pressure, the number of molecules present is constant for a particular volume. So when water molecules (vapor) are introduced to the air, the number of air molecules must reduce by the same number in a given volume, without the pressure or temperature increasing. Hence the mass per unit volume of the gas (its density) decreases. The density of humid air may be calculated as a mixture of ideal gases. In this case, the partial pressure of water vapor is known as the vapor pressure. Using this method, error in the density calculation is less than 0.2% in the range of −10°C to 50°C.

77.   **ORGANIZE AND PLAN** (a) We use the equation for linear thermal expansion $\dfrac{\Delta L}{L} = \alpha \Delta T$ and calculate the increase in length at 38°C, corresponding to the gap necessary. (b) We calculate by how much the rails will contract at −15°C, and add that distance to the gap, since the shrinking of a given rail contributes to two gaps.

SOLVE

(a)

$$\Delta L = \alpha \; \Delta TL = \left(1.2 \times 10^{-5} \text{°C}^{-1}\right) \times \left(38\text{°C} - 15\text{°C}\right) \times \left(20 \text{ m}\right) = 0.00552 \text{ m}$$

The gap has to be made to be 5 mm.

(b)

$$\Delta L = \alpha \Delta TL = \left(1.2 \times 10^{-5} \text{°C}^{-1}\right) \times \left(-20\text{°C} - 15\text{°C}\right) \times \left(20 \text{ m}\right) = 0.0084 \text{ m}$$

The gap will be 13 mm.

REFLECT   Many solid materials will expand evenly in all three directions, but this is not true for all. Graphite for example has a pronounced layer structure and the expansion in the direction perpendicular to the layers is quite different from that in the layers. In general the proper description of the thermal expansion of a solid must therefore include its symmetry. For cubic materials a single expansion coefficient suffices, but for a material with triclinic symmetry six parameters must be distinguished, three for each of the three axes (a,b,c) and three for the change in the angles (α,β,γ) between them. An excellent way of measuring the entire expansion tensor is to perform powder diffraction on the material during a heating or cooling run and monitor the position of its diffraction peaks.

79.   **ORGANIZE AND PLAN**   We use the equation for linear thermal expansion $\dfrac{\Delta L}{L} = \alpha \Delta T$.

(a) The actual length of the beam will be the measured length plus the length the tape will shrink.

(b) The actual length of the beam at 33°C is the length from part (a) plus the length it expanded. The tape will measure a distance at 33°C that is the actual length plus the length the tape expands with respect to its calibration temperature.

SOLVE

(a)

$$\Delta L = \alpha \Delta TL = \left(1.2 \times 10^{-5} \text{°C}^{-1}\right) \times \left(22\text{°C} + 5\text{°C}\right) \times \left(19.357 \text{ m}\right) = 0.006 \text{ m}$$

The actual distance is 19.357 m + 0.006 m = 19.363 m.

(b)

$$\Delta L = \alpha \Delta TL = \left(1.2 \times 10^{-5} \text{°C}^{-1}\right) \times \left(33\text{°C} + 5\text{°C}\right) \times \left(19.363 \text{ m}\right) = 0.008 \text{ m}$$

The actual length of the beam will be 19.363 m + 0.008 m = 19.372 m.

$$\Delta L = \alpha \Delta TL = \left(1.2 \times 10^{-5} \text{°C}^{-1}\right) \times \left(33\text{°C} - 22\text{°C}\right) \times \left(19.363 \text{ m}\right) = 0.003 \text{ m}$$

The tape will measure 19.363 m + 0.003 m = 19.366 m.

REFLECT   All errors in the measurements are well below 1%.

81.   **ORGANIZE AND PLAN**   We use Gay-Lussac's law that at constant volume and amount of gas, the temperature is proportional to the pressure.

SOLVE $\dfrac{P_{initial}}{T_{initial}} = \dfrac{P_{final}}{T_{final}}$

We set the initial conditions as 440 kPa and 276.15 K, and use the final temperature of 323.15 K to obtain for the maximum allowed pressure:

$$P_{final} = \frac{P_{initial}}{T_{initial}} T_{final} = \frac{(440 \text{ kPa})}{(276.15° \text{ K})} \times (323.15° \text{ K}) = 514.9 \text{ kPa}$$

REFLECT   The maximum allowed pressure in the can corresponds to about 5 atm!

83.   **ORGANIZE AND PLAN**   We first calculate the number of moles of oxygen gas based on the conditions at 1 atm of pressure using the ideal gas law. Subsequently, we use the number of moles of O_2 gas to calculate the pressure inside the cylinder, for which we have to determine the volume.

SOLVE   The number of moles of O_2 gas is:

$$n(O_2) = \frac{PV}{RT} = \frac{(101325 \text{ Pa}) \times (0.165 \text{ m}^3)}{(8.314 \text{ J K}^{-1} \text{ mol}^{-1}) \times (293.15 \text{ K})} = 6.8596 \text{ mol}$$

The pressure inside the cylinder is then:

$$P(O_2) = \frac{nRT}{V} = \frac{nRT}{\pi r^2 h} = \frac{(6.8596 \text{ mol}) \times (8.314 \text{ J K}^{-1} \text{ mol}^{-1}) \times (293.15 \text{ K})}{\pi (3.4 \times 10^{-2} \text{ m})^2 \times (0.28 \text{ m})} = 1.644 \times 10^7 \text{ Pa} = 162.3 \text{ atm}$$

85.   **ORGANIZE AND PLAN**   We use the equation for the rms speed as $v_{rms} = \sqrt{\dfrac{3k_B T}{m}}$, where we find the mass of one UF_6 molecule by dividing by Avogadro's number, N_A.

SOLVE

$$v_{rms}(^{238}U) = \sqrt{\frac{3k_B T}{(M/N_A)}} = \sqrt{\frac{3(1.38 \times 10^{-23} \text{ J K}^{-1}) \times (298.15 \text{ K})}{\left(352 \times 10^{-3} \text{ kg mol}^{-1} \middle/ 6.022 \times 10^{23} \text{ mol}^{-1}\right)}} = 145.32 \text{ m s}^{-1}$$

$$v_{rms}(^{235}U) = \sqrt{\frac{3k_B T}{(M/N_A)}} = \sqrt{\frac{3(1.38 \times 10^{-23} \text{ J K}^{-1}) \times (298.15 \text{ K})}{\left(349 \times 10^{-3} \text{ kg mol}^{-1} \middle/ 6.022 \times 10^{23} \text{ mol}^{-1}\right)}} = 145.94 \text{ m/s}^{-1}$$

The ratio of the rms speeds is:

$$\frac{v_{rms}(^{238}U)}{v_{rms}(^{235}U)} = \frac{145.32 \text{ m/s}^{-1}}{145.94 \text{ m/s}^{-1}} = 0.9957$$

REFLECT   The difference in rms speeds is only about 0.5 %! Nevertheless, this difference is sufficient to enrich uranium by diffusion.

87.   **ORGANIZE AND PLAN**   We use the equation for linear thermal expansion, $\dfrac{\Delta L}{L} = \alpha \Delta T$, to estimate the rise of the sea level.

SOLVE $\Delta L = \alpha \Delta T L = (2.1 \times 10^{-4} \text{°C}^{-1}) \times (1\text{°C}) \times (3.8 \times 10^3 \text{ m}) = 0.798 \text{ m}$

REFLECT   Current sea level rise has occurred at a mean rate of 1.8 mm per year for the past century, and more recently at rates estimated near 2.8 ± 0.4 to 3.1 ± 0.7 mm per year (1993–2003). Current sea level rise is due partly to human-induced global warming, which will increase sea level over the coming century and longer periods.

Increasing temperatures result in sea level rise by the thermal expansion of water and through the addition of water to the oceans from the melting of continental ice sheets. Thermal expansion, which is well quantified, is currently the primary contributor to sea level rise and is expected to be the primary contributor over the course of the next century. Glacial contributions to sea-level rise are less important and are more difficult to predict and quantify. Values for predicted sea level rise over the course of the next century typically range from 90 to 880 mm, with a central value of 480 mm. Based on an analog to the deglaciation of North America at 9,000 years before present, some scientists predict sea level rise of 1.3 meters in the next century. However, models of glacial flow in the smaller present-day ice sheets show that a probable maximum value for sea level rise in the next century is 80 centimeters, based on limitations on how quickly ice can flow below the equilibrium line altitude and to the sea.

13

HEAT

CONCEPTUAL QUESTIONS

1. **SOLVE** Heat is a transfer of energy, whereas temperature is a measure of the internal energy of an object.
    REFLECT It's easy to see how these two concepts might be confused. An object with a high temperature feels "hot" when you touch it because heat flows from it to your hand.

3. **SOLVE** Heat is not energy. It's the flow of energy. You don't say that a glass contains so much "water flow." It contains water, which can flow out.
    REFLECT Notice that heat is not like other quantities that have a set amount. You can put a hot brick in a room and heat will flow out of the brick and into the room until the brick and the room are at the same temperature. Just because no more heat is flowing doesn't mean the brick is "out of heat." If you pick up the brick and put it in the refrigerator, heat will start flowing again because there will be a temperature difference between the brick and the fridge.

5. **SOLVE** We can invert Equation 13.2 to solve for the temperature: $\Delta T = Q/mc$. From Table 13.1, the specific heat of aluminum is $900 \text{ J/kg}^\circ\text{C}$, whereas for iron it's $449 \text{ J/kg}^\circ\text{C}$. Therefore, for the same heat transfer and mass, the iron will have the greater temperature increase.
    REFLECT What does it mean to say that aluminum has a larger heat capacity than iron? In one sense, it takes more heat to change the temperature of aluminum than it does for an equal amount of iron. Or equivalently, the same amount of heat has less effect on aluminum than it does on iron.

7. **SOLVE** The specific heat of water is very high, about 200 times that of air (see Tables 13.1 and 13.2). Consequently, it can take longer (i.e., more hot, sunny days) for the lake to absorb enough heat to increase its temperature. This works in the other direction as well: it can take longer for the water to lose enough heat to decrease its temperature once the summer is over.
    REFLECT This is sometimes described as "thermal inertia." A large lake has so much thermal inertia that it's very hard to "move" its temperature. Whereas the air temperature may fluctuate from day to night and from summer to winter, lake and ocean temperatures remain relatively stable.

9. **SOLVE** The object with more mass has the higher heat capacity, so its temperature will change less.
    REFLECT Imagine adding a drop of hot water to a bucket of cold water. It's rather obvious that when the combination reaches equilibrium, the temperature will be more cold than hot. The bucket's water doesn't change temperature by very much, whereas the drop's water changes a lot.

11. **SOLVE** Falling below 0°C would be damaging for the fruit, but spraying water can prevent that from happening. Imagine a blanket of water around the fruit. See figure below.

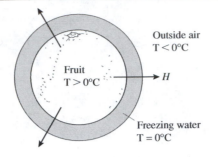

If the outside air falls below freezing, the water will remain at 0°C while it turns to ice. Therefore, the fruit's temperature will not fall below zero degrees.

REFLECT   If the water blanket freezes completely, then the ice will begin to drop below zero degrees. Even so, the ice will slow down the loss of heat from the fruit. This can buy enough time for the Sun to come out and start warming back up the air. As long as the night doesn't get too cold, the fruit will likely avoid damage.

13.   As you can see from Table 13.2, the molar specific heat of all the gases is higher for constant pressure than for constant volume. That means the temperature will climb faster at constant volume.

REFLECT   Heating at constant pressure means that the gas will be able to expand and therefore do work on its container (see Figure 13.7). As a consequence, less of the heat goes to increasing the internal energy of the gas, which is proportional to the temperature. By contrast, when heating at constant volume, all the heat goes into the internal energy of the gas, so the temperature rises faster.

15.   **SOLVE**   The runner sweats, drinks, breathes, and radiates. Sweating cools the runner through evaporation (the sweat absorbs heat from the skin as it vaporizes). Drinking brings the internal organs in contact with a cool liquid, which induces heat conduction out of the body and into the liquid. Breathing cools through convection: the air breathed in is cooler than the air breathed out, so there's a net transfer of heat out of the body. Lastly, the runner's temperature is typically higher than his surroundings so he will radiate heat in accordance with the Stefan-Boltzman law (Equation 13.8).

REFLECT   You can get a sense of how important each of these effects is by regarding the following: Problem 13.73 (sweating), Problem 13.74 (drinking), Problem 13.100 (breathing), and Example 13.13 (radiation).

17.   **SOLVE**   The two panes trap air, which has a much lower thermal conductivity than glass (see Table 13.3). That means less heat escapes through a double-paned window than a single pane. But if the gap is too wide, convection cells can develop between the windows (see Figure 13.2). The churning air of convection cells is more effective at passing heat than stationary air.

REFLECT   The optimum air gap is about half an inch. Less than that, and you don't have enough of an air buffer. More than that, and you start to have convection through air movement.

19.   **SOLVE**   The curtains themselves are usually thin, but they hold air in front of the windows. This confined air acts as insulation, since it has a low thermal conductivity. The curtains also reduce the amount of convection. Without curtains, air currents in the room bring warm air in contact with the windows, increasing the amount of heat the room is losing. Curtains partially block the air currents, so less warm air is cycled by the windows.

REFLECT   You might think opening the curtains would be better, since it would let in some sunlight that could help heat the room. In fact, the curtains do bring in the energy of the Sun into the room, but it's indirectly. The combination of curtain and window creates a kind of greenhouse effect, where the sunlight warms the curtain material and the glass window prevents the energy radiating from the curtain from escaping.

21.   **SOLVE**   Your tongue exchanges heat with the bulb, where most of the mercury is. After several seconds, the mercury in the bulb is the same temperature as your tongue (which is an accurate representation of your body's temperature). In response to this temperature increase, the mercury in the bulb expands and pushes a small amount of mercury out along the thin column where the measurement is made. The mercury in the column will not be exactly the same temperature as your body, but this liquid is only a small fraction of the total mercury, so it has a negligible effect on the measurement.

REFLECT   Most liquids expand in a similar way as mercury when they are heated. The advantage of mercury is that it remains liquid over a wide range of temperatures: from –39°C to 358°C (see Table 13.3). Some thermometers use ethanol, which is liquid between –114°C and 78°C.

MULTIPLE-CHOICE PROBLEMS

23.   **SOLVE**   The change in gravitational potential energy is equal to:

$$U = mgh = (60 \text{ kg})(9.80 \text{ m/s}^2)(1200 \text{ m}) = 706 \text{ kJ} \left[\frac{1 \text{ Cal}}{4186 \text{ J}} \right] = 169 \text{ Cal}$$

The answer is (a).

REFLECT   Of course, the hike will take more energy than this, since humans are not perfectly efficient machines. Some energy is spent on normal bodily functions.

25.   **SOLVE**   From Equation 13.2 and Table 13.1:

$$\Delta T = \frac{Q}{mc} = \frac{(100 \text{ cal})}{(25 \text{ g})(1.00 \text{ cal/g}°\text{C})} = 4°\text{C}$$

The answer is (b).

REFLECT   The calorie was defined to be the energy needed to raise 1 g of water by 1°C.

27.   **SOLVE**   The hot water and cold water transfer heat to each other in equal and opposite amounts: $Q_h = -Q_c$. Using Equation 13.2, this becomes:

$$m_h c \Delta T_h = -m_c c \Delta T_c$$

Where $\Delta T_h = T_f - 45°\text{C}$ and $\Delta T_c = T_f - 10°\text{C}$. Solving for the final temperature gives:

$$T_f = \frac{(2.0 \text{ kg})(45°\text{C}) + (1.2 \text{ kg})(10°\text{C})}{(2.0 \text{ kg}) + (1.2 \text{ kg})} = 32°\text{C}$$

The answer is (a).

REFLECT   The final temperature is basically like a weighted average of the two temperatures.

29.   **SOLVE**   We can determine the mass of the ice from Equation 13.5: $m = Q_f / L_f$. We can then plug into Equation 13.6 to find the energy to vaporize water with the same mass:

$$Q_v = mL_v = Q_f \frac{L_v}{L_f} = (6.5 \text{ kJ}) \frac{(2.26 \times 10^6 \text{ J/kg})}{(3.33 \times 10^5 \text{ J/kg})} = 44 \text{ kJ}$$

The answer is (d).

REFLECT   It takes almost 7 times more energy to boil water than melt an equal amount of ice. This trend of greater boiling than melting energy is true of all the materials in Table 13.3.

31.   **SOLVE**   The surface area of a sphere is: $A = 4\pi r^2$. Using the Stefan-Boltzmann law (Equation 13.8), the rate of energy radiation from the star is:

$$P = e\sigma A T^4 = (1)(5.67 \times 10^{-8} \text{ W/m}^2\text{K}^4)\left(4\pi(1.62 \times 10^9 \text{ m})^2\right)(7200 \text{ K})^4 = 5 \times 10^{27} \text{ W}$$

The answer is (a).

REFLECT   This is approximately 10 times the power of the Sun (see Problem 13.82). That's because this star is slightly bigger and slightly hotter.

PROBLEMS

33.   **ORGANIZE AND PLAN**   We are again asked to convert food calories to Joules: 1 Cal $= 4186$ J.
SOLVE   The candy bar has:

$$280 \text{ Cal} \left[\frac{4186 \text{ J}}{1 \text{ Cal}} \right] = 1.2 \times 10^6 \text{ J}$$

REFLECT   A million Joules in a single wrapper! Note that this is 10% of a normal athletic person's caloric intake.

35.   **ORGANIZE AND PLAN**   As we did before, we will assume that the kinetic energy of the truck is transferred to thermal energy in the brakes when it stops.
Known: $m = 34,000$ kg, $v = 60$ mph.
SOLVE   In the previous problem, we calculated the velocity in SI units. Plugging this into the kinetic energy equation, we obtain

$$E_{th} = K = \tfrac{1}{2} mv^2 = \tfrac{1}{2} (34,000 \text{ kg})(27 \text{ m/s})^2 = 1.2 \times 10^7 \text{ J}$$

REFLECT   Truckers have to be careful about overheating their brakes, so they often use engine braking to slow down. In this case, the engine and the drive-train absorb the thermal energy associated with stopping.

37.   **ORGANIZE AND PLAN**   We're told that the energy expended per stride is equal to the kinetic energy ($K = \tfrac{1}{2} mv^2$), but notice that we are not given how fast the runner is running. We'll have to determine it from the time it takes her to run the given distance.
Known: $m = 65$ kg, $\Delta x = 1000$ m, $\Delta t = 5$ min.
SOLVE   The velocity of the runner is:

$$v = \frac{\Delta x}{\Delta t} = \frac{(1000 \text{ m})}{(5 \text{ min})} \left[\frac{1 \text{ min}}{60 \text{ s}} \right] = 3.3 \text{ m/s}$$

Using this to find the kinetic energy:

$$K = \tfrac{1}{2} mv^2 = \tfrac{1}{2} (65 \text{ kg})(3.3 \text{ m/s})^2 = 350 \text{ J}$$

We are told to assume this is the energy expended per stride, so the total energy is this times the number of strides, which is the total distance divided by the stride-length:

$$E = K \cdot N_{stride} = (350 \text{ J}) \left(\frac{1000 \text{ m}}{1.5 \text{ m}} \right) \left[\frac{1 \text{ Cal}}{4186 \text{ J}} \right] = 56 \text{ Cal}$$

REFLECT   The answer is reasonable, if a bit on the small side.

39.   **ORGANIZE AND PLAN**   During each repetition, you expend energy in increasing the gravitational potential of the barbell ($U_{grav} = mgh$). We'll assume that you don't expend energy letting the barbell fall back to the starting position.
Known: $m = 75$ kg, $h = 1.9$ m, $N_{reps} = 20$.
SOLVE   The barbell's gravitational potential changes during each rep by:

$$U_{grav} = mgh = (75 \text{ kg})(9.80 \text{ m/s}^2)(1.9 \text{ m}) = 1400 \text{ J}$$

The total energy expended in Joules and Calories:

$$E = U_{grav} \cdot N_{reps} = (1400 \text{ J})(20) = 28,000 \text{ J}$$

$$= 28,000 \text{ J} \left[\frac{1 \text{ Cal}}{4186 \text{ J}} \right] = 6.7 \text{ Cal}$$

REFLECT   This matches data on how much energy is burned during 1 minute-worth of weight training.

41. **ORGANIZE AND PLAN** From the preceding problem, we have the conversion: 1 Btu $= 1055$ J. Power is energy divided by time.

Known: $\Delta E = 92$ therms, $\Delta t = 1$ mo.

SOLVE The power in watts (J/s) is:

$$P = \frac{\Delta E}{\Delta t} = \frac{(92 \text{ therms})}{(1 \text{ mo})}\left[\frac{10^5 \text{ Btu}}{1 \text{ therm}}\right]\left[\frac{1055 \text{ J}}{1 \text{ Btu}}\right]\left[\frac{1 \text{ mo}}{30 \cdot 24 \cdot 60 \cdot 60 \text{ s}}\right] = 3700 \text{ W}$$

REFLECT This is equivalent to having 37 100-Watt lightbulbs on all the time.

43. **ORGANIZE AND PLAN** The definition of the heat capacity is from Equation 13.1: $Q = C\Delta T$. The value for C found in part (a) should remain valid for the temperature increase in part (b).

Known: $Q = 2.48$ kJ, $\Delta T = 25°C$ for part (a); $\Delta T = 200°C$ for part (b).

SOLVE (a) From the definition of heat capacity:

$$C = \frac{Q}{\Delta T} = \frac{2.48 \text{ kJ}}{25°C} = 99.2 \text{ J/°C}$$

(b) The heat capacity is a constant of the material, so we can use it to find the heat absorbed for other temperature increases:

$$Q = C\Delta T = (99.2 \text{ J/°C})(200°C) = 19.8 \text{ kJ}$$

REFLECT The temperature change increases by a factor of 8 between part (a) and part (b). The same 8-fold increase should be seen in the heat absorbed and it is: $(19.8 \text{ kJ})/(2.48 \text{ kJ}) = 8$.

45. **ORGANIZE AND PLAN** This problem is similar to Example 13.4, but in this case we don't know the final temperature. All we can say, then, is that the hotter water changes temperature by: $\Delta T_{hot} = T_f - 25°C$, while the colder water changes temperature by: $\Delta T_{cold} = T_f - 2.0°C$. We will be able to solve for T_f using Equation 2.2, and the fact that the heat lost by the hot water is gained by the cold water: $Q_{hot} = -Q_{cold}$, assuming of course that no heat is lost to the surroundings.

Known: $m_{hot} = 18$ kg, $m_{cold} = 6$ kg.

SOLVE The equal but opposite heat exchange implies:

$$Q_{hot} = -Q_{cold} \quad \Rightarrow \quad m_{hot}c\Delta T_{hot} = -m_{cold}c\Delta T_{cold}$$

Solving for the final temperature:

$$T_f - 25°C = -\frac{6 \text{ kg}}{18 \text{ kg}}(T_f - 2.0°C) \quad \Rightarrow \quad T_f = 19°C$$

REFLECT The final temperature is closer to the hotter temperature, which makes sense since there is three times more hot water than cold water.

47. **ORGANIZE AND PLAN** We will follow the argument of Example 13.4. Heat is exchanged between the hot coffee and the cold water, but the whole system does not lose or receive heat. Therefore, $Q_{hot} + Q_{cold} = 0$. The temperature of the coffee drops, while that of the added water rises:

$$\Delta T_{hot} = 49°C - 55°C = -6°C$$
$$\Delta T_{cold} = 49°C - 10°C = 39°C$$

Known: $m_{hot} = 300$ g

SOLVE The heat exchange is written:

$$Q_{hot} + Q_{cold} = m_{hot}c\Delta T_{hot} + m_{cold}c\Delta T_{cold} = 0$$

The specific heat of coffee is the same as water, so the c's will cancel out of the equation. Solving for the cold water mass:

$$m_{cold} = -m_{hot}\frac{\Delta T_{hot}}{\Delta T_{cold}} = -(300 \text{ g})\frac{(-6°C)}{(39°C)} = 46 \text{ g}$$

REFLECT   The small amount of cold water makes sense, since the coffee temperature is only dropping a few degrees. Note that 46 g of water is 46 mL, or about 9 teaspoons.

49.   **ORGANIZE AND PLAN**   We'll need the equation for the gravitational potential ($U = m_{weight}gh$) and equate it to the heat needed to raise the given water temperature ($Q = m_{water}c\Delta T$).

Known: $h = 1$ m, $m_{water} = 1$ kg, $\Delta T = 1$ K.

SOLVE   Equating the gravitational potential and the heat, we can solve for the mass of the weight:

$$m_{weight} = \frac{m_{water}c\Delta T}{gh} = \frac{(1 \text{ kg})(4186 \text{ J/kg°C})(1 \text{ K})}{(9.80 \text{ m/s}^2)(1 \text{ m})} = 427 \text{ kg}$$

REFLECT   Note that we have used the fact that for temperature changes, a Kelvin is the same as a degree Celsius ($1 \text{ K} = 1°C$). The calculated mass is equivalent to 940 pounds, which is slightly smaller than the 1,000 pounds we found in Problem 8.48 (b) for a mass falling 1 foot to raise the temperature of 1 pound of water by 1°F.

51.   **ORGANIZE AND PLAN**   Water boils at 100°C, so we'll use Equation 13.2 ($Q = mc\Delta T$) to calculate how much heat is needed to bring the water up to that temperature. The time it takes to reach boiling is just the heat delivered divided by the kettle's power: $t = Q/P$, where recall that W = J/s.

Known: $P = 1250$ W, $V_{H2O} = 1.0$ L, $T_i = 20°C$, $T_f = 100°C$.

SOLVE   The temperature change in the water is: $\Delta T = T_f - T_i = 80°C$, and given that the density of water is 1.00 g/mL, the mass of 1.0 L is 1.0 kg. The heat needed to bring this much water to boil is:

$$Q = mc\Delta T = (1.0 \text{ kg})(4186 \text{ J/kg°C})(80°C) = 3.3 \times 10^5 \text{ J}$$

The time it takes the kettle to deliver this much heat is:

$$t = \frac{Q}{P} = \frac{(3.3 \times 10^5 \text{ J})}{(1250 \text{ W})} = 268 \text{ s}$$

REFLECT   This is about 4 and half minutes, which seems reasonable for bringing that much water to boil. Notice that the time could be longer because some of the heat may escape.

53.   **ORGANIZE AND PLAN**   We know that the sum of the heat lost by the material and gained by the water is zero: $Q_M + Q_W = 0$, so we'll use that to solve for the unknown specific heat: c_M. The temperature changes for the material and the water are:

$$\Delta T_M = 22.9°C - 34.5°C = -11.6°C$$
$$\Delta T_W = 22.9°C - 18°C = 4.9°C$$

Known: $m_M = 25.0$ g, $m_W = 125$ g.

SOLVE   Using the heat exchange and Equation 13.2:

$$c_M = -\frac{m_W c_W \Delta T_W}{m_M \Delta T_M} = -\frac{(0.125 \text{ kg})(4186 \text{ J/kg°C})(4.9°C)}{(0.0250 \text{ kg})(-11.6°C)} = 8841 \text{ J/kg°C}$$

REFLECT   Looking through Table 13.1, there's no material that matches this specific heat. But of course this list is not exhaustive, so we shouldn't be concerned.

55.   **ORGANIZE AND PLAN**   This is a straightforward use of Equation 13.2: $Q = mc\Delta T$, where the specific heat of mercury is from Table 13.1: $c = 140$ J/kg°C. The one thing we will need is the density of liquid mercury from Table 10.1: $\rho = 13,600$ kg/m³.

Known: $V = 2.30$ mL, $\Delta T = 100$°C.

SOLVE   Plugging the mass of mercury ($m = \rho V$) into Equation 13.2:

$$Q = \rho V c \Delta T = (0.0136 \text{ kg/mL})(2.30 \text{ mL})(140 \text{ J/kg}°\text{C})(100°\text{C}) = 438 \text{ J}$$

REFLECT   This heat causes the mercury to expand slightly, which results in the liquid rising inside the thermometer. Because this rise is uniform, we can use it to measure the temperature. In this way, the thermometer works simply by absorbing heat (or losing heat) to the environment.

57.   **ORGANIZE AND PLAN**   The nitrogen starts off colder, so it will gain heat from the helium: ($Q_N = -Q_{He}$). We'll assume that the gases are combined under fixed pressure, so that the heat gained or lost will come from Equation 8.4: $Q = nc_p\Delta T$. The molar specific heats can be taken from Table 13.2 for nitrogen ($c_N = 29.1$ J/mol°C) and for helium ($c_{He} = 20.8$ J/mol°C). We'll need to convert the given masses into moles, and write the temperature change for the nitrogen as: $\Delta T_N = T_f - 25$°C, and the helium as: $\Delta T_{He} = T_f - 45$°C.

Known: $m_N = 56$ g, $m_{He} = 12$ g.

SOLVE   The equal but opposite heat exchange implies:

$$Q_N = -Q_{He} \Rightarrow n_N c_N \Delta T_N = -m_{He} c_{He} \Delta T_{He}$$

The molar masses are 28 g/mol for nitrogen gas and 4 g/mol for helium gas, so the number of moles are 2 mol of nitrogen and 3 mol of helium. Solving for the final equilibrium temperature:

$$T_f - 25°\text{C} = -\frac{(3 \text{ mol})(20.8 \text{ J/mol}°\text{C})}{(2 \text{ mol})(29.1 \text{ J/mol}°\text{C})}(T_f - 45°\text{C}) \Rightarrow T_f = 35°\text{C}$$

REFLECT   The answer makes sense, since the final temperature is halfway between the initial temperatures of the nitrogen and the helium. If you assumed that the gases were mixed with constant volume, the result would be practically the same: $T_f = 34$°C. This is because the ratio of the molar specific heats (c_{He}/c_N) is practically the same for constant volume and constant pressure.

59.   **ORGANIZE AND PLAN**   We'll assume a constant volume of air in the house. We can find the total number of moles using the ideal gas law ($PV = nRT$). We could then find the number of moles of nitrogen and the number of moles of oxygen using the percentages given. But they are both diatomic gases, so they have essentially the same molar specific heat. In Table 13.3, the molar specific heat of air is given ($c_V = 20.8$ J/mol°C), so we'll use that to find the energy needed to raise the air temperature by one degree Celsius (Equation 13.3).

Known: $A = 190$ m², $h = 2.3$ m, $\Delta T = 1$°C, $r = 16$¢/kWh.

SOLVE   (a) The volume is the area times the height ($V = Ah$), and we assume that the pressure is 1 atm (or 101,325 Pa in SI units) and the room temperature is 20°C (or 293 K). So the total number of moles is:

$$n = \frac{PV}{RT} = \frac{(101,325 \text{ Pa})(190 \text{ m}^2)(2.3 \text{ m})}{(8.315 \text{ J/mol} \cdot \text{K})(293 \text{ K})} = 18,200 \text{ mol}$$

You could also come to a similar result by using the fact that the molar volume of an ideal gas is 22.4 L/mol at standard temperature and pressure, but that would be assuming the house is a very chilly 0°C to begin with. Plugging the total moles into Equation 13.3:

$$Q = nc_V\Delta T = (18,200 \text{ mol})(20.8 \text{ J/mol}°\text{C})(1°\text{C}) = 3.79 \times 10^5 \text{ J}$$

(b) The cost of this much energy is:

$$C = rQ = (16¢/\text{kWh})(3.79 \times 10^5 \text{ J})\left[\frac{1 \text{ W} \cdot \text{s}}{1 \text{ J}}\right]\left[\frac{1 \text{ kWh}}{(1000 \text{ W})(60 \cdot 60 \text{ s})}\right] = 1.7¢$$

REFLECT   The heating system won't be 100% efficient, and we've neglected the energy needed to heat the furniture, the floors, the walls, etc. But assuming this cost is approximate, it is interesting to consider how often a heating system has to rewarm the air to counter heat loss from the house. Imagine the heating bill for one month is $150, then we could think of that as the heating system turning on almost 9,000 times (or once every 5 minutes) to reheat the air by one degree.

61.   **ORGANIZE AND PLAN**   This is like the previous problem, except that the ice first has to warm to 0°C before it starts to melt. For this initial warming, we use Equation 13.2: $Q = mc\Delta T$, with the specific heat of ice from Table 13.1: $c = 2090$ J/kg°C.
Known: $m = 120$ g, $\Delta T = 25°$C.
SOLVE   The energy needed to bring the ice to the melting point is:

$$Q = mc\Delta T = (0.120 \text{ kg})(2090 \text{ J/kg°C})(25°\text{C}) = 6.27 \times 10^3 \text{ J}$$

In the previous problem, we found the energy for the melting $Q = 4.00 \times 10^4$ J, so the total energy is the sum:

$$Q_{tot} = 6.27 \times 10^3 \text{ J} + 4.00 \times 10^4 \text{ J} = 4.63 \times 10^4 \text{ J}$$

REFLECT   Notice that the majority of the energy is the melting. The heat required for the temperature change is a small fraction of the total.

63.   **ORGANIZE AND PLAN**   The copper is in liquid form, so energy must be removed from it to cause it to solidify. This loss of heat is from Equation 13.5: $Q = -mL_f$, where we have included a negative sign to signify that this is heat taken away from the copper. The latent heat of fusion for copper from Table 13.3 is: $L_f = 2.05 \times 10^5$ J/kg. Once it turns completely solid, the copper temperature will be at its melting point: $T_i = 1084°$C from Table 13.3. As it cools to 600°C, the heat removed from the copper will be $Q = mc\Delta T$, where the specific heat of copper from Table 13.1 is: $c = 385$ J/kg°C. In the end we will sum these two energies.
Known: $m = 15$ g, $\Delta T = 600°$C $- 1084°$C $= -484°$C.
SOLVE   The energy removed while the copper is solidifying is:

$$Q = -mL_f = -(0.015 \text{ kg})(2.05 \times 10^5 \text{ J/kg}) = -3100 \text{ J}$$

The energy removed while it is cooling is:

$$Q = mc\Delta T = (0.015 \text{ kg})(385 \text{ J/kg°C})(-484°\text{C}) = -2800 \text{ J}$$

The total energy removed is:

$$Q_{tot} = -3100 \text{ J} - 2800 \text{ J} = -5900 \text{ J}$$

REFLECT   The energies are all negative, as they should be, because they represent heat loss from the copper.

65.   **ORGANIZE AND PLAN**   The copper needs to first be heated to its melting point of 1084°C. The energy required will be a sum of $Q = mc\Delta T$ and $Q = mL_f$, where $c = 385$ J/kg°C and $L_f = 2.05 \times 10^5$ J/kg. We will divide this energy by the furnace power to find the time.
Known: $m = 52$ kg, $\Delta T = 1084°$C $- 20°$C $= 1064°$C, $P = 120$ kW.
SOLVE   The energy needed to heat and then melt the copper ingot is:

$$Q = mc\Delta T = (52 \text{ kg})(385 \text{ J/kg°C})(1064°\text{C}) = 2.1 \times 10^7 \text{ J}$$
$$Q = mL_f = (52 \text{ kg})(2.05 \times 10^5 \text{ J/kg}) = 1.1 \times 10^7 \text{ J}$$

The time to supply all this heat is:

$$t = \frac{Q_{tot}}{P} = \frac{2.1 \times 10^7 \text{ J} + 1.1 \times 10^7 \text{J}}{120,000 \text{ J/s}} = 270 \text{ s}$$

Where we have used the definition: 1 kW = 1,000 J/s.

REFLECT   This is about 4 and half minutes for a large chunk of metal. Mighty impressive, but realize that the furnace here is about 30 times more powerful than a household oven.

67.   **ORGANIZE AND PLAN**   This will require three steps: (1) lower the water temperature from 11°C to 0°C; (2) freeze the water; (3) lower the ice temperature from 0°C to –14 °C. We'll need the specific heat of water: $c_w = 4186$ J/kg°C, the latent heat of fusion of water: $L_f = 3.33 \times 10^5$ J/kg and the specific heat of ice: $c_I = 2090$ J/kg°C. The water changes temperature by $\Delta T = -11°C$, while the ice changes temperature by $\Delta T = -14°C$.

Known: $m = 410$ g.

SOLVE   Equating the heat gain and loss ($Q_{ice} = -Q_{water}$) allows us to solve for the ice mass:

$$Q_1 = mc_w\Delta T = (0.41 \text{ kg})(4186 \text{ J/kg°C})(-11°C) = -1.9 \times 10^4 \text{ J}$$
$$Q_2 = -mL_f = -(0.41 \text{ kg})(3.33 \times 10^5 \text{ J/kg}) = -1.4 \times 10^5 \text{ J}$$
$$Q_3 = mc_I\Delta T = (0.41 \text{ kg})(2090 \text{ J/kg°C})(-14°C) = -1.2 \times 10^4 \text{ J}$$

Notice that we included a minus sign for the freezing step, since it corresponds to a heat loss, just as the other steps. The total energy removed from the water is:

$$Q_{tot} = Q_1 + Q_2 + Q_3 = -1.7 \times 10^5 \text{ J}$$

REFLECT   The value seems reasonable. The majority of energy is needed to do the freezing.

69.   **ORGANIZE AND PLAN**   Let's first imagine what the plot will look like. There are four main steps: (1) the temperature will start off rising from –10°C to 0°C; (2) then it will stay at 0°C until the ice has melted; (3) next it will rise from 0°C to 100°C; (4) lastly it will stay at 100°C until all the water has boiled away. What we have to determine is how long each step is, which means finding the energy needed in each step and then dividing by the power. We'll need the specific heat of ice ($c_I = 2090$ J/kg°C), the latent heat of fusion ($L_f = 3.33 \times 10^5$ J/kg), the specific heat of water ($c_w = 4186$ J/kg°C), and the latent heat of vaporization ($L_v = 2.26 \times 10^6$ J/kg).

Known: $m = 0.50$ kg, $\Delta T_I = 10°C$, $\Delta T_w = 100°C$, $P = 1000$ W.

SOLVE   The four steps each have an associated energy, which we can turn into a time by dividing by the power:

$$\Delta t_1 = \frac{mc_I\Delta T_I}{P} = \frac{(0.50 \text{ kg})(2090 \text{ J/kg°C})(10°C)}{1000 \text{ W}} = 10 \text{ s}$$
$$\Delta t_2 = \frac{mL_f}{P} = \frac{(0.50 \text{ kg})(3.33 \times 10^5 \text{ J/kg})}{1000 \text{ W}} = 170 \text{ s}$$
$$\Delta t_3 = \frac{mc_w\Delta T_w}{P} = \frac{(0.50 \text{ kg})(4186 \text{ J/kg°C})(100°C)}{1000 \text{ W}} = 210 \text{ s}$$
$$\Delta t_4 = \frac{mL_v}{P} = \frac{(0.50 \text{ kg})(2.26 \times 10^6 \text{ J/kg})}{1000 \text{ W}} = 1130 \text{ s}$$

These are the duration of each step. To plot the temperature as a function of time, we'll need to know when each step starts and ends: (1) from 0 s to 10 s the temperature rises from –10°C to 0°C; (2) from 10 s to 180 s the temperature stays at 0°C; (3) from 180 s to 390 s the temperature rises from 0°C to 100°C; (4) from 390 s to 1520 s the temperature stays at 100°C. See figure below.

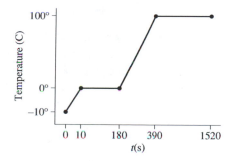

REFLECT   The whole process takes about 25 minutes, but 19 of those minutes are devoted to boiling the water. This agrees with our experience. If we put an ice cube in a pan and put it on the stove, it will melt quickly and soon reach a boil, but for all the water to disappear into vapor takes a long time.

71.   ORGANIZE AND PLAN   There are two steps here: (1) the steam condenses into water and (2) the two quantities of water equilibrate in temperature. See figure below.

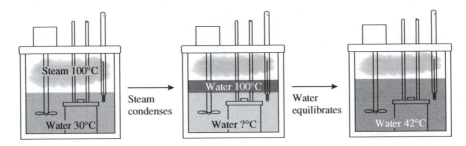

We don't know the temperature of the initial water after the first step when the steam condenses, but we won't need it. We only need to determine the total heat gained by the water in the calorimeter, which is easy to find using Equation 13.2: $Q_W = m_W c \Delta T_W$ and the full temperature change: $\Delta T_W = 42°C - 30°C = 12°C$. This heat is equal to the total heat lost by the steam as it both condenses ($Q_{cond} = -m_S L_v$) and cools to the final temperature ($Q_{cool} = m_S c \Delta T_S$). In this case, the steam-supplied water starts off at 100°C and cools by:
$\Delta T_S = 42°C - 100°C = -58°C$.

Known: $m_W = 150$ g.

SOLVE   We assume no heat is lost out of the calorimeter, so the sum of the heat exchanges should be zero:

$$Q_W + Q_{cond} + Q_{cool} = 0$$

Plugging in the parameters for each of these energies, we can solve for the unknown mass:

$$m_S = \frac{-m_W c \Delta T_W}{-L_v + c \Delta T_S} = \frac{-(150 \text{ g})(4186 \text{ J/kg}°\text{C})(12°\text{C})}{-(2.26 \times 10^6 \text{ J/kg}) + (4186 \text{ J/kg}°\text{C})(-58°\text{C})} = 3.0 \text{ g}$$

REFLECT   This seems like a small amount of steam, but we have to remember that a lot of energy is given up by condensing steam, seeing as the latent heat of vaporization is so large.

73.   ORGANIZE AND PLAN   As described in the section on "Evaporative Cooling," the body loses heat to the evaporation of sweat according to: $Q = -m_S L_v$, where $L_v = 2.4 \times 10^6$ J/kg is the latent heat of vaporization of water at the body temperature 37°C. The body responds to this heat loss by dropping in temperature: $Q = m_B c \Delta T$, where the specific heat of the body is: $c = 3500$ J/kg°C.

Known: $m_B = 90$ kg, $\Delta T = -1°C$.

SOLVE   Solving for the mass of sweat:

$$m_S = \frac{m_B c \Delta T}{-L_v} = \frac{(90 \text{ kg})(3500 \text{ J/kg}°\text{C})(-1°\text{C})}{-(2.4 \times 10^6 \text{ J/kg})} = 0.13 \text{ kg}$$

REFLECT   This is equal to 130 mL of water or a little less than half a cup. That seems like a reasonable amount if you consider that it would be spread over a large portion of the body.

75.   ORGANIZE AND PLAN   The heat conduction is given in Equation 13.7: $H = kA\Delta T / \Delta x$. For iron, the thermal conductivity is: $k = 52$ W/°C·m, from Table 13.5. The heat flows along the length of the cylinder, through the round face ($A = \pi r^2$). The temperature difference is: $\Delta T = 250°C - 20°C = 230°C$.

Known: $\Delta x = 25.0$ cm, $r = 1.0$ cm.

SOLVE   Plugging the values into the heat conduction equation:

$$H = kA\frac{\Delta T}{\Delta x} = (52 \text{ W/}^\circ\text{C}\cdot\text{m})(\pi(0.01 \text{ m})^2)\frac{(230^\circ\text{C})}{(0.25 \text{ m})} = 15 \text{ W}$$

REFLECT   For a metal, iron is not especially good at conducting heat. This is why you often find steel pots with copper bottoms, since the thermal conductivity of copper is almost 8 times that of iron.

77.   ORGANIZE AND PLAN   As worked out in the Example, a wooden house loses heat by: $H = k_w A\Delta T / \Delta x_w$ (Equation 13.7), where the thermal conductivity of wood is: $k_w = 0.12 \text{ W/}^\circ\text{C}\cdot\text{m}$, from Table 13.5. We now assume a layer of Styrofoam is added to the walls. Like in the previous problem, we can show that the heat flow becomes:

$$H = \frac{A\Delta T}{(\Delta x_w / k_w + \Delta x_s / k_s)}$$

The thermal conductivity of Styrofoam is: $k_s = 0.024 \text{ W/}^\circ\text{C}\cdot\text{m}$. Once we have this, we can find the heat lost in a day: $Q = Ht$, and how much it costs assuming a utility rate of $r = \$0.10$ per kWh.
Known: $A = 275 \text{ m}^2$, $\Delta x_w = 1.0 \text{ cm}$, $\Delta x_s = 6.0 \text{ cm}$, $\Delta T = 18^\circ\text{C}$.
SOLVE   If the wooden walls have Styrofoam insulation, the heat flow becomes:

$$H = \frac{A\Delta T}{(\Delta x_w / k_w + \Delta x_s / k_s)} = \frac{(275 \text{ m}^2)(18^\circ\text{C})}{(0.01/0.12 + 0.06/0.024)(\text{m}^2 \,^\circ\text{C/W})} = 1.92 \text{ kW}$$

The heat lost in a day and the corresponding cost are:

$$Q = Ht = (1{,}920 \text{ W})(24\cdot 60\cdot 60 \text{ s}) = 1.66\times 10^8 \text{ J}$$

$$C = Qr = (1.66\times 10^8 \text{ J})\left[\frac{1 \text{ kWh}}{3.6\times 10^6 \text{ J}}\right](\$0.10/\text{kWh}) = \$4.61$$

REFLECT   The Styrofoam reduces the daily cost by a factor of 30. Even a modest investment in insulation can have a big effect on heating bills.

79.   ORGANIZE AND PLAN   From Table 13.4, the thermal conductivities are $0.8 \text{ W/}^\circ\text{C}\cdot\text{m}$ for glass, $0.12 \text{ W/}^\circ\text{C}\cdot\text{m}$ for wood, and $0.024 \text{ W/}^\circ\text{C}\cdot\text{m}$ for Styrofoam.
Known: $\Delta x = 3.2 \text{ mm}$.
SOLVE   (a) For glass:

$$R = \frac{\Delta x}{k} = \frac{0.0032 \text{ m}}{0.8 \text{ W/}^\circ\text{C}\cdot\text{m}} = 0.004 \text{ m}^2\,^\circ\text{C/W}$$

(b) For wood:

$$R = \frac{\Delta x}{k} = \frac{0.0032 \text{ m}}{0.12 \text{ W/}^\circ\text{C}\cdot\text{m}} = 0.027 \text{ m}^2\,^\circ\text{C/W}$$

(c) For Styrofoam:

$$R = \frac{\Delta x}{k} = \frac{0.0032 \text{ m}}{0.024 \text{ W/}^\circ\text{C}\cdot\text{m}} = 0.133 \text{ m}^2\,^\circ\text{C/W}$$

REFLECT   As we might imagine, the Styrofoam provides the best resistance (i.e., insulation) to heat flow.

81. **ORGANIZE AND PLAN** We have shown that the R-value of a combination of materials is the sum of their respective R-values. For a double-pane window, there are three materials: glass, air, and glass again. The R-value for 3.2 mm of glass was calculated in Problem 13.79: $R_{glass} = 0.004$ m² °C/W. To compute the R-value of the air gap, we'll need the thermal conductivity of air: $k_{air} = 0.026$ W/°C·m. We will then compare the sum to the standard R-19 wall construction, which is defined as:

$$R_{wall} = 19 \text{ ft}^2 \cdot {}^\circ F \cdot h/Btu$$

Where a Btu is equal to 1055 J (see Problem 13.40). We will convert our answer into these customary units for comparison.

SOLVE For a 2.0-mm-wide air gap, the R-value is:

$$R_{air} = \frac{\Delta x_{air}}{k_{air}} = \frac{0.002 \text{ m}}{0.026 \text{ W/}^\circ\text{C}\cdot\text{m}} = 0.077 \text{ m}^2 \, {}^\circ\text{C/W}$$

For the double-pane window, then the R-value in SI and traditional units is:

$$R_{dp} = R_{glass} + R_{air} + R_{glass} = (0.004 + 0.077 + 0.004) \text{ m}^2 \, {}^\circ\text{C/W} = 0.085 \text{ m}^2 \, {}^\circ\text{C/W}$$

$$= 0.085 \text{ m}^2 \, {}^\circ\text{C s/J} \left[\frac{3.28 \text{ ft}}{1 \text{ m}}\right]^2 \left[\frac{1.8 \, {}^\circ\text{F}}{{}^\circ\text{C}}\right]\left[\frac{1 \text{ h}}{3600 \text{ s}}\right]\left[\frac{1055 \text{ J}}{1 \text{ Btu}}\right] = 0.48 \text{ ft}^2 \, {}^\circ\text{F h/Btu}$$

REFLECT In customary units, the double-pane window has R-0.48, which is 40 times less insulating than a standard R-19 wall.

83. **ORGANIZE AND PLAN** We found the Sun's power in the previous problem. When that radiation reaches the Earth, the power is uniformly distributed over a sphere whose radius is equal to the Earth-Sun distance: $r_{AU} = 1.496 \times 10^{11}$ m. The power per unit area, or what's called the intensity, is then:

$$I = \frac{P_{sun}}{A} = \frac{P_{sun}}{4\pi r_{AU}^2}$$

Known: $P_{sun} = 3.91 \times 10^{26}$ W.

SOLVE (a) The power per unit area felt at the Earth's orbit is:

$$I = \frac{P_{sun}}{4\pi r_{AU}^2} = \frac{3.91 \times 10^{26} \text{ W}}{4\pi(1.496 \times 10^{11} \text{ m})^2} = 1,390 \text{ W/m}^2$$

(b) The sunlight power hitting the panels in IA_{panel}, but the panels only convert 20% into electricity power: $P_{panel} = 0.2IA_{panel}$. Solving for the area needed for 1.0 GW of electricity.

$$A_{panel} = \frac{P_{panel}}{0.2I} = \frac{1.0 \times 10^9 \text{ W}}{0.2(1,390 \text{ W/m}^2)} = 3.6 \times 10^6 \text{ m}^2$$

REFLECT The area is approximately a square mile. This seems like a large thing to be putting into space, but we should compare it to the 1.3 million tons of coal or the 5 million barrels of oil that a 1GW-power-plant on Earth requires every year. The advantage of having solar panels in space is that there's no night and no clouds, so the panels would work all the time at their maximum efficiency. Still, putting up a square mile of solar panels is not easy and neither is figuring out how to transmit the electricity down to Earth.

85. **ORGANIZE AND PLAN** In Example 13.10, you have a cup with 300 g of coffee at 85°C. You add 400 g ice. It's not clear that this much ice is going to melt completely. We will start by calculating how much energy it takes to melt the ice: $Q_i = m_i L_f$, where $L_f = 3.33 \times 10^5$ J/kg. We will compare this to the energy it takes to cool the coffee from 85°C to 0°C: $Q_c = m_c c |\Delta T|$, where we assume the coffee's specific heat is that of water: $c = 4186$ J/kg°C. If $Q_i < Q_c$, then the ice melts completely and we proceed like in Example 13.10. If $Q_i > Q_c$, then the coffee cools to

0°C before all the ice melts. That's because the coffee and ice are at the same temperature, so no more heat is exchanged between them. If this is the case, then we will figure out much ice remains in the cup.

Known: $m_c = 300$ g, $m_i = 400$ g.

SOLVE   First the energy it would take to melt the ice:

$$Q_i = m_i L_f = (0.400 \text{ kg})(3.33 \times 10^5 \text{ J/kg}) = 1.33 \times 10^5 \text{ J}$$

Then the energy it would take to cool the coffee to the freezing point:

$$Q_c = m_c c |\Delta T| = (0.300 \text{ kg})(4186 \text{ J/kg}^\circ\text{C})(85^\circ\text{C}) = 1.07 \times 10^5 \text{ J}$$

So $Q_i > Q_c$, and therefore the ice only partially melts. The coffee can only transfer Q_c-worth of energy before it reaches 0°C, which is the same temperature as the ice. The amount of ice that does end up melting is:

$$m = \frac{Q_c}{L_f} = \frac{(1.07 \times 10^5 \text{ J})}{(3.33 \times 10^5 \text{ J/kg})} = 321 \text{ g}$$

That means there will remain 79 g of ice along with the 0°C-coffee.

REFLECT   Note that we dealt only with the positive values of heat transfer just for ease of comparison. Technically, Q_c is a negative quantity, since it's the heat lost by the coffee. It is perhaps helpful to ask what would have happened if you had followed the same exact steps in Example 13.10? You would have calculated the energy to melt the ice: $Q_i = m_i L_f$, and then you would have used this to find the change in the coffee's temperature. $\Delta T_c = -Q_i / m_c c$. If you did that, you would have found $\Delta T_c = -106^\circ\text{C}$, which doesn't make physical sense since it implies that the coffee cools to below zero. This would signal to you that there's more ice than the coffee is capable of melting.

87.   **ORGANIZE AND PLAN**   The key is whether the energy to melt the ice: $Q_i = m_i L_f$ is more or less than the energy to cool the water to 0°C: $Q_w = m_w c(25^\circ\text{C})$. If $Q_i < Q_w$, then there is more than enough energy to melt all the ice, but if $Q_i > Q_w$, then the water cools to 0°C before all the ice melts. Once the water reaches 0°C, it is the same temperature as the ice and no more melting can occur.

Known: $m_i = m_w = 250$ g.

SOLVE   (a) The energies we want to compare are:

$$Q_i = m_i L_f = (0.25 \text{ kg})(3.33 \times 10^5 \text{ J/kg}) = 8.33 \times 10^4 \text{ J}$$
$$Q_w = m_w c(25^\circ\text{C}) = (0.25 \text{ kg})(4186 \text{ J/kg}^\circ\text{C})(25^\circ\text{C}) = 2.62 \times 10^4 \text{ J}$$

$Q_i > Q_w$, which means that we are left with a water-ice mixture.

(b) The water cools to 0°C and releases Q_w into the ice to melt a portion of it. The mass of melted ice is:

$$m = \frac{Q_w}{L_f} = \frac{(2.62 \times 10^4 \text{ J})}{(3.33 \times 10^5 \text{ J/kg})} = 78.7 \text{ g}$$

This melted ice increases the amount of water to 328.7 g, whereas the amount of ice drops to 171.3 g.

REFLECT   Since you are starting with equal amounts of water and ice, an interesting question is at temperature would the water need to start at in order to just barely melt all the ice (i.e., $Q_i = Q_w$). If you work it out, you'll find that the water would have to start out at 80°C. To reiterate what that means: the energy that it takes to chill 80-degree water to the freezing point is the same as the energy it takes to completely melt ice of the same mass.

89.   **ORGANIZE AND PLAN**   The skillet absorbs heat according to $Q = mc\Delta T$, where $c = 449$ J/kg$^\circ$C is the specific heat of iron from Table 13.1. The time for the burner to produce this temperature change is: $t = Q/P$, where P is the stove's power output.

Known: $m = 3.4$ kg, $\Delta T = 130^\circ\text{C} - 20^\circ\text{C} = 110^\circ\text{C}$, $P = 2.0$ kW.

SOLVE   (a) The energy needed to heat to warm the skillet is:

$$Q = mc\Delta T = (3.4 \text{ kg})(449 \text{ J/kg}^\circ\text{C})(110^\circ\text{C}) = 1.7 \times 10^5 \text{ J}$$

(b) The time to supply all this heat is:

$$t = \frac{Q}{P} = \frac{1.7 \times 10^5 \text{ J}}{2,000 \text{ J/s}} = 85 \text{ s}$$

Where we have used the definition: 1 kW=1,000 J/s.

REFLECT   An empty skillet should heat up fairly fast. According to our calculations, this skillet will take just over a minute to reach high temperatures.

91.   **ORGANIZE AND PLAN**   From Table 13.3, uranium melts at 1133°C with a latent heat of fusion: $L_f = 8.28 \times 10^4$ J/kg. We first need to find the total heat needed to melt the core: $Q = mL_f$, and then we'll divide this by the rate at which the uranium is producing heat.

Known: $m = 2.5 \times 10^5$ kg, $P = 120$ MW.

SOLVE   The energy needed to melt the core is:

$$Q = mL_f = (2.5 \times 10^5 \text{ kg})(8.28 \times 10^4 \text{ J/kg}) = 2.07 \times 10^{10} \text{ J}$$

The time for this to occur is:

$$t = \frac{Q}{P} = \frac{2.07 \times 10^{10} \text{ J}}{120 \times 10^6 \text{ J/s}} = 173 \text{ s}$$

REFLECT   This is less than 3 minutes, which isn't a lot of time to react. We've neglected the time it takes to reach the melting point, so that should give a little more time but not much. One way to stop this is by quenching the core with water (see Problem 13.106).

93.   **ORGANIZE AND PLAN**   We're given the thickness of the stovetop and we can figure out its area from the dimensions. We're also given the temperature difference between the inside and outside surfaces. The thermal conductivity of steel is found in Table 13.4: $k = 40$ W/°C·m. That should allow us to calculate the stovetop's heat conduction: $H = kA\Delta T / \Delta x$.

For part (b), we're told that the rate at which the stove heats the room is 3 times that of the heat flow from part (a): $Q / \Delta t = 3H$. The room will rise in temperature at a rate of:

$$\frac{\Delta T}{\Delta t} = \frac{Q}{nc_V \Delta t} = \frac{3H}{nc_V}$$

Here, the molar specific heat is for air at constant volume: $c_V = 20.8$ J/mol°C from Table 13.2. We'll need to determine the number of moles from the ideal gas law: $(PV = nRT)$.

Known:

For the stove: $A = 90 \text{ cm} \times 40 \text{ cm} = 0.36 \text{ m}^2$, $\Delta x = 0.45$ cm, $\Delta T = 310°\text{C} - 308°\text{C} = 2°\text{C}$.

For the room: $V = 8.6 \text{ m} \times 6.5 \text{ m} \times 2.8 \text{ m} = 156 \text{ m}^3$, $P = 1$ atm, $T = 20°\text{C}$.

SOLVE   (a) The heat flow through the stove top is:

$$H = kA\frac{\Delta T}{\Delta x} = (40 \text{ W/°C} \cdot \text{m})(0.36 \text{ m}^2)\frac{(2°\text{C})}{(0.0045 \text{ m})} = 6.4 \text{ kW}$$

(b) In the room, the pressure is 1atm (or 101,325 Pa in SI units) and the room temperature is 20°C (or 293K). By the ideal gas law tells us that the number of moles of air in the room are:

$$n = \frac{PV}{RT} = \frac{(101,325 \text{ Pa})(156 \text{ m}^3)}{(8.315 \text{ J/mol} \cdot \text{K})(293 \text{ K})} = 6,500 \text{ mol}$$

We can plug this into the equation that we derived for the rate of temperature change:

$$\frac{\Delta T}{\Delta t} = \frac{3H}{nc_V} = \frac{3 \cdot (6,400 \text{ J/s})}{(6,500\text{mol})(20.8 \text{ J/mol°C})} = 0.14°\text{C/s}$$

REFLECT The result is unreasonably high. The room is not going to heat by one degree every 7 seconds. The air next to the stove will get hot very fast, but heat will have a harder time reaching other parts of the room. We've also not considered the heat escaping the room, which will limit how fast it warms up.

95. **ORGANIZE AND PLAN** All we need is the thermal conductivity of concrete from Table 13.4: $k_C = 1.28$ W/°C·m.
    Known: $A = 8.0$ m×12 m = 96 m², $\Delta x = 23$ cm, $\Delta T = 18°C - 10°C = 8°C$
    SOLVE From Equation 13.7, the heat loss is:

$$H = kA\frac{\Delta T}{\Delta x} = (1.28 \text{ W/}°\text{C}\cdot\text{m})(96 \text{ m}^2)\frac{(8°\text{C})}{(0.23 \text{ m})} = 4.3 \text{ kW}$$

REFLECT We typically think the heat loss in a house is out of the ceiling or through the walls and windows. But if the ground is cold, it can pull heat out through the floors as well. To help reduce this heat loss, one can install some carpeting, ceramic tiles, or a wooden floor (see previous problem).

97. **ORGANIZE AND PLAN** When the car stops, it's kinetic energy $(K = \frac{1}{2}m_{car}v^2)$ is dissipated into the brakes. We'll assume the energy is distributed equally to the four brakes: $Q = K/4$, so the temperature change will be: $\Delta T = K/(4m_{brake}c)$. For the specific heat of steel, we use $c = 500$ J/kg°C from Table 13.1.
    Known: $m_{car} = 1500$ kg, $v = 32$ m/s, $m_{brake} = 5.0$ kg
    SOLVE The kinetic energy of the car before stopping is:

$$K = \frac{1}{2}m_{car}v^2 = \frac{1}{2}(1500 \text{ kg})(32 \text{ m/s})^2 = 7.68\times10^5 \text{ J}$$

When this energy is deposited into the brakes, they will rise in temperature by:

$$\Delta T = \frac{K}{4m_{brake}c} = \frac{(7.68\times10^5 \text{J})}{4(5.0 \text{ kg})(500 \text{ J/kg}°\text{C})} = 77°\text{C}$$

REFLECT The answer makes sense, since car brakes tend to get extremely hot from use.

99. **ORGANIZE AND PLAN** In the "Earth's Climate" application, we are told that the Earth receives on average 240 W of sunlight per square meter. The Earth must emit this same power per unit area (P/A), otherwise it would continue to heat up. By the Stefan-Boltzmann law (Equation 13.8), the temperature of the Earth must then be $T = \sqrt[4]{P/e\sigma A}$ to be in equilibrium.
    Known: $P/A = 240$ W/m², $e = 1$, $\sigma = 5.67\times10^{-8}$ W/m²K⁴.
    SOLVE Substituting the values into the Stefan-Boltzmann law:

$$T = \sqrt[4]{\frac{P/A}{e\sigma}} = \sqrt[4]{\frac{240 \text{ W/m}^2}{(1)(5.67\times10^{-8} \text{ W/m}^2\text{K}^4)}} = 255 \text{ K} = -18°\text{C}$$

REFLECT This temperature is obviously not correct! The global average temperature is more like 15°C, which is 33°C higher than we calculated. The reason for this discrepancy is the greenhouse effect. Our planet's atmosphere lets sunlight in to warm up the Earth's surface. The warm ground emits heat just as the Stefan-Boltzman law dictates, but certain gases in the atmosphere, like carbon dioxide and methane, trap some of this heat before it can escape into space. This causes the temperature of the Earth to be higher than it would if there were no atmosphere. In fact, the moon and Mars are good examples of how cold it would be if Earth didn't have its atmosphere.

101. **ORGANIZE AND PLAN** The lake absorbs power that is equal to the given intensity multiplied by the area: $P = IA$. The energy needed to raise the temperature by the given amount is: $Q = mc\Delta T$, and the time it takes the sun to deliver this much energy is $t = Q/P$. To find the mass of the water in the lake, we'll need the volume $(V = A \cdot d)$ and the density of water ($\rho = 1000$ kg/m³).
    Known: $r = 1.0$ km, $d = 10$ m, $I = 200$ W/m², $\Delta T = 15°C$.

SOLVE   By combining the above equations, we can solve for the time:

$$t = \frac{Q}{P} = \frac{(\rho A d)c\Delta T}{IA} = \frac{\rho d c \Delta T}{I}$$

$$= \frac{(1000 \text{ kg/m}^3)(10 \text{ m})(4186 \text{ J/kg°C})(15°C)}{(200 \text{ W/m}^2)} = 3.14 \times 10^6 \text{ s}$$

REFLECT   Notice that the area cancels out of the calculation. The answer implies that the lake warms up in 36 days, which seems reasonable, at least, for the top layer of the lake. The bottom of the lake will probably take longer to warm up.

103.   **ORGANIZE AND PLAN**   This is a problem in heat conduction. The body is producing heat, which inevitably escapes through the sleeping bag. For the temperature to remain stable inside the sleeping bag, the heat flow ($H = kA\Delta T/\Delta x$) through the sleeping bag must be equal to the 100 W that the body generates. We're told the body's surface area, so we'll assume that sleeping bag has roughly the same surface area. (Don't be concerned that the surface in this case is not flat.) We have the thickness of the down in the sleeping bag, and we can find the thermal conductivity of goose down in Table 13.4: $k = 0.043 \text{ W/°C·m}$. All we need to find is the temperature difference that develops between the inside and outside of the sleeping bag, given the heat that the body is producing.
Known: $H = 100 \text{ W}$, $A = 1.5 \text{ m}^2$, $\Delta x = 4.0 \text{ cm}$.
SOLVE   Solving for the temperature difference:

$$\Delta T = \frac{H\Delta x}{kA} = \frac{(100 \text{ W})(0.04 \text{ m})}{(0.043 \text{ W/°C·m})(1.5 \text{ m}^2)} = 62°C$$

If the body temperature is to remain above 37°C, then the outside cannot dip below: $37°C - 62°C = -25°C$.
REFLECT   If the outside temperature fell below –25°C, then the heat flow from inside to outside the sleeping bag would become greater than 100 W. This would draw more heat from the body than it is able to generate, so the body temperature would drop.

105.   **ORGANIZE AND PLAN**   We're told that basically all of the reactor's thermal power output goes into heating the water. Therefore, the time to achieve the desired temperature change will be:

$$\Delta t = \frac{Q}{P} = \frac{mc\Delta T}{P}$$

We'll assume the specific heat of the water is 4186 J/kg°C, even though it will be different under the reactor's high pressures.
Known: $m = 5.4 \times 10^6 \text{ kg}$, $\Delta T = 350°C - 10°C = 340°C$, $P = 1.42 \text{ GW}$.
SOLVE   The time for the water to reach the needed temperature is:

$$\Delta t = \frac{mc\Delta T}{P} = \frac{(5.4 \times 10^6 \text{ kg})(4186 \text{ J/kg°C})(340°C)}{1.42 \times 10^9 \text{ W}} = 5400 \text{ s}$$

REFLECT   This says that it takes only 90 minutes to super-heat 5 million liters of water. That's *if* you have over a billion-watts-worth of power at your disposal.

107.   **ORGANIZE AND PLAN**   The water and the heat are both flowing continuously, but it might to think of the heat transfer in discrete steps. Therefore, imagine that over a given time period, Δt, the reactor is in contact with a certain mass of water, m, from the river. You might picture this water being collected temporarily in a huge tank that surrounds the reactor. However the contact is made, the water absorbs a certain amount of waste heat, Q, from the reactor. See figure below.

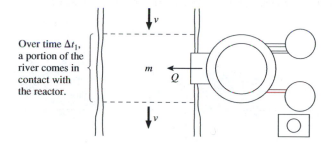

Over time Δt_1, a portion of the river comes in contact with the reactor.

The amount of waste heat is just the rate of reactor waste output multiplied by the time period: $Q = P\Delta t$. Likewise, the amount of mass is the mass flow rate of the river multiplied by the time period: $m = F\Delta t$. This mass flow rate is: $F = \rho A v$, where ρ is the density of water, A is the cross-sectional area of the river, and v is the speed of the river flow.

You might wonder what the time period is, but it doesn't actually matter. We are looking for the temperature increase of the water, which is: $\Delta T = Q/mc$. The time period cancels out of the equation.

Known: $P = 2.0$ GW, $\rho = 1000$ kg/m^3, $A = 250$ m$\times 3.0$ m, $v = 1.5$ m/s, $c = 4186$ J/kg$^\circ$C.

SOLVE   The water temperature rises by

$$\Delta T = \frac{P}{\rho A v c} = \frac{(2.0 \times 10^9 \text{ W})}{(1000 \text{ kg/m}^3)(750 \text{ m}^2)(1.5 \text{ m/s})(4186 \text{ J/kg}^\circ\text{C})} = 0.42\,^\circ\text{C}$$

REFLECT   This is sometimes called thermal pollution because it is believed that temperature increases like this can have negative effects on the ecosystem of the river. Engineers can reduce the temperature change by using cooling towers or cooling ponds to let the water release some of its heat to the atmosphere before going back into the lake.

14

THE LAWS OF THERMODYNAMICS

CONCEPTUAL QUESTIONS

1. **SOLVE** The work done is the area under a pressure-versus-volume graph.
   (a) For the constant pressure case, the area is simply a rectangle $P\Delta V$.
   (b) For the adiabatic case, the pressure increases when the volume decreases, keeping the quantity PV^γ constant. Consequently, the work is larger than in the constant pressure case.
   (c) For the isothermal case, the pressure also increases when the volume decreases, keeping the product PV constant to satisfy the ideal gas law. Consequently, the work is larger than in the constant pressure case, but it is smaller than in the adiabatic case because the pressure increase is larger in the adiabatic case than in the isothermal case.
   REFLECT The pressure increase is larger in the adiabatic case than in the isothermal case because $\gamma > 1$ for all gases.

3. **SOLVE** (a) Yes, because both heat flowing into the gas and work done on the gas increase the internal energy of the gas, and an increase in internal energy means an increase in temperature.
   (b) Not necessarily, because heat flowing into the gas increases the internal energy while the expansion means that the gas does work, i.e., decreases its internal energy.
   REFLECT Whether the temperature goes up or down in (b) can be solved by knowing the specific heat of the gas, how much heat goes into the gas, and how much the gas expands and through which process.

5. **SOLVE** The temperature of your body doesn't drop because heat is released in chemical reactions; essentially the heats of transformation of various chemical reactions heat the body.
   REFLECT Your body is not a gas, so it doesn't obey the law of physics for a gas stating that a drop in the internal energy means a drop in temperature.

7. **SOLVE** Air conditioners work by moving heat from one location (indoors) to another (outdoors). If an air conditioning unit was placed in the middle of the room it would just move the heat from the air on one side of the room to the air of the other side of the room — the air would soon mix and no cooling would be achieved.
   REFLECT It takes work (electrical energy) to move the heat. If the air-conditioning unit was placed in the middle of the room, not only would no heat be removed from the room, but the electrical energy used by the air conditioner is converted into heat which would end up heating the room — placing an air conditioner in the middle of a room is a dumb idea for two reasons!

9. **SOLVE** The reason is the same as the reason why an air conditioner placed in the middle of a room doesn't cool the room (see Problem 14.7). Refrigerators work by moving heat from one location (inside the refrigerator) to another (the room the refrigerator is in). If the door was left open, the refrigerator would just move the heat from the air inside the open refrigerator to the air of the room — the air would soon mix and no cooling would be achieved.
   REFLECT It takes work (electrical energy) to move the heat. If the refrigerator door was left open, not only would no heat be removed from the room, but the electrical energy used by the air conditioner is converted into heat which would end up heating the room — leaving a refrigerator door open with the hope of cooling a room is a dumb idea twice over.

11.   **SOLVE**   Entropy is a measure of disorder. The molecules in a solid are more ordered than those in a liquid, so the entropy is larger in liquid silver than in solid silver.

REFLECT   If you add heat to the solid silver, its entropy increases by

$$\Delta S = \frac{Q}{T}$$

If you add enough heat the silver melts, so it makes sense that liquid silver has higher entropy than solid silver.

13.   **SOLVE**   No, this is just less likely than 50 molecules on either side, and if you wait a little longer, the most likely configuration you will see is that with 50 molecules on either side again.

REFLECT   If you know the rate with which the molecules swap sides (and you can estimate this rate from the size of the box divided by the average speed of a molecule) you can work out how long, on average, you would have to wait before you see a specific configuration. Some configurations, like all 100 molecules being on the same side, may be so unlikely that you will have to wait (or expect to wait) longer than the lifetime of the universe before you see it.

15.   **SOLVE**   Water in a glass freezing to ice by dumping its heat into the surrounding room.

Carbon dioxide and various paper debris spontaneously forming a ready-to-use fire work.

All the dust and dirt on the floor in your room upstairs forming a neat pile outside your house.

REFLECT   Think about any process you are familiar with that you know cannot reverse itself as if time had run backwards. Very frequently this will be a process that conserves energy but which violates the second law of thermodynamics.

MULTIPLE-CHOICE PROBLEMS

17.   **SOLVE**   For a monatomic gas the internal energy is:

$$U = \tfrac{3}{2}PV = \tfrac{3}{2}(1.0\,\text{atm})(23.0\,\text{L}) = 3500\,\text{J}$$

The correct answer is (d).

REFLECT   This is an easy calculation to do in your head, without a calculator.

19.   **SOLVE**   The temperature increases, so the correct answer is (a).

REFLECT   From the ideal gas law, temperature is proportional to the product of pressure and volume $T \propto PV$. For an adiabatic process, the quantity PV^γ is constant, so

$$T \propto PV = \frac{1}{V^\gamma} V = \frac{1}{V^{\gamma-1}}$$

and the specific heat ratio $\gamma > 1$.

21.   **SOLVE**   When the gas expands, the gas does work. To compress the gas, work is done on the gas. Consequently, the work done on the gas is positive if the compression is done at a higher pressure than the expansion, and that requires that the path is traversed counterclockwise. The correct answer is (b).

REFLECT   The gas does work if the path is traversed clockwise.

23.   **SOLVE**   The entropy change is the heat added to the water divided by its temperature. The heat can be calculated from the heat of transformation from Table 13.3. Consequently, the entropy change is:

$$\Delta S = \frac{Q}{T} = -\frac{mL_f}{T} = -\frac{(40\,\text{g})(3.33\times10^5\,\text{J/kg})}{(0°\text{C})} = -49\,\text{J/K}$$

The correct answer is (b).

REFLECT   In this case heat is not added to but removed from the water, so the entropy change of the water is negative. However, there must be an entropy increase somewhere else that more than makes up for this entropy loss.

25. **SOLVE** The coefficient of performance of a refrigerator is heat removed divided by work (electrical energy) used, so the heat removed per unit time is:

$$\frac{Q_C}{t} = \text{COP} \cdot \frac{W}{t} = \text{COP} \cdot P = (3.5)(800 \text{ W}) = 2800 \text{ W}$$

The correct answer is (c).

REFLECT It's important to be comfortable going from energy to power and vice versa. Power is energy per unit time. All the formulae you've learned in this chapter for various heats and energies also apply to power by simply dividing all heats and energies by time.

PROBLEMS

27. **ORGANIZE AND PLAN** Helium is a monatomic gas so the internal energy difference when warmed is $\Delta U = \frac{3}{2} N k_B \Delta T$. The number of atoms in 1.25 L is given by the ideal gas law.
Known: $V = 1.25$ L; $\Delta T = 100°C$.
SOLVE From the ideal gas law $PV = N k_B T$ we find that internal energy increases by:

$$\Delta U = \tfrac{3}{2} N k_B \Delta T = \tfrac{3}{2} PV \frac{\Delta T}{T}$$

Insert known values:

$$\Delta U = \tfrac{3}{2}(1 \text{ atm})(1.25 \text{ L}) \frac{(100°C)}{(20°C)} = \tfrac{3}{2}(101 \text{ kPa})(1.25 \times 10^{-3} \text{ m}^3) \frac{(100 \text{ K})}{(293 \text{ K})} = 64.8 \text{ J}$$

REFLECT We will assume room temperature and atmospheric pressure unless stated otherwise (or unless such an assumption appears obviously erroneous). Note that it's important to convert the room temperature to Kelvin because it's not a temperature difference.

29. **ORGANIZE AND PLAN** Heat, internal energy, and work are related through the first law of thermodynamics. We will divide the quantities in this law by time to get powers and heat flow rates.
Known: $\Delta U / \Delta t = 45$ W; $P = -165$ W.
SOLVE The first law of thermodynamics divided by time is:

$$\frac{\Delta U}{\Delta t} = \frac{\Delta Q}{\Delta t} + P$$

which we can rewrite to get an expression for the heat flow into the system:

$$\frac{\Delta Q}{\Delta t} = \frac{\Delta U}{\Delta t} - P = (45 \text{ W}) - (-165 \text{ W}) = 210 \text{ W}$$

REFLECT The power is negative because the system is doing work.

31. **ORGANIZE AND PLAN** We will use the first law of thermodynamics to calculate the work, given the heat added and the internal energy change of a monatomic gas.
Known: $n = 4.0$ mol; $T = 293$ K; $p = 1$ atm; $Q = 1830$ J; $\Delta T_a = 45°C$; $\Delta T_c = 15°C$.
SOLVE (a) The internal energy change in a monatomic gas is:

$$\Delta U_a = \tfrac{3}{2} n R \Delta T_a = \tfrac{3}{2}(4.0 \text{ mol})(8.31 \text{ J}/(\text{mol} \cdot \text{K}))(45°C) = 2.2 \text{ kJ}$$

The work done on the krypton is given by the first law of thermodynamics:

$$W_a = \Delta U_a - Q = (2.2 \text{ kJ}) - (1830 \text{ J}) = 0.41 \text{ kJ}$$

(b) The work done on the gas is positive so the gas volume is decreasing.

(c) The internal energy change for a smaller temperature increase of 15°C is:

$$\Delta U_c = \tfrac{3}{2} nR\Delta T_c = \tfrac{3}{2}(4.0 \text{ mol})(8.31 \text{ J/(mol·K)})(15°C) = 0.75 \text{ kJ}$$

The work done on the krypton is:

$$W_c = \Delta U_c - Q = (0.75 \text{ kJ}) - (1830 \text{ J}) = -1.1 \text{ kJ}$$

The work done on the gas is negative so the gas volume is increasing.

REFLECT   The expanding gas in part (c) is doing work.

33.   **ORGANIZE AND PLAN**   The internal energy equals the thermal energy of the atoms in the gas. We can use the ideal gas law to calculate a new temperature when the pressure and volume double, and use the new temperature to calculate the new internal energy.

Known: $n = 2.0 \text{ mol}$; $T = 273 \text{ K}$; $p = 1 \text{ atm}$.

SOLVE   The internal energy equals the thermal energy:

$$U = \tfrac{3}{2} nRT = \tfrac{3}{2}(2.0 \text{ mol})(8.31 \text{ J/(mol·K)})(273 \text{ K}) = 6.8 \text{ kJ}$$

From the ideal gas law $PV = nRT$ we see that the temperature quadruples when both pressure and volume double. Consequently, the internal energy must quadruple as well, to $U_{new} = 4U = 4(6.8 \text{ kJ}) = 27 \text{ kJ}$. The change in internal energy is:

$$\Delta U = U_{new} - U = (27 \text{ kJ}) - (6.8 \text{ kJ}) = 20 \text{ kJ}$$

REFLECT   From the result in Problem 14.32 we immediately see that the internal energy quadruples when both pressure and volume double.

35.   **ORGANIZE AND PLAN**   We can use the first law of thermodynamics to calculate the work done on the gas. Nitrogen is a diatomic gas.

Known: $V = 8.5 \text{ L}$; $Q = 860 \text{ J}$; $\Delta T = 82°C$.

SOLVE   From the ideal gas law $PV = Nk_B T$ we find that internal energy of the diatomic increases by:

$$\Delta U = \tfrac{5}{2} Nk_B \Delta T = \tfrac{5}{2} PV \frac{\Delta T}{T}$$

Insert known values:

$$\Delta U = \tfrac{5}{2}(1 \text{ atm})(8.5 \text{ L})\frac{(82°C)}{(20°C)} = \tfrac{5}{2}(101 \text{ kPa})(8.5 \times 10^{-3} \text{ m}^3)\frac{(82 \text{ K})}{(293 \text{ K})} = 0.60 \text{ kJ}$$

From the first law of thermodynamics we can calculate the work done on the gas:

$$W = \Delta U - Q = (0.60 \text{ kJ}) - (860 \text{ J}) = -0.26 \text{ kJ}$$

REFLECT   The negative value means that the gas does work. Very likely this work is done in the form of expanding the elastic skin of the balloon.

37.   **ORGANIZE AND PLAN**   We can use the ideal gas law to calculate the temperature change. When the heat is added the balloon has already been compressed, so the heat is added in a constant volume process.

Known: $V = 2.5 \text{ L}$; $p = 1.0 \text{ atm}$; $\Delta V = -0.10V$.

SOLVE   (a) From the ideal gas law we see that when the pressure is constant, temperature is proportional to volume, so when the volume decreased by 10% the temperature also decreased by 10%:

$$\Delta T_a = -0.10T = -0.10(293 \text{ K}) = -29 \text{ K}$$

(b) For a constant volume process the work done is zero. This means that the heat that must be added equals the increase in internal energy that brings the temperature in the gas back up to room temperature. Oxygen is diatomic, so we have:

$$Q = \Delta U = \tfrac{5}{2} nR\Delta T_b$$

where $\Delta T_b = -\Delta T_a$. Rewrite this expression using the ideal gas law:

$$Q = \tfrac{5}{2} nR\Delta T_b = \tfrac{5}{2} PV \frac{\Delta T_b}{T} = \tfrac{5}{2}(1.0\ \text{atm})(2.5\ \text{L})\frac{(29\ \text{K})}{(293\ \text{K})} = 63\ \text{J}$$

REFLECT   It doesn't matter if you use the original values or the compressed values for V and T in the final expression, because the ratio is the same. We are using V and T here only to calculate the number of moles the gas contains.

39.   **ORGANIZE AND PLAN**   The temperature is still constant, so this is still an isothermal process where pressure varies inversely with volume.
Known: $n = 0.30$ mol; $T = 300$ K; $p_i = 100$ kPa; $p_f = 75$ kPa.
SOLVE   (a) The volume increases by a factor:

$$\frac{V_f}{V_i} = \frac{P_i}{P_f} = \frac{(100\ \text{kPa})}{(75\ \text{kPa})} = 1.3$$

(b) The work done on the gas is:

$$W = nRT \ln\left(\frac{V_i}{V_f}\right) = (0.30\ \text{mol})(8.31\ \text{J}/(\text{mol}\cdot\text{K}))(300\ \text{K})\ln\left(\frac{1}{1.3}\right) = -0.22\ \text{kJ}$$

The work done by the gas is positive 0.22 kJ.

REFLECT   It is from the ideal gas law we see that pressure varies inversely with volume in an isothermal process.

41.   **ORGANIZE AND PLAN**   In an adiabatic process pressure times volume raised to the power of the adiabatic exponent is constant. That allows us to calculate the new pressure. Then we can calculate the work done using the expression for an adiabatic process. The temperature can be calculated from the ideal gas law.
Known: $V_i = 0.60$ L; $P_i = 1.0$ atm; $T_i = 293$ K; $V_f = V_i/20$; $\gamma = \tfrac{7}{5}$.
SOLVE   (a) In an adiabatic process PV^γ is a constant, so the final pressure is:

$$P_f = P_i\left(\frac{V_i}{V_f}\right)^\gamma = (1.0\ \text{atm})(20)^{\frac{7}{5}} = 66\ \text{atm} = 6.7\ \text{MPa}$$

(b) The work done in an adiabatic process is:

$$W = \frac{P_f V_f - P_i V_i}{\gamma - 1}$$

Insert known values:

$$W = \frac{\tfrac{1}{20}P_f - P_i}{\gamma - 1} V_i = \frac{\tfrac{1}{20}(6.7\ \text{MPa}) - (1.0\ \text{atm})}{\tfrac{7}{5} - 1}(0.60\ \text{L}) = 0.35\ \text{kJ}$$

(c) The final temperature is calculated from the ideal gas law:

$$T_f = \frac{P_f V_f}{nR} = T_i \frac{P_f V_f}{P_i V_i} = \left(293\ \text{K}\right)\frac{\left(6.7\ \text{MPa}\right)}{\left(1.0\ \text{atm}\right)20} = 9.7 \times 10^2\ \text{K}$$

REFLECT   When a gas is compressed adiabatically, both the pressure and the temperature rise.

43.   **ORGANIZE AND PLAN**   In a constant volume process there is no work done, so the heat added equals the change in internal energy. Helium is a monatomic gas.
Known: $m = 8.0$ g; $T = 273$ K; $\Delta T = T$; $\mu_{\text{He}} = 4.0$ g/mol.
SOLVE   The number of moles in 8.0 g of helium is:

$$n = \frac{m}{\mu_{\text{He}}} = \frac{\left(8.0\ \text{g}\right)}{\left(4.0\ \text{g/mol}\right)} = 2.0\ \text{mol}$$

The change in internal energy is:

$$\Delta U = \tfrac{3}{2}nR\Delta T = \tfrac{3}{2}\left(2.0\ \text{mol}\right)\left(8.31\ \text{J/}\left(\text{mol}\cdot\text{K}\right)\right)\left(273\ \text{K}\right) = 6.8\ \text{kJ}$$

REFLECT   You could also solve this problem using Equation 13.3 and the molar specific heat of helium from Table 13.2:

$$Q = nc_V\Delta T = \left(2.0\ \text{mol}\right)\left(12.5\ \text{J/}\left(\text{mol}\cdot\text{K}\right)\right)\left(273\ \text{K}\right) = 6.8\ \text{kJ}$$

45.   **ORGANIZE AND PLAN**   The volume can be calculated using the ideal gas law. The pressure and work done can be calculated using our equations for adiabatic processes.
Known: $n = 0.50$ mol; $T_i = 273$ K; $P_i = 1.0$ atm; $V_f = 2V_i$.
SOLVE   (a) The initial volume is calculated from the ideal gas law:

$$V_i = \frac{nRT_i}{P_i} = \frac{\left(0.50\ \text{mol}\right)\left(8.31\ \text{J/}\left(\text{mol}\cdot\text{K}\right)\right)\left(273\ \text{K}\right)}{\left(1.0\ \text{atm}\right)} = 11\ \text{L}$$

(b) In an adiabatic process PV^γ is a constant, so the final pressure is:

$$P_f = P_i\left(\frac{V_i}{V_f}\right)^\gamma = \left(1.0\ \text{atm}\right)\left(\frac{1}{2}\right)^{\frac{7}{5}} = 0.38\ \text{atm} = 38\ \text{kPa}$$

The work done in an adiabatic process is:

$$W = \frac{P_f V_f - P_i V_i}{\gamma - 1}$$

Insert known values:

$$W = \frac{2P_f - P_i}{\gamma - 1}V_i = \frac{2\left(38\ \text{kPa}\right) - \left(1.0\ \text{atm}\right)}{\frac{7}{5} - 1}\left(11\ \text{L}\right) = -0.67\ \text{kJ}$$

REFLECT   The work done on the gas is negative when the gas expands.

47.   **ORGANIZE AND PLAN**   The work done in each step is the area under the curve in the pressure-versus-volume graph. The work done is negative when the gas expands, positive when the gas compresses.
Known: pressure-versus-volume diagram.

SOLVE   (a) The work done in step AB (going from point A to point B in the pressure-versus-volume graph) is:

$$W_{AB} = -(300 \text{ kPa})(3 \text{ m}^3) = -900 \text{ kJ}$$

The work done in step BC is zero, because there is no area under the vertical line from point B to point C. The work done in step CD is:

$$W_{CD} = (100 \text{ kPa})(3 \text{ m}^3) = 300 \text{ kJ}$$

The work done in step DA is zero, because there is no area under the vertical line from point D to point A.
(b) The net work done over the cycle is:

$$W_{net} = W_{AB} + W_{BC} + W_{CD} + W_{DA} = (-900 \text{ kJ}) + (0) + (300 \text{ kJ}) + (0) = -600 \text{ kJ}$$

REFLECT   The work done is always the area under the curve, even for much more complicated relationships between pressure and volume. For example, the expressions for work done in isothermal and adiabatic processes are derived using calculus to calculate the area under the curve in the pressure-versus-volume graphs for these processes.

49.   **ORGANIZE AND PLAN**   We can use the ideal gas law to determine whether the temperature increases or decreases, and internal energy will change similarly. The work is the area under the curve.
Known: pressure-versus-volume diagram.
SOLVE   (a) The product of pressure times volume decreases when going from point A to point B, so according to the ideal gas law, the temperature must decrease as well. This means the internal energy decreases.
(b) The work done is the area under the curve. The area consists of one triangular and one rectangular part. The work is positive because the gas is compressed. The work is:

$$W_{AB} = \tfrac{1}{2}(1 \text{ atm})(10 \text{ L}) + (1 \text{ atm})(10 \text{ L}) = 1.5 \text{ kJ}$$

REFLECT   Because both pressure and volume are halved, the temperature decreases to one-fourth its original value.

51.   **ORGANIZE AND PLAN**   In an adiabatic process pressure times volume raised to the power of the adiabatic exponent is constant. That allows us to calculate the new pressure. The temperature can be calculated from the ideal gas law.
Known: $P_i = 1.0$ atm; $T_i = 290$ K; $V_f = V_i/10$; $\gamma = \tfrac{7}{5}$.
SOLVE   (a) In an adiabatic process PV^γ is a constant, so the final pressure is:

$$P_f = P_i \left(\frac{V_i}{V_f}\right)^\gamma = (1.0 \text{ atm})(10)^{\frac{7}{5}} = 25 \text{ atm} = 2.5 \text{ MPa}$$

(b) The final temperature is calculated from the ideal gas law:

$$T_f = \frac{P_f V_f}{nR} = T_i \frac{P_f V_f}{P_i V_i} = (290 \text{ K}) \frac{(2.5 \text{ MPa})}{(1.0 \text{ atm})10} = 7.3 \times 10^2 \text{ K}$$

REFLECT   When a gas is compressed adiabatically, both the pressure and the temperature rise.

53.   **ORGANIZE AND PLAN**   The heat lost is the heat of vaporization of 120 g of perspiration, which we can approximate with 120 g of water. The heat of vaporization of water is listed in Table 13.3. The work done is the internal energy change minus the heat lost.
Known: $\Delta U = 80$ kcal; $m = 120$ g; $L_v = 2.26 \times 10^6$ J/kg.

SOLVE   (a) The heat lost is given by Equation 13.6:

$$Q = mL_v = (120 \text{ g})(2.26 \times 10^6 \text{ J/kg}) = 271 \text{ kJ}$$

(b) The work done by the runner is:

$$W = \Delta U - Q = (80 \text{ kcal}) - (271 \text{ kJ}) = 64 \text{ kJ}$$

REFLECT   The heat of vaporization for water has a slight dependency on temperature. The value in Table 13.3 is the heat of vaporization at the boiling point. At body temperature, the heat of vaporization of water is slightly larger, 2.4×10^6 J/kg, but the value used above is an acceptable approximation.

55.   **ORGANIZE AND PLAN**   The entropy change is the heat flow divided by the temperature.
Known: $T = 25°C$; $Q_a = 1000$ J; $Q_b = -1000$ J.
SOLVE   (a) The entropy change when the water absorbs heat is:

$$\Delta S_a = \frac{Q_a}{T} = \frac{(1000 \text{ J})}{(25°C)} = \frac{(1000 \text{ J})}{(298 \text{ K})} = 3.35 \text{ J/K}$$

(a) The entropy change when the water loses heat is:

$$\Delta S_b = \frac{Q_b}{T} = \frac{(-1000 \text{ J})}{(25°C)} = \frac{(-1000 \text{ J})}{(298 \text{ K})} = -3.35 \text{ J/K}$$

REFLECT   Remember to convert the temperature to Kelvin before calculating an entropy change!

57.   **ORGANIZE AND PLAN**   The entropy change is the heat added to the ice or the water divided by the temperature. The heat added can be calculated using the heats of transformation from Table 13.3.
Known: $m = 100$ g; $T_f = 0°C$; $L_f = 3.33 \times 10^5$ J/kg; $T_v = 100°C$; $L_v = 2.26 \times 10^6$ J/kg.
SOLVE   When the ice melts the entropy change is:

$$\Delta S_f = \frac{Q_f}{T_f} = \frac{mL_f}{T_f} = \frac{(100 \text{ g})(3.33 \times 10^5 \text{ J/kg})}{(0°C)} = 122 \text{ J/K}$$

When the water boils the entropy change is:

$$\Delta S_v = \frac{Q_v}{T_v} = \frac{mL_v}{T_v} = \frac{(100 \text{ g})(2.26 \times 10^6 \text{ J/kg})}{(100°C)} = 606 \text{ J/K}$$

REFLECT   The entropy change is greater when water boils than when ice melts.

59.   **ORGANIZE AND PLAN**   The entropy change of the water is the heat removed from the water divided by the temperature of the water. The entropy change of the freezer is the heat added to the freezer divided by the temperature of the freezer.
Known: $m = 150$ g; $T_{water} = 0°C$; $L_f = 3.33 \times 10^5$ J/kg; $T_{freezer} = -4°C$
SOLVE   The heat removed from the water in order to freeze it is:

$$Q = -mL_f = -(150 \text{ g})(3.33 \times 10^5 \text{ J/kg}) = -50.0 \text{ kJ}$$

The entropy change of water is:

$$\Delta S_{water} = \frac{Q}{T_{water}} = \frac{(-50.0 \text{ kJ})}{(0°C)} = -183 \text{ J/K}$$

The entropy change of the freezer is:

$$\Delta S_{freezer} = \frac{-Q}{T_{freezer}} = \frac{(50.0 \text{ kJ})}{(-4°C)} = 186 \text{ J/K}$$

The net entropy change is:

$$\Delta S_{water} + \Delta S_{freezer} = (-183 \text{ J/K}) + (186 \text{ J/K}) = 2.72 \text{ J/K}$$

REFLECT The net entropy is always positive.

61. **ORGANIZE AND PLAN** The entropy change of the nitrogen is the heat added to the nitrogen divided by the boiling temperature of nitrogen.
    Known: $m = 100$ g; $T = -196°C$; $L_v = 1.96 \times 10^5$ J/kg.
    SOLVE The heat added to the nitrogen in order to vaporize it is:

$$Q = mL_v = (100 \text{ g})(1.96 \times 10^5 \text{ J/kg}) = 19.6 \text{ kJ}$$

The entropy change of the nitrogen is:

$$\Delta S = \frac{Q}{T} = \frac{(19.6 \text{ kJ})}{(-196°C)} = 254 \text{ J/K}$$

REFLECT Remember to convert the temperature to Kelvin before calculating an entropy change!

63. **ORGANIZE AND PLAN** The efficiency is the ratio between work done and heat used.
    Known: $W = 650$ J; $Q_H = 1270$ J.
    SOLVE The heat engine's efficiency is:

$$e = \frac{W}{Q_H} = \frac{(650 \text{ J})}{(1270 \text{ J})} = 0.512.$$

REFLECT In most applications 51.2% would be a very good efficiency.

65. **ORGANIZE AND PLAN** The efficiency is the ratio between work done and the thermal energy removed from the uranium fuel. The work done is the difference between the thermal energy and the waste heat.
    Known: $Q_H = 1700$ MJ; $Q_C = 1100$ MJ; $t = 1$ s.
    SOLVE (a) The work done is:

$$W = Q_H - Q_C = (1700 \text{ MJ}) - (1100 \text{ MJ}) = 600 \text{ MJ}$$

The efficiency is the power plant is:

$$e = \frac{W}{Q_H} = \frac{(600 \text{ MJ})}{(1700 \text{ MJ})} = 0.353$$

(b) The rate of electrical energy produced, i.e., the electrical power is:

$$P = \frac{W}{t} = \frac{(600 \text{ MJ})}{(1 \text{ s})} = 600 \text{ MW}$$

REFLECT If the nuclear power plant is close to a city, the waste heat could instead be used to heat thousands of homes.

67. **ORGANIZE AND PLAN** The maximum efficiency is that of a Carnot engine, one minus the temperature ratio between the cold and hot reservoirs.
    Known: $T_C = 0°C$; $T_H = 100°C$.

SOLVE   The maximum efficiency is:

$$e_{Carnot} = 1 - \frac{T_C}{T_H} = 1 - \frac{(0°C)}{(100°C)} = 1 - \frac{(273 \text{ K})}{(373 \text{ K})} = 0.268$$

REFLECT   Remember to convert all temperatures to Kelvin before calculating the temperature ratio!

69.   **ORGANIZE AND PLAN**   The maximum efficiency is that of a Carnot engine, one minus the temperature ratio between the cold and hot reservoirs. The energy produced is the work done and equals the efficiency times the heat removed from the surface water.
Known: $T_C = 4°C$; $T_H = 25°C$; $P = 1000$ MW; $t = 1$ day.
SOLVE   (a) The maximum efficiency is:

$$e_{Carnot} = 1 - \frac{T_C}{T_H} = 1 - \frac{(4°C)}{(25°C)} = 0.070$$

(b) The desired energy production each day is:

$$W = Pt = (1000 \text{ MW})(1 \text{ day}) = 8.64 \times 10^{13} \text{ J}$$

The amount of heat that must be removed from the surface water per day is:

$$Q_H = \frac{W}{e} = \frac{(8.64 \times 10^{13} \text{ J})}{(0.070)} = 1.2 \times 10^{15} \text{ J}$$

REFLECT   At a theoretical maximum efficiency of 7%, it would be very challenging to make this a viable energy source.

71.   **ORGANIZE AND PLAN**   The coefficient of performance of a refrigerator is the heat removed divided by the required work.
Known: COP $= 3.8$; $P = 600$ W.
SOLVE   The refrigerator can remove heat at a rate:

$$\frac{Q_C}{t} = \text{COP} \frac{W}{t} = \text{COP} \times P = (3.8)(600 \text{ W}) = 2.3 \text{ kJ/s}$$

REFLECT   The higher the COP the larger the amount of heat removed.

73.   **ORGANIZE AND PLAN**   The coefficient of performance of a refrigerator is the heat removed divided by the required work. The maximum COP is that of a Carnot cycle.
Known: $T_C = 77$ K; $T_H = 22°C$.
SOLVE   The maximum COP is given in Example 14.11:

$$\text{COP}_{Carnot} = \frac{1}{\dfrac{T_H}{T_C} - 1} = \frac{1}{\dfrac{(22°C)}{(77 \text{ K})} - 1} = 0.35$$

REFLECT   Remember to convert the temperatures to Kelvin before calculating the temperature ratio!

75.   **ORGANIZE AND PLAN**   The coefficient of performance of a heat pump is the heat delivered divided by the work (electrical energy) used. Everything else in this problem is unit conversions.
Known:
$Q_H/t = 250$ kBtu/h; COP $= 3.0$; $c_{electricity}/W = \$0.09/$kWh; $t = 1$ day; $c_{oil}/V = \$4.20/$gal; $Q_{oil}/V = 40$ kWh/gal; $e_{oil} = 87\%$.

SOLVE   (a) The required amount of heat in one day is:

$Q_H = \dfrac{Q_H}{t} t = (250 \text{ kBtu/h})(1 \text{ day}) = (250 \times 10^3 \text{ Btu/h})(1055 \text{ J/Btu})(1 \text{ day})(24 \text{ h/day}) = 6.33 \times 10^9 \text{ J}$. The required

electrical energy to supply this amount of heat using a heat pump with COP = 3.0 is:

$$W = \frac{Q_H}{\text{COP}} = \frac{(6.33 \times 10^9 \text{ J})}{(3.0)} = 2.11 \times 10^9 \text{ J}$$

The daily cost of this electrical energy is:

$$c_{\text{electricity}} = \frac{c_{\text{electricity}}}{W} W = (\$0.09 / \text{kWh})(2.11 \times 10^9 \text{ J}) = \$52.8$$

(b) Expressed in terms of energy, the required amount of oil to burn each day is:

$$Q_{\text{oil}} = \frac{Q_H}{e_{\text{oil}}} = \frac{(6.33 \times 10^9 \text{ J})}{(0.87)} = 7.3 \times 10^9 \text{ J}$$

In volume, the required amount of oil to burn each day is:

$$V = \frac{Q_{\text{oil}}}{\dfrac{Q_{\text{oil}}}{V}} = \frac{(7.3 \times 10^9 \text{ J})}{(40 \text{ kWh/gal})} = 51 \text{ gal}$$

The daily cost of this oil is:

$$c_{\text{oil}} = \frac{c_{\text{oil}}}{V} V = (\$4.20 / \text{gal})(51 \text{ gal}) = \$2.1 \times 10^2$$

The daily monetary savings is:

$$\Delta c = c_{\text{oil}} - c_{\text{electricity}} = (\$2.1 \times 10^2) - (\$52.8) = \$1.6 \times 10^2$$

or roughly 160 dollars per day.

REFLECT   The savings per year is over $58,000. The owner of this office building probably recovered the cost of investing in a heat pump very quickly.

77.   **ORGANIZE AND PLAN**   We will sum all probabilities.

Known: Figure in solution to Problem 76.

SOLVE   (a) The sum of all probabilities are:

$$\frac{1}{32} + \frac{5}{32} + \frac{10}{32} + \frac{10}{32} + \frac{5}{32} + \frac{1}{32} = \frac{32}{32} = 1$$

(b) The sum of all probabilities are:

$$\frac{1}{64} + \frac{6}{64} + \frac{15}{64} + \frac{20}{64} + \frac{15}{64} + \frac{6}{64} + \frac{1}{64} = \frac{64}{64} = 1$$

REFLECT   If you sum the ratios rather than the rounded-off decimal numbers you are guaranteed to get a sum that equals 1.

79.   **ORGANIZE AND PLAN**   First we will calculate how many ways we can choose n molecules *in a specific order* out of the N molecules. Then we will calculate how many ways we can rearrange the order of n molecules. Finally we will combine these two results to calculate how many ways we can choose n molecules without regards to order (which equals the number of microstates).

Known: N; n.

SOLVE   We are looking for a number x which is the number of ways one can choose n molecules out of N molecules without regard to the order between these n molecules. However, consider for the moment the number of ways we can choose n molecules *in a specific order* out of the N molecules. This number must be:

$$N \cdot (N-1) \cdot (N-2) \cdot ... \cdot (N-n+1)$$

because for our first molecule we have N choices, for our second molecule we have $(N-1)$ choices remaining, for our third molecule we have $(N-2)$ choices remaining, and so on.

Similarly, the number of ways we can rearrange the order of these n molecules is:

$$n \cdot (n-1) \cdot (n-2) \cdot ... \cdot 2 \cdot 1 = n!$$

Consider now a different approach to choosing n molecules *in a specific order* out of the N molecules. We could do this by first choosing n molecules *without* regards to order (there are x ways to do this), and then once we have these n molecules we can rearrange them (there are $n!$ ways to do this). These two numbers must multiply to get the result above. Consequently, our number x must fulfill:

$$x \cdot n! = N \cdot (N-1) \cdot (N-2) \cdot ... \cdot (N-n+1)$$

The right-hand side of this equation can be rewritten using factorials as:

$$N \cdot (N-1) \cdot (N-2) \cdot ... \cdot (N-n+1) = \frac{N!}{(N-n)!}$$

Solve for x and we find the number of microstates for an N-molecule gas with n molecules on one side of the box:

$$x = \frac{N!}{n!(N-n)!}$$

REFLECT   Try the reasoning above using the established cases (two, three, four, five, six molecules).

81.   **ORGANIZE AND PLAN**   We will use the formula we derived in Problem 14.79.

Known: $N = 52$; $n = 5$.

SOLVE   The number of different poker hands is:

$$x = \frac{N!}{n!(N-n)!} = \frac{52!}{5!(52-5)!} = 2,589,960$$

REFLECT   If we were talking about molecules instead of cards, we would say that there are 2,589,960 microstates (poker hands) for placing 5 out of 52 molecules (cards) in one box (dealt to one player) and the other 47 molecules in another box (left in the deck).

83.   **ORGANIZE AND PLAN**   The internal energy of the gas does not change since the temperature remains constant. This means that the gas does as much work as the heat it absorbs. We can calculate how many moles the gas has by using the formula for work done on a gas in an isothermal process.

Known: $V_f = 10V_i$; $T = 440$ K; $W = -3.3$ kJ.

SOLVE   (a) The amount of heat absorbed by the gas is $Q = -W = -(-3.3\,\text{kJ}) = 3.3\,\text{kJ}$.

(b) The work done on a gas in an isothermal process is:

$$W = nRT \ln\left(\frac{V_i}{V_f}\right)$$

Rewrite this equation to calculate the number of moles in the gas:

$$n = \frac{W}{RT \ln\left(\dfrac{V_i}{V_f}\right)} = \frac{(-3.3\,\text{kJ})}{(8.31\,\text{J/(mol}\cdot\text{K)})(440\,\text{K}) \ln\left(\dfrac{1}{10}\right)} = 0.39\,\text{mol}$$

REFLECT   It does not matter in an isothermal process whether the gas is monatomic or diatomic.

85.   ORGANIZE AND PLAN   In an adiabatic process, the quantity PV^γ remains constant.
Known: $V_1 = V_0/2; P_1 = 2.55P_0$.
SOLVE   The quantity PV^γ before the compression equals the quantity PV^γ after the compression:

$$P_0 V_0^\gamma = P_1 V_1^\gamma$$

If we insert our known ratios between the two volumes and the two pressures, we get an expression we can solve for γ:

$$P_0 V_0^\gamma = P_1 V_1^\gamma$$

$$P_0 V_0^\gamma = 2.55 P_0 \left(\frac{V_0}{2}\right)^\gamma$$

$$1 = 2.55 \left(\frac{1}{2}\right)^\gamma$$

$$\ln\left(\frac{1}{2.55}\right) = \gamma \ln\left(\frac{1}{2}\right)$$

$$\gamma = \frac{\ln\left(\dfrac{1}{2.55}\right)}{\ln\left(\dfrac{1}{2}\right)} = \frac{\ln(2.55)}{\ln(2)} = 1.35$$

REFLECT   This value is slightly smaller than the value for diatomic gas (7/5), so the answer seems reasonable.

87.   ORGANIZE AND PLAN   The pressure can be calculated from the quantity PV^γ remaining constant. The work for each segment can be calculated using the formulas for an adiabatic, constant volume, and constant pressure process, respectively. The net work is the sum of the work of the three segments.
Known: $\gamma = 1.4; P_A = P_C = 60\,\text{kPa}; V_A = 5\,\text{L}; V_B = V_C = 1\,\text{L}$.
SOLVE   (a) The pressure at B equals:

$$P_B = P_A \left(\frac{V_A}{V_B}\right)^\gamma = (60\,\text{kPa})\left(\frac{(5\,\text{L})}{(1\,\text{L})}\right)^{1.4} = 0.57\,\text{MPa}$$

(b) The work done on segment AB is the work done in an adiabatic process:

$$W_{AB} = \frac{P_B V_B - P_A V_A}{\gamma - 1} = \frac{(0.57\,\text{MPa})(1\,\text{L}) - (60\,\text{kPa})(5\,\text{L})}{1.4 - 1} = 0.68\,\text{kJ}$$

The work done on segment BC is zero, because the volume of the gas does not change.

The work done on segment CA is the work done in a constant pressure process:

$$W_{CA} = P_C (V_C - V_A) = (60 \text{ kPa})((1 \text{ L}) - (5 \text{ L})) = -0.24 \text{ kJ}$$

The net work done on the gas is:

$$W_{net} = W_{AB} + W_{BC} + W_{CA} = (0.68 \text{ kJ}) + (0) + (-0.24 \text{ kJ}) = 0.44 \text{ kJ}$$

REFLECT   The net work done on the gas is larger for the adiabatic compression than for the isothermal compression in the previous problem. Similarly, had the arrows in the graph been in the other direction the gas would have done larger work in an adiabatic expansion than in an isothermal expansion.

89.   **ORGANIZE AND PLAN**   We will assume that the power plant operates at the maximum thermodynamic efficiency, i.e., the efficiency of a Carnot cycle, where the efficiency is one minus the temperature ratio between the cold and hot reservoirs. The power produced is proportional to the efficiency, so from knowing the winter production and both the winter and summer efficiencies we can calculate the summer production.
Known: $T_H = 310°C$; $P_{winter} = 650 \text{ MW}$; $T_{C,winter} = 0°C$; $T_{C,summer} = 38°C$.
SOLVE   (a) The theoretical maximum winter efficiency is:

$$e_{Carnot, winter} = 1 - \frac{T_{C,winter}}{T_H} = 1 - \frac{(0°C)}{(310°C)} = 0.532$$

(b) The theoretical maximum summer efficiency is:

$$e_{Carnot, summer} = 1 - \frac{T_{C,summer}}{T_H} = 1 - \frac{(38°C)}{(310°C)} = 0.466$$

The production is proportional to efficiency, so comparing the winter and summer numbers we must have:

$$\frac{P_{winter}}{e_{winter}} = \frac{P_{summer}}{e_{summer}}$$

If we assume that the efficiencies are the Carnot efficiencies, we have:

$$\frac{P_{winter}}{e_{Carnot, winter}} = \frac{P_{summer}}{e_{Carnot, summer}}$$

$$P_{summer} = \frac{e_{Carnot, summer}}{e_{Carnot, winter}} P_{winter} = \frac{(0.466)}{(0.532)}(650 \text{ MW}) = 570 \text{ MW}$$

REFLECT   The actual efficiency of a power plant is nowhere near the maximum thermodynamic efficiency, so our answer in this particular problem may not apply to a real world situation. One thing we can say for sure, however, is that the summer production would be smaller than the winter production.

91.   **ORGANIZE AND PLAN**   The pressure can be calculated from the quantity PV^γ remaining constant. The work can be calculated using the formulas for an adiabatic process.
Known: $\gamma = 1.4$; $V_i = 5.0 \text{ L}$; $P_i = 100 \text{ kPa}$; $V_f = 2.5 \text{ L}$.
SOLVE   (b) The final pressure equals:

$$P_f = P_i \left(\frac{V_i}{V_f}\right)^\gamma = (100 \text{ kPa})\left(\frac{(5.0 \text{ L})}{(2.5 \text{ L})}\right)^{1.4} = 0.26 \text{ MPa}$$

(a) The work done is the work done in an adiabatic process:

$$W = \frac{P_f V_f - P_i V_i}{\gamma - 1} = \frac{(0.26 \text{ MPa})(2.5 \text{ L}) - (100 \text{ kPa})(5.0 \text{ L})}{1.4 - 1} = 0.40 \text{ kJ}$$

REFLECT The net work done on the gas is larger for the adiabatic compression than for the isothermal compression in the previous problem. Similarly, had the arrows in the graph been in the other direction the gas would have done larger work in an adiabatic expansion than in an isothermal expansion.

93. **ORGANIZE AND PLAN** Because no heat is exchanged with the surroundings, this is an adiabatic process. We can calculate the final temperature from the quantity PV^γ remaining constant and the ideal gas law. Air is predominantly a diatomic gas, so the specific heat ratio $\gamma = \frac{7}{5}$.

Known: $P_i = 62.0$ kPa; $T_i = -11.0°C$; $P_f = 86.5$ kPa; $V_i = 1.00$ m³; $\gamma = \frac{7}{5}$.

SOLVE (a) Take the quantity PV^γ and the ideal gas law and divide the final quantities with the initial quantities:

$$\frac{P_f V_f^\gamma}{P_i V_i^\gamma} = 1$$

$$\frac{P_f V_f}{P_i V_i} = \frac{nRT_f}{nRT_i} = \frac{T_f}{T_i}$$

Combine these two equations to relate a change in pressure to a change in temperature for an ideal gas in an adiabatic process:

$$\frac{T_f}{T_i} = \frac{P_f V_f}{P_i V_i} = \frac{P_f}{P_i}\left(\frac{P_i}{P_f}\right)^{1/\gamma} = \left(\frac{P_f}{P_i}\right)^{1-1/\gamma}$$

and solve for the final temperature of the air once it's reached Calgary:

$$T_f = T_i \left(\frac{P_f}{P_i}\right)^{1-1/\gamma} = (-11.0°\,\text{C})\left(\frac{(86.5\text{ kPa})}{(62.0\text{ kPa})}\right)^{1-5/7} = 15.2°C$$

(b) The final volume of what was initially a cubic meter of air is:

$$V_f = V_i\left(\frac{P_i}{P_f}\right)^{1/\gamma} = (1.00\text{ m}^3)\left(\frac{(62.0\text{ kPa})}{(86.5\text{ kPa})}\right)^{5/7} = 0.788\text{ m}^3$$

The work done on this volume is:

$$W = \frac{P_f V_f - P_i V_i}{\gamma - 1} = \frac{(86.5\text{ kPa})(0.788\text{ m}^3) - (62.0\text{ kPa})_i(1.00\text{ m}^3)}{\frac{7}{5} - 1} = 15.5\text{ kJ}$$

REFLECT This type of hot wind is called a foehn wind and exists above other mountain ranges as well, where moist air is forced up a mountain by a strong wind, then dumps its moisture content as rain or snow near the top of the mountain. This sets up the conditions described in the problem text.

95. **ORGANIZE AND PLAN** The coefficient of performance (COP) of a refrigerator equals the heat removed divided by the electrical energy used. The heat removed for a given mass of water can be calculated from water's specific heat. The maximum possible COP is that of a Carnot cycle.

Known:

$V = 4.0$ L; $T_i = 9.0°C$; $P = 130$ W; $t = 4.0$ min; $T_f = T_C = 1.0°C$; $T_H = 25°C$; $\rho = 1000$ kg/m³; $c = 4186$ J/$(\text{kg} \cdot °C)$.

SOLVE The COP of the refrigerator is:

$$\text{COP} = \frac{Q_C}{W} = \frac{-mc\Delta T}{W} = \frac{-\rho V c \Delta T}{W} = \frac{-\rho V c (T_f - T_i)}{W} = \frac{-\rho V c (T_f - T_i)}{Pt}$$

Insert our known values to calculate the COP:

$$\text{COP} = \frac{-(1000 \text{ kg/m}^3)(4.0 \text{ L})(4186 \text{ J/}(\text{kg}\cdot{}^\circ\text{C}))((1.0{}^\circ\text{C})-(9.0{}^\circ\text{C}))}{(130 \text{ W})(4.0 \text{ min})} = 4.3$$

The maximum COP is given in Example 14.11:

$$\text{COP}_{\text{Carnot}} = \frac{1}{\dfrac{T_H}{T_C}-1} = \frac{1}{\dfrac{(25{}^\circ\text{C})}{(1{}^\circ\text{C})}-1} = 11.4$$

REFLECT   The actual COP is less than the theoretical maximum —as it should be—and at approximately two and a half times smaller than the theoretical maximum, this value for the actual COP is quite reasonable.

97.   **ORGANIZE AND PLAN**   The maximum efficiency is that of a Carnot engine, one minus the temperature ratio between the cold and hot reservoirs.
Known: $T_H = 1050{}^\circ\text{C}$; $T_M = 590{}^\circ\text{C}$; $T_C = 42{}^\circ\text{C}$.
SOLVE   (a) The maximum efficiency of the steam cycle alone is:

$$e_{\text{Carnot}} = 1 - \frac{T_C}{T_M} = 1 - \frac{(42{}^\circ\text{C})}{(590{}^\circ\text{C})} = 0.635$$

(b) The maximum efficiency of the combined cycle is:

$$e_{\text{Carnot}} = 1 - \frac{T_C}{T_H} = 1 - \frac{(42{}^\circ\text{C})}{(1050{}^\circ\text{C})} = 0.762$$

REFLECT   The internal mechanism of the power plant does not affect the theoretical maximum efficiency; only the temperatures of the hot and cold reservoirs matter. However, the internal mechanisms very much will affect how close to a Carnot cycle the power plant operates, and consequently will affect the actual efficiency.

99.   **ORGANIZE AND PLAN**   The pressure at B can be calculated from the quantity PV^γ remaining constant. The pressure at C can be found using the ideal gas law. The work for each segment can be calculated using the formulas for an adiabatic, a constant volume, and an isothermal process, respectively. The net work is the sum of the work of the three segments.
Known: $\gamma = \frac{5}{3}$; $V_A = 1.00 \text{ m}^3$; $P_A = 250 \text{ kPa}$; $V_B = V_C = 3V_A$; $T_A = T_C$.
SOLVE   (a) The pressure at B equals:

$$P_B = P_A \left(\frac{V_A}{V_B}\right)^\gamma = (250 \text{ kPa})\left(\frac{1}{3}\right)^{5/3} = 40.0 \text{ kPa}$$

(b) The pressure at C equals:

$$P_C = \frac{nRT_C}{V_C} = \frac{nRT_A}{3V_A} = \frac{P_A}{3} = \frac{(250 \text{ kPa})}{3} = 83.3 \text{ kPa}$$

(c) The work done on segment AB is the work done in an adiabatic process:

$$W_{AB} = \frac{P_B V_B - P_A V_A}{\gamma - 1} = V_A \frac{3P_B - P_A}{\gamma - 1} = (1.00 \text{ m}^3)\frac{3(40.0 \text{ kPa}) - (250 \text{ kPa})}{\frac{5}{3}-1} = -195 \text{ kJ}$$

The work done on segment BC is zero, because the volume of the gas does not change.

The work done on segment CA is the work done in an isothermal process:

$$W_{CA} = nRT_A \ln\left(\frac{V_C}{V_A}\right) = P_A V_A \ln\left(\frac{V_C}{V_A}\right) = (250 \text{ kPa})(1.00 \text{ m}^3) \ln(3) = 275 \text{ kJ}$$

The net work done on the gas is:

$$W_{net} = W_{AB} + W_{BC} + W_{CA} = (-195 \text{ kJ}) + (0) + (275 \text{ kJ}) = 79.9 \text{ kJ}$$

REFLECT   The pressure is higher when we are compressing the gas than when it expands, resulting in a net positive work done on the gas.